Applied Electromagnetism and Materials

Applied Electromagnetism and Materials

André Moliton
Université de Limoges
Limoges, France

Springer

André Moliton
Laboratory Xlim-MINACOM
Faculte des Sciences et Techniques
123 Avenue Albert Thomas
87060 Limoges Cedex
France
andre.moliton@unilim.fr

ISBN-13: 978-1-4419-2283-0
e-ISBN-10: 0-387-38064-7
e-ISBN-13: 978-0-387-38064-3

9 8 7 6 5 4 3 2 1

springer.com

Preface

This book describes the applications of the fundamental interactions of electromagnetic waves and materials as described in the preceding volume, "*Basic Electromagnetism and Materials*". It is addressed to students studying masters or doctorate courses in electronics, electromagnetism, applied physics, materials physics, or chemical physics. In particular, this volume analyzes the behavior of materials in the presence of an electromagnetic field and related applications in the fields of electronics, optics, and materials physics. The study of the fundamental processes due to electromagnetic fields is detailed and includes processes of radiation, intense fields (including the electrooptical effect), and confined media and of those related to particle mechanics that are used in large-scale apparatuses.

The first three chapters are dedicated to the description of materials placed under a varying electric field. The material is treated particularly from the viewpoint of a dielectrician, with the classic representations (sometimes termed as in the complex plane) of Debye, Cole–Cole, and Cole–Davidson. The origins of the relaxation mechanisms in the Hertzian domain due to slow-moving charges, such as space charges, bound charges associated with permanent dipoles, and electrons trapped in insulators or semiconductors are treated. A study of the latter material makes possible a determination of the energy associated with trap depths, an important value for electricians, semiconductor optoelectricians, dielectricians, and opticians (through fluorescence studies).

The second chapter details the relations between components for conductivity and dielectric permittivity. They show, among other things, that a material is never an ideal (perfect) conductor or insulator. For a material subject to an electromagnetic wave, there are simple relations between these two components written in a complex form to take into account the realities of the materials. Spectroscopic analysis is notably detailed.

The Kramers–Krönig relations are then established in Chapter 3 by means of a formal physical treatment of signals. They show that the knowledge of a spectrum of the real component of the dielectric permittivity (also true for an optical index or for the magnetic permeability) permits a deduction of the imaginary component for a given frequency and *vice versa*. The spectra of dielectrics thus are completed for the infrared and optical frequencies by a study of the ionic and above all electronic polarization. For each degree of bonding of electrons (valence electrons in internal or external layers) it is shown that there is a corresponding absorption frequency, which in turn gives rise to different absorption peaks and dispersion domains. The range of various phenomena is grouped together in a single figure, which summarizes the responses of a dielectric to waves throughout the electromagnetic spectrum. For further details on infrared or optical waves, the reader would obviously have to turn to a course on quantum mechanics, but the physical origin of optical losses that intervene in guided propagation is well described using the presented classical electromagnetic theory.

Chapters 4 and 5 detail studies of waves in more semiconductor-type solids. Initially, the behavior of a wave associated with electrons in periodic medium is described to show that it is the establishment of stationary waves for particular values of the wave vector that generates forbidden bands. These bands make possible a distinction between insulators, semiconductors, and metals for which the population of the bands is described as simply by introducing state density functions. The electromagnetic properties, and in particular in the optical domain, of semiconductors are then depicted. These include, notably, reflectivity and absorption coefficients and the establishment of the relation between the size of the gap and the dielectric permittivity. The origin of forbidden transitions and radiative transitions is given, and Chapter 4 is completed with a determination of the equation for the levels of absorption and transmission in semiconductors, which controls their opto-electronic properties. Chapter 5 describes the electrical and magnetic properties of homogeneous and inhomogeneous semiconductors. Point by point, the practical characteristics of conductivity (along with the physical significance of the squared resistance and a discussion on abuses of its use) and the thermoelectric effect are dealt with (in the form of problems). The characterizations of magnetoresistances are introduced and the Hall effect with two types of carriers is studied through the use of set problems. Finally, the origin of the Gunn effect is presented along with its application to the generation of microwaves.

Chapters 6 and 7 go beyond the electromagnetic phenomena introduced in the preceding chapters. It is here that the nonlinear response of a material to an electromagnetic wave is brought under examination, in particular by looking at the Pockels effect underlying the operation of electrooptical modulators. The effects of cavities and microcavities on the behavior of waves then is introduced, taking into account its importance in present-day applications, especially in optics where the laser effect is widely used in material physics (for optical characterizations). The ablation by laser of materials has given rise to deposition methods, which here are limited to the application of ion beams in the last three chapters.

Thus Chapters 8 and 9 study and apply particle mechanics in the presence of an electromagnetic field. The classic trajectories are first detailed, and they include electromagnetic deviation, cycloid-type trajectories in E x B fields and associated with the magnetron effect (used in vacuum techniques), and trajectories associated with electronic or ionic optics in large-scale apparatuses such as electron microscopes and ion accelerators. A description of such a machine thus is proposed, with a detailed appraisal of an ion accelerator equipped with an ion source [termed a cyclotronic resonance, ECR], ionic optics, and a mass filter, the operation of which all rely on the laws of electromagnetism (for example, the penetration of microwaves into a plasma to generate ECR-type ions). Chapter 10 is used to study ion–material interactions that govern implantation mechanisms in semiconductors, material surface treatments (cleaning, engraving, and densification), and their fabrication (pulverization using a single- or double-beam configuration). Finally, a detailed description of ion sources is given, each of which use the generation of a plasma distributed in a specific manner by the electromagnetic interaction of charged particles with the atomic element to be ionized.

ACKNOWLEDGMENTS. I would like to offer my special thanks to the translator of this text, Dr. Roger C. Hiorns. Dr. Hiorns is following postdoctorial studies in the synthesis of polymers for electroluminescent and photovoltaic applications at the Laboratoire de Physico-Chimie des Polymères (Université de Pau et des Pays de l'Adour, France).

Contents

Chapter 1

Dielectrics under Varying Regimes: Phenomenological Study of Dielectric Relaxation

1.1. Definitions for dielectric permittivities and dielectric conductivity and classification of dielectric phenomena

1.1.1. Absolute permittivity

The absolute permittivity (ε) of an isotropic material can be defined in general terms as the quotient of the electrical induction (D) and the electric field (E), i.e., $\varepsilon = \frac{D}{E}$.

Given that the field and the induction are zero in a metal at equilibrium (where V = constant and E_{metal} = - grad V = 0 and D_{metal} = 0), the equation for continuity at an interface between a metallic electrode and a dielectric means that for the normal component $[D_n]_{diel}$ (= D in the present configuration) to the dielectric:

$$[D_n]_{diel} - [D_n]_{metal} = D - 0 = D = \sigma_{real}, \text{ from which } \varepsilon = \frac{\sigma_{real}}{E}.$$

For two electrodes of equal surface (S) separated by a dielectric so that they are a distance (d) apart, having a potential difference (V) that gives rise to a charge (Q), the preceding ratio means that:

$$\varepsilon = \frac{\sigma_{real}}{E} = \frac{Q/S}{V/d} = \frac{Cd}{S},$$

where C is the capacity.

As a consequence, in the MKS unit system, the absolute permittivity is expressed as Farad meter^{-1}. If the "material" under question is a vacuum, then

$$\varepsilon_0 = \frac{1}{36\pi}10^{-9} \text{ F m}^{-1}.$$

However, in the UES–CGS system, $\varepsilon_0 = 1$.

1.1.2. Relative permittivity

It is more often practical to use relative permittivity (ε_r)—commonly and abusively cited as simply "permittivity"—which is defined by the equation $\varepsilon = \varepsilon_r \varepsilon_0$, where ε_r is a number without dimensions. This number in the UES system is equal to the absolute permittivity (or the dielectric constant) which explains the confusion often found between the two magnitudes.

If an alternating tension, given by $V = V_0 e^{j\omega t}$, is applied to the terminals of a planar condenser (that has a capacitance denoted by C_0 when the dielectric is a vacuum), the intensity of the current ($\underline{I}$) circulating between the electrodes when there is a supposedly perfect dielectric between the electrodes, which does not give rise to leak currents and has a permittivity denoted by ε_r, is given by $\underline{I} = \frac{dQ}{dt}$.

Knowing that $\varepsilon_0 = \frac{C_0 d}{S}$, we thus find that $\varepsilon_r = \frac{\varepsilon}{\varepsilon_0} = \frac{Cd/S}{C_0 d/S} = \frac{C}{C_0}$. For the alternating tension, $Q = CV = \varepsilon_r C_0 V_0 e^{j\omega t}$, from which $\underline{I} = j\omega\varepsilon_r C_0 V$.

1.1.3. Complex relative permittivity

The reality is that the dielectrics generally used are not perfect and in fact provide for a wide variety of currents, which will be detailed later on. Nevertheless, the different causes can result in similar effects, for example, free or bound carriers can result both in heating and dielectric losses. Such currents are entirely due to the dielectric material, and they can be characterized as an imaginary component with a relative permittivity:

$$\underline{\varepsilon}_r = \varepsilon'_r - j\varepsilon''_r.$$

In order to simplify the notation used, the indices (r) are often omitted, and it is up to the reader to know whether the text deals with relative or absolute permittivities.

The current intensity in the condenser is now stated therefore as:

$$\underline{I} = \omega\varepsilon_r'' C_0 V + j\omega\varepsilon_r' C_0 V = I_R + jI_C.$$

The second term, corresponding to a dephasing of $\frac{\pi}{2}$ between tension and current, is pure capacitance, and ε_r' thus only characterizes the capacitance (insulating) of the dielectric. The first term, due to the fact that the tension (V) and the intensity are in phase, corresponds to the resistive part of the dielectric which is characterized by ε_r''.

The power dissipated by the Joule effect thus is given by

$$P_J = \frac{1}{2} V_0 I_{R0} = \frac{1}{2} C_0 \varepsilon_r'' \omega V_0^2 ,$$

where $I_R = \omega\varepsilon_r'' C_0 V_0 e^{j\omega t} = I_{R0} e^{j\omega t}$, and the value ε_r'' is called the dielectric absorption as it intervenes in the equation for the electrical energy converted into heat in the medium (therefore absorbed by the dielectric) and then lost into the electric circuit.

The quantity given by:

$$\tan\delta = \frac{|I_R|}{|I_c|} = \frac{\omega\varepsilon'' C_0 V}{\omega\varepsilon' C_0 V} = \frac{\varepsilon''}{\varepsilon'}$$

is the (dielectric) tangential loss and makes it possible to define the loss angle (δ) which represents the dephasing between the resultant current and the "ideal capacitance" (I_c) for this current. The value given by $Q = \dfrac{1}{\tan\delta}$ is the quality factor of the condenser and increases as $\tan\delta$ decreases (Figure 1.1.).

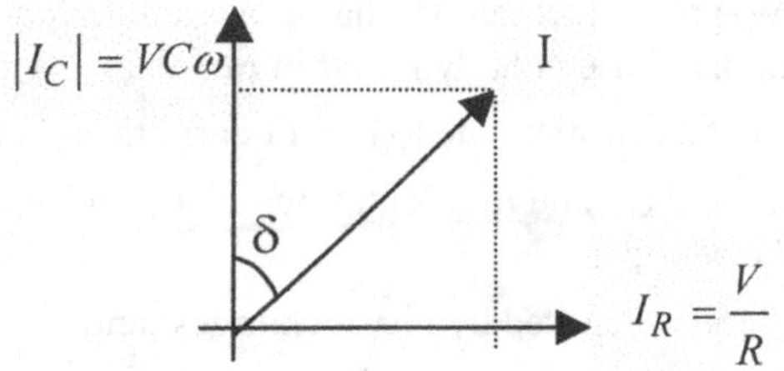

Figure 1.1. *Definition of the loss angle δ.*

1.1.4. Limited permittivity

The permittivity limited to low frequencies (ε_s) can be absolute or—occasionally—relative with the "r" being omitted by convention. At zero frequency, the field is in effect static. There is also a permittivity limited to high frequencies (ε_∞) (infinite frequency).

1.1.5. Dielectric conductivity

By plugging the equations E = V/d and $C_0 = \varepsilon_0 S/d$ into the for $\underline{I}$, we obtain:

$$\underline{I} = \omega\varepsilon_r''\varepsilon_0 \frac{S}{d} Ed + j\omega\varepsilon_r'\varepsilon_0 \frac{S}{d} Ed .$$

It is possible to write the complex equation by using the equations $J = \gamma E$ and $I = \iint J\,dS = J\,S$, so that: $\underline{\gamma} = \frac{J}{E} = \frac{\underline{I}}{ES} = \omega\varepsilon_0\varepsilon''_r + j\omega\varepsilon_0\varepsilon'_r$.

The real component of the conductivity is in this case called the dielectric conductivity, and is thus

$$\gamma_d = \omega\varepsilon_0\varepsilon''_r.$$

1.1.6. Classification of diverse dielectric phenomena

As was above indicated, the introduction of an imaginary term into the dielectric permittivity (in ε'') to take into account a current that is in phase with the tension does not make any reference to the origin of that current, its conduction, and any associated losses. In fact, a large number of phenomena can be responsible for this dielectric absorption:

I. absorption due to free charge conduction, for example thermal carriers generated in the permitted band present at very low densities in insulators due to their large forbidden band limiting their generation, and ions giving rise to ionic conduction at low frequencies;

II. the Maxwell–Wagner effect due to charge accumulation at discontinuities in the dielectric, for example, those notably found in powders;

III. the Debye dipolar absorption due to bound carriers and electric dipoles;

IV. anomalies in ε' and ε'' due to orientations changed by impurities such as water at the surface of the solid;

V. anomalies in ε' and ε'' caused by phase changes; and

VI. absorptions due to resonances.

Each phenomenon can be classed as a function of its appearance at a particular frequency or temperature range as indicated in Freymann's representation, which is shown in Figure 1.2. For a more classical representation of ε' and ε'' as functions of frequency only, please see Chapter 3.

Phenomenon I is caused by leak currents of free charges alone, the number of which can be increased by the introduction of impurities as they insert trapping levels into the large forbidden band of the dielectric. Examples include ZnS and CdS doped with copper in a fashion similar to semiconductors. It is the electrons from the impurities, which, under the action of a sufficiently high field, reach high enough energies so that their collisions with the lattice of the material result in an ionization of atoms. This facilitates the ejection of electrons to the conduction band and results in what is called an intrinsic breakdown of the dielectric.

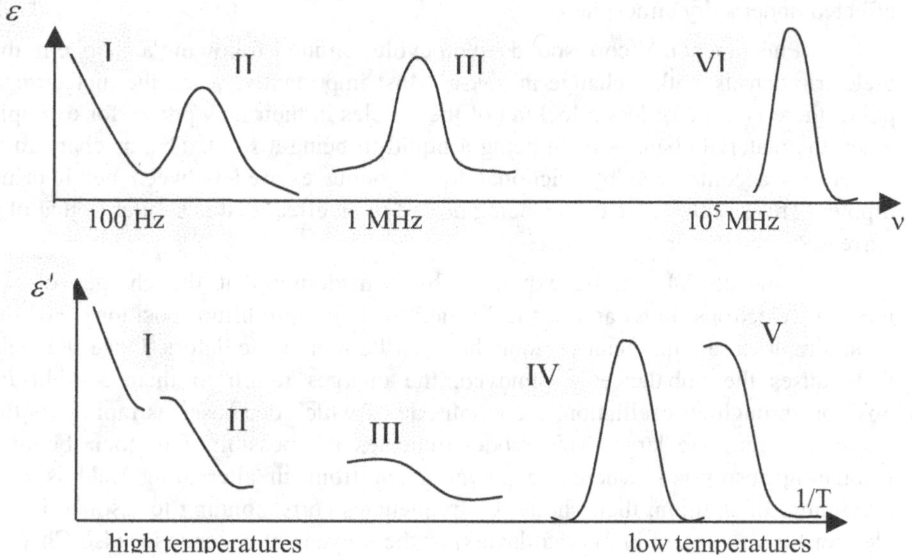

Figure 1.2. *Representation of ε' and ε'' as a function of 1/T.*

Phenomenon II corresponds to an accumulation of charges such as electrons around discontinuities in the dielectric. This problem can be modeled as a condenser consisting of several layers of dielectric.

Phenomenon III, the Debye dipolar absorption (DDA), is generally the result of two neighboring and indissolubly tied dipole charges (+ and -), otherwise called bound carriers. It is caused by the dielectric relaxation associated with the orientation of dipoles excited by an applied electric field. This orientation is delayed by a relaxation time due to an inertia in the movement of the dipole caused by viscous frictions in the material. Section 1.2 details this mechanism and furthermore shows how a single charge jumping over a potential barrier can be characterized as the same type of relaxation.

Phenomenon IV can correspond to a superficial conductivity caused by "semiconducting" materials such as dust, soiling, water (which has a very high dielectric constant of around 80 and is a conductor when impure), etc., deposited on the surface of the insulator. This is the reason why dielectric surfaces should be well cleaned prior to study (notably with alcohol or acetone to remove any humidity and then with a nonpolar solvent such as benzene so as to remove all polar molecules and grease that may perturb the measurements) and why measurements should be effected under a dry atmosphere.

Phenomenon V corresponds to an evolution in ε' following a change in the dielectric density with a change in phase. Most importantly, when the dielectric is polar, there is more or less a locking of the dipoles in their new phase, for example when the material changes from being a liquid to being a solid. This mechanism is sometimes accompanied by frictional forces being exerted between neighboring dipoles. This results in the dipole being heated in an effect that is similar to that of a current.

Phenomenon VI can be explained by considering that the charges of the medium (electrons, ions) are elastically tied to their equilibrium positions. For the most simple cases, they can be thought of as harmonic oscillators. Once the field that causes the imbalance is removed, the charges return to their equilibrium position through an oscillation, the amplitude of which decreases as rapidly as the damping forces are large. This model indicates the possibility of their being a resonating absorption, where the power taken from an alternating field is at a maximum. In addition, their should be frequencies corresponding to resonations of electrons (more or less in internal layers) or the movement of ions (see also Chapter 8 of Volume 1, and also Chapter 3 in this volume).

This chapter will look specifically at phenomena I, II, and II, which are normally observed in the Hertzian domain (very approximately in the range from continuous to 10^{11} Hz). Chapter 3 establishes the general theory for the effect of variable fields on a linear material, which gives rise to the Kramers-Krönig equations. These in turn make it possible to obtain the values for ε' from a spectrum of ε'' and visa versa for a given frequency. In addition, there is a summary of the various behaviors exhibited by dielectrics throughout the electromagnetic spectrum.

1.2. Classic study of the Debye dipolar absorption (DDA)

Dielectric relaxation phenomena are associated with the orientation of permanent dipoles subject to an excitation due to an electric field. The orientation is delayed by a frictional resistance of the material and is characterized by a relaxation time.

1.2.1. The form of the polarization under a continuous (stationary) regime

When a field (E) is applied to a dielectric, the electrics and then the nuclei move almost immediately as their inertia (mass) is very small and the deformation polarization ($P_E + P_A$) is established nearly instantaneously. However, things are not quite the same for permanent dipoles, as whole molecules have to be orientated while interacting with their neighbors. To summarize, after a very long time (t), for the stationary regime, we can state that the static polarization is the sum of:

- a quasi-instantaneous polarization denoted by P_∞ (which is equal to the sum of electronic and atomic polarizations and exists whatever the frequency of the applied field); and
- a stationary dipolar polarization $P_{S(dipole)}$.

Therefore, $P_{S(total)} = P_\infty + P_{S(dipole)}$.

Given that the general expression for polarization is $P = (\varepsilon - \varepsilon_0)E$, then

$$P_\infty = (\varepsilon_\infty - \varepsilon_0)E \text{ and } P_S(total) = (\varepsilon_s - \varepsilon_0)E,$$

from which can be deduced that:

$$P_{S(dipole)} = P_{S(total)} - P_\infty = (\varepsilon_s - \varepsilon_0)E - (\varepsilon_\infty - \varepsilon_0)E = (\varepsilon_s - \varepsilon_\infty)E.$$

1.2.2. Dipolar polarization as a function of time

If E is applied at an instant t = 0, the polarization given by $P_\infty = P_E + P_A$ thus will disappear instantaneously. Given that with time the dipoles will undergo an orientation, the polarization at the instant t is given by

$P(t) = P_\infty + P_{(dipole)} = (\underline{\varepsilon} - \varepsilon_0)E(t)$ from which

$$P_{(dipole)} = (\underline{\varepsilon} - \varepsilon_\infty)E.$$

If it is accepted that the variation in dipolar polarization is proportional to its difference from the equilibrium value, which also means accepting that the dipolar polarization varies with time even more that the difference from its final value, we can state that:

$$\frac{dP_{(dipole)}}{dt} = \frac{1}{\tau}\left(P_{S(dipole)} - P_{(dipole)}\right),$$

where τ is the relaxation time.

As $P_{S(dipole)} = (\varepsilon_s - \varepsilon_\infty)E$, we have

$$\frac{dP_{(dipole)}}{dt} = \frac{1}{\tau}\left[(\varepsilon_s - \varepsilon_\infty)E - P_{(dipole)}\right].$$

If we apply an alternating sinusoidal field, we thus have:

$$\tau\frac{dP_{(dipole)}}{dt} + P_{(dipole)} = (\varepsilon_s - \varepsilon_\infty)E_0 e^{j\omega t}.$$

The integration of the equation without its second term gives:

$$P_{(dipole)} = C\exp\left(-\frac{t}{\tau}\right).$$

By applying the varying constant method, we find that by substituting C exp(- t/τ) into the differential equation, such that:

$$C'(t) = \frac{1}{\tau}(\varepsilon_s - \varepsilon_\infty)E_0 e^{\frac{1+j\omega\tau}{\tau}t} \quad \text{from which} \quad C = \frac{(\varepsilon_s - \varepsilon_\infty)E_0}{1+j\omega\tau} e^{\frac{1+j\omega\tau}{\tau}t} + K.$$

The consequence is that $P_{(dipole)} = Ke^{-\frac{t}{\tau}} + \frac{(\varepsilon_s - \varepsilon_\infty)E_0}{1+j\omega\tau} e^{j\omega t}$.

The first term is characteristic of a transitional regime which tends toward zero when $t \to \infty$. The second represents a permanent regime for which:

$$P_{(dipole)} = \frac{(\varepsilon_s - \varepsilon_\infty)E_0}{1+j\omega\tau} e^{j\omega t}.$$

1.2.3. Debye equations and the Argand diagram

1.2.3.1. Debye equations

Identification with P(dipôle) = $(\underline{\varepsilon} - \varepsilon_\infty)E$ where $\underline{\varepsilon} = \varepsilon' - j\varepsilon''$, gives:

$$\underline{\varepsilon} = \varepsilon_\infty + \frac{\varepsilon_s - \varepsilon_\infty}{1+j\omega\tau} = \varepsilon_\infty + \frac{(\varepsilon_s - \varepsilon_\infty)(1-j\omega\tau)}{1+\omega^2\tau^2}.$$

Identification of the real and imaginary parts yields the Debye equations, which are also plotted in Figure 1.3.

Via the use of an equivalent circuit, it is possible to see that the Debye equations can determine relaxation spectra.

$$\varepsilon' = \varepsilon_\infty + \frac{\varepsilon_s - \varepsilon_\infty}{1+\omega^2\tau^2}$$

$$\varepsilon'' = \frac{(\varepsilon_s - \varepsilon_\infty)\omega\tau}{1+\omega^2\tau^2}$$

$$\tan\delta = \frac{\varepsilon''}{\varepsilon'} = \frac{(\varepsilon_s - \varepsilon_\infty)\omega\tau}{\varepsilon_s + \varepsilon_\infty\omega^2\tau^2}$$

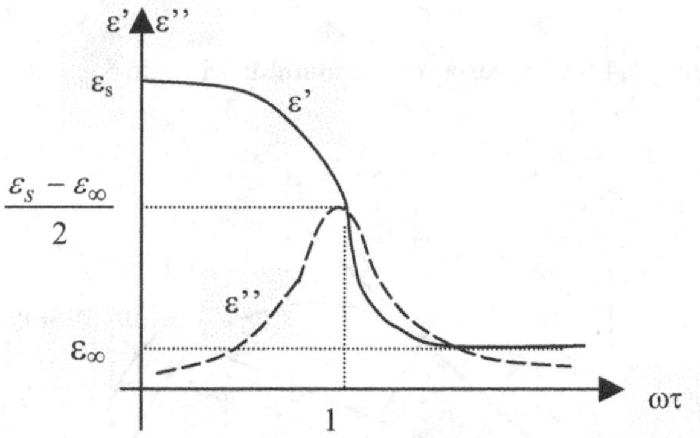

Figure 1.3. *Debye relaxation plots.*

1.2.3.2. The Argand diagram (Cole–Cole representation)

By suitably rearranging Debye's equations, Cole and Cole showed that a circle represents ε'' as a function of ε'.

In effect, these equations make it possible to write: $(\varepsilon' - \varepsilon_\infty)^2 + (\varepsilon'')^2 = \dfrac{(\varepsilon_s - \varepsilon_\infty)^2}{1 + \omega^2\tau^2}$

from which can be pulled: $(\varepsilon')^2 - \varepsilon'(\varepsilon_s + \varepsilon_\infty) + \varepsilon_s\varepsilon_\infty + (\varepsilon'')^2 = 0$.

When $\omega = 0$ and $\omega = \infty$, we know that $\varepsilon'' = 0$ and, respectively, $\varepsilon' = \varepsilon_s$ and $\varepsilon' = \varepsilon_\infty$.

When $\omega\tau = 1$, $\varepsilon'' = \dfrac{\varepsilon_s - \varepsilon_\infty}{2}$ and $\varepsilon' = \dfrac{\varepsilon_s + \varepsilon_\infty}{2}$.

In a system with axes $x = \varepsilon'$ and $y = \varepsilon''$, the coordinate points are $(\varepsilon_s, 0)$, $(\varepsilon_\infty, 0)$, and $(\dfrac{\varepsilon_s + \varepsilon_\infty}{2}, \dfrac{\varepsilon_s - \varepsilon_\infty}{2})$. All of these points are distributed along a curve defined by the preceding equation, which can be rearranged in the form:

$$\left[x - \frac{(\varepsilon_s + \varepsilon_\infty)}{2}\right]^2 + y^2 = \left[\frac{(\varepsilon_s - \varepsilon_\infty)}{2}\right]^2 .$$

This line is in fact a circle and has as its center $(\frac{\varepsilon_s + \varepsilon_\infty}{2}, 0)$ and a radius given by $\frac{\varepsilon_s - \varepsilon_\infty}{2}$. Figure 1.4 thus gives a representation, with $\sin\delta_{max} = \frac{\varepsilon_s - \varepsilon_\infty}{\varepsilon_s + \varepsilon_\infty}$.

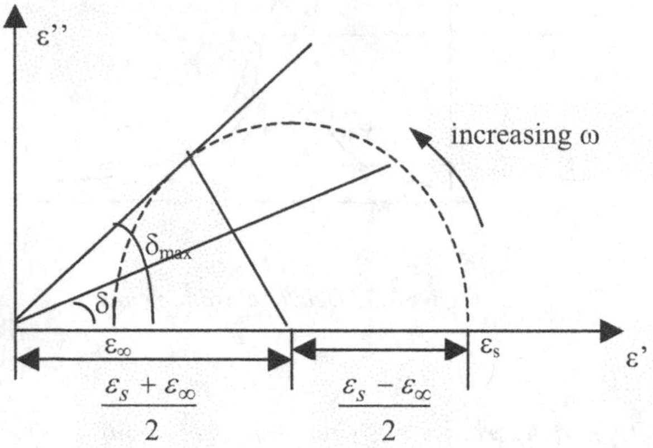

Figure 1.4. *Argand's diagram for $\varepsilon'' = f(\varepsilon')$.*

1.2.4. Practical representations

In practical terms, it is quite rare to actually see a semicircular plot of $\varepsilon'' = f(\varepsilon')$ directly from the Debye equations. This is because in most dielectrics there are actually several types of dipoles each with its own relaxation time and indeed a number of different relaxation mechanisms (related to the different equilibrium positions that the dipole can take up). The upshot is that there is a distribution of relaxation times and a representation of $\varepsilon'' = f(\varepsilon')$ that is not quite semicircular. In reality, there are two types of diagrams that can be observed.

1.2.4.1. Cole and Cole's flattened half-sphere (Figure 1.5 and Problem 1.4.2.)

With $u = (\varepsilon' - \varepsilon_\infty) - i\varepsilon''$

$v = (\varepsilon_s - \varepsilon') + i\varepsilon''$

and with an angle of $\frac{\pi}{2}(1\text{-}h)$ between u and v, we can show that:

$$\underline{\varepsilon} = \varepsilon_\infty + \frac{\varepsilon_s - \varepsilon_\infty}{1 + (j\omega\tau)^{(1-h)}}.$$

The parameter h introduced in the above equation characterizes the relaxation time. If h = 0, then semicircular Debye form is found, whereas if $h \to 1$, then an infinite relaxation time is indicated.

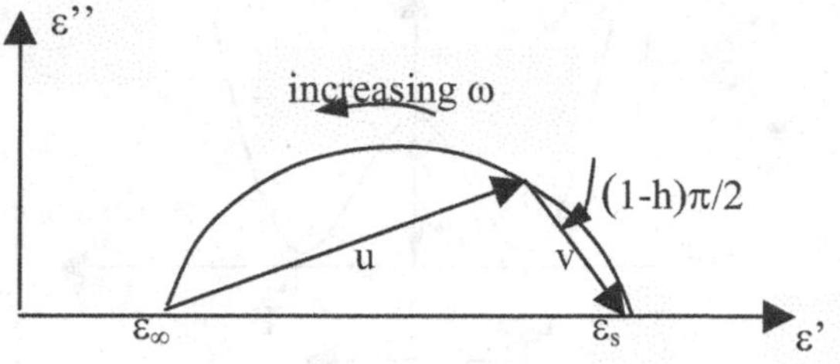

Figure 1.5. *Flattened half-circle after Cole–Cole.*

1.2.4.2. Cole–Davidson's oblique arc (see also Problem 1.4.3)

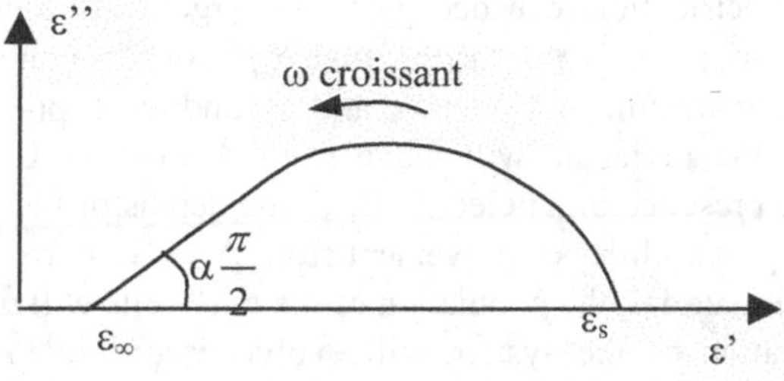

Figure 1.6. *The Cole–Davidson asymmetric arc.*

The plot shown in Figure 1.6 represents a distribution of relaxation times with an "excess" absorption at high frequencies. This behavior can be represented by the following equation:

$$\frac{\varepsilon - \varepsilon_\infty}{\varepsilon_s - \varepsilon_\infty} = \frac{1}{(1 + i\omega\tau)^\alpha}, \text{ where } 0 < \alpha \le 1.$$

It is possible to analytically show that $\frac{d\varepsilon''}{d\varepsilon'} = tg\frac{\pi\alpha}{2}$, where geometric definition of α in Figure 1.6 represents the Cole–Davidson arc.

1.3. The double-well potential model: physical representations

1.3.1. Introduction

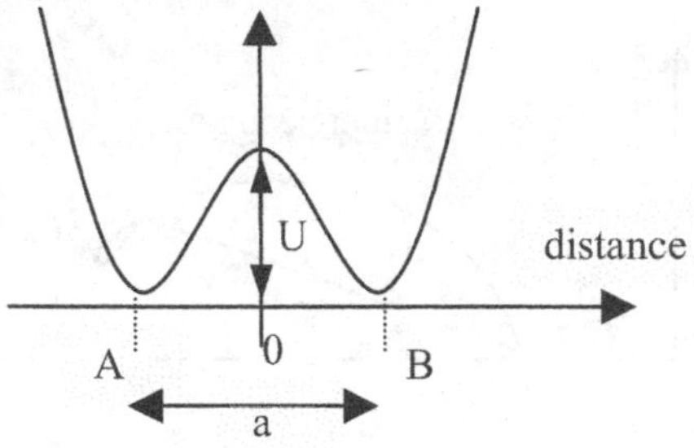

Figure 1.7. *The double-well potential model.*

A system that has two equilibrium positions, each separated by a potential barrier of a certain height (U), can be represented by a potential model based on two wells, as shown in Figure 1.7. With this in mind, it is possible to see how a molecule in the absence of an electric field can occupy two energetically identical sites A and B separated by U. If there is no barrier, then $\nu_0 = \omega_0/2\pi$ represents the oscillation frequency for a molecule between A and B and $\nu_0 \exp(-U/kT)$ represents the probability that the molecule will move from A to B or B to A per unit time. However, in the presence of an electric field, the depths of the potential wells differ by ΔU, and the probabilities of movement from A to B or from B to A also will be unequal. For the material, the population of the two wells will no longer be the same and the polarization of the system will evolve according to an exponential law toward an equilibrium value.

The potential model based on two wells also can be used to represent the displacement of an electron between two localized trapping levels in a forbidden band of an insulator or semiconductor. The A and B correspond to two trap levels between which the electron may slip via the conduction band. The U represents the depth of the trapping levels (the energies of which may be modified by applying an electric field) with respect to the conduction band. From the induced polarization it is possible to determine U for the traps. The following sections will detail this problem further before returning to look at dipolar relaxation. How to obtain an empirical value of U for these potential barriers then will be described.

1.3.2. Polarization associated with the displacement of electrons between two positions separated by a potential barrier

1.3.2.1. Trap levels and phosphorescence

The existence of trap levels in solids was demonstrated while following the optical processes in class II–VI semiconductors. Figure 1.8 shows an example of the electronic levels found in an inorganic phosphor based on ZnS(Cu).

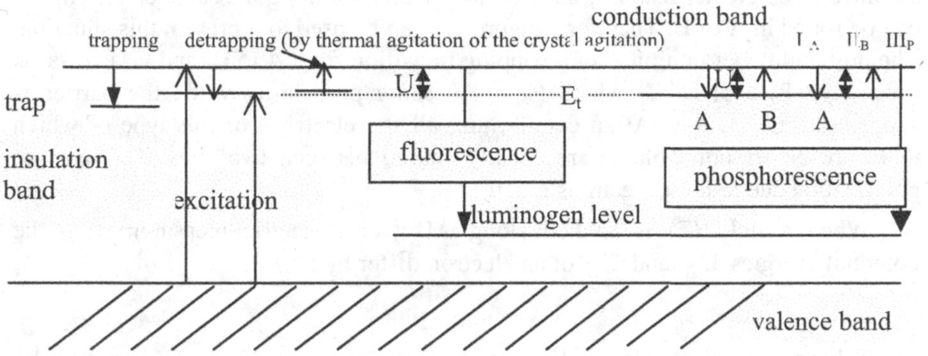

Figure 1.8. *Fluorescence and phosphorescence mechanisms.*

Phosphorescence is a radiation that follows, after a certain delay, a luminous excitation. The latter excites electrons toward the conduction band (CB), which then move through intermediate levels due to traps in the forbidden band. The delay is caused by this stepwise movement. It is the relaxation of the electrons from their excited state to lower states that is delayed with a probability given by the law of Mott, Randall, and Wilkins. This law, which can be written as:

$$p = s \exp(-U/kT),$$

where s is a constant for a group of traps, describes the process whereby there is a delay and a persistence of the emission. The latter gives rise to the phosphorescence persistence. In contrast, fluorescence comes from an emission that is instantaneous with respect to the excitation, i.e., the electrons have not passed through trap levels. $\tau = 1/p$ is equivalent to the average lifetime of an electron in a trap.

For phosphorescence, in physical terms, the electrons trapped at a level A that are subject to a thermal energy (kT) (from phonons) can pass (transition I_A in Figure 1.8) through to another energetic position B by crossing (transition II_B) a potential barrier of a given height (U). This barrier is at a trapping level denoted by E_t with respect to the bottom of the conduction band. The group of transitions I_A, II_B, III_P describe the mechanism of phosphorescence. In addition, different levels of traps can exist (U, U_2, etc.).

1.3.2.2. Transposition to a dielectric possessing trap levels and the electric field effect on transitions between them

In the absence of an electric field, for an electron (of charge –q) the two equilibrium positions A and B separated by a given distance (a) and a potential barrier (U) assumed to be greater than kT, there is statistically speaking the same chance that it can be found in A or B. Therefore, Figure 1.7 can be used to represent this situation. The probability per unit time for a hopping transition from A to B (and visa versa) is given by $P_0 \exp(-U/kT)$ where P_0 is the same probability when the barrier is suppressed, i.e., U = 0. At an equilibrium, all the electrons of this type of which there are N per unit volume are equally shared between two sites. The average polarization due to these electrons is zero.

When a field (E) is applied along AB (for dielectric measurements), the potential energies U_A and U_B of an electron differ by :

$$U_A - U_B = qaE\,.$$

Taking the orientation of E into account, we have $V_A < V_B$, so that by consequence (where U = - q V and $|V| = |-\int E d\ell| = |Ea|$), U_A is greater than U_B by

$$\Delta U = qaE,$$

if the origin of the potentials is taken with respect to B, as in Figure 1.9. This destroys the symmetry of the system and implies that there is a probability of presence of an electron at B greater than that at A. This assumes that we use the Boltzmann function, which is proportional to exp(-U/kT) to describe the distribution function.

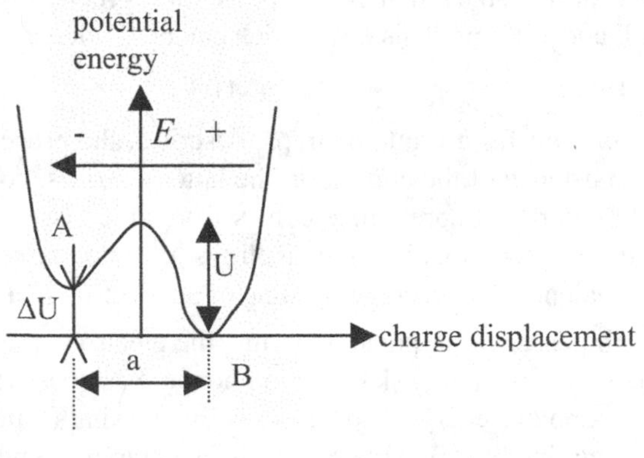

Figure 1.9. *Effect of applying an electric field to a potential barrier.*

(i) Calculation of transition probabilities

Initially, it is assumed that the electrons oscillate (through thermal energy) about their equilibrium positions at a frequency given by $\omega_0/2\pi$. In the absence of a barrier, the probability (P_0) that an electron will carry out a transition from A to B or from B to A in one second is thus $P_0 = \omega_0/2\pi$.

However, with a barrier of a given height (U) and in the presence of an electric field as in Figure 1.9, the probability for transitions per second from A to B (P_{AB}) or B to A (P_{BA}) must take the apparent height of the barrier into account, that is (U – qaE) on the A side and U on the B side. So, now:

$$P_{AB} = \frac{\omega_0}{2\pi}\exp(-\frac{U - qaE}{kT}) \text{ and } P_{BA} = \frac{\omega_0}{2\pi}\exp(-\frac{U}{kT}) \qquad (1').$$

If we assume that qaE << kT, as is generally the case for dielectric measurements carried out for weak fields of around 1 V cm^{-1}, we can state that:

$$P_{AB} = \frac{\omega_0}{2\pi}e^{-\frac{U}{kT}}(1 + \frac{qaE}{kT}) = P_{BA}(1 + \frac{qaE}{kT}). \qquad (1)$$

(ii) Calculation for the number of particles in the A and B states at a given instant

If at a given time (t) there are N_A electrons in A and N_B in B, then there are N_AP_{AB} electrons making the transition A→B and N_BP_{BA} undergoing the transition B→A. From this, we can state that

$$\frac{dN_A}{dt} = -N_AP_{AB} + N_BP_{BA} \quad \text{and} \quad \frac{dN_B}{dt} = N_AP_{AB} - N_BP_{BA}. \qquad (2')$$

With the total number of electrons, given by $N = N_A + N_B$, being constant and by taking the difference between these two equations [by their addition and the removal of $(N_AP_{BA} - N_BP_{AB})$ to make apparent $(P_{BA} + P_{AB})$ and $(P_{AB} - P_{BA})$], we obtain:

$$\frac{d(N_B - N_A)}{dt} = -(P_{BA} + P_{AB})(N_B - N_A) + (P_{AB} - P_{BA})N. \qquad (2)$$

Using Eq. (1) we have (with qaE << kT):

$$P_{AB} + P_{BA} = P_{BA}(1 + \frac{qaE}{kT}) + P_{BA} \approx 2P_{BA}, \qquad (3)$$

and

$$P_{AB} - P_{BA} \approx \frac{qaE}{kT}P_{BA}. \qquad (4)$$

By substituting Eqs. (3) and (4) into Eq. (2), we obtain:

$$\frac{d(N_B - N_A)}{dt} = -2P_{BA}(N_B - N_A) + 2P_{BA}\frac{qaE}{2kT}N.$$

The integration of the differential equation without the second term yields: $N_B - N_A = C\exp(-2P_{BA}t)$. By transferring this into the differential equation and by varying the constant denoted C, we obtain:

$$C = \frac{qaE}{2kT}N\exp(2P_{BA}t) + K.$$

By using the limiting condition, as in when t = 0, $N_A = N_B = N/2$, it is possible to deduce that $K = -\frac{qaE}{2kT}N$, and from which:

$$N_B - N_A = \frac{N}{2}\frac{qaE}{kT}(1 - \exp[-2P_{BA}t]). \qquad (5)$$

1.3.2.3. Equation for the polarization at an instant t following electron movement

1.3.2.3.1. Displacement of charges of a given concentration

The movement of a charge (q_i) by a distance (δ_i) is effectively the same as superimposing a dipole given by $\delta\mu_i = q_i\delta_i$. This is schematized in Figure 1.10.

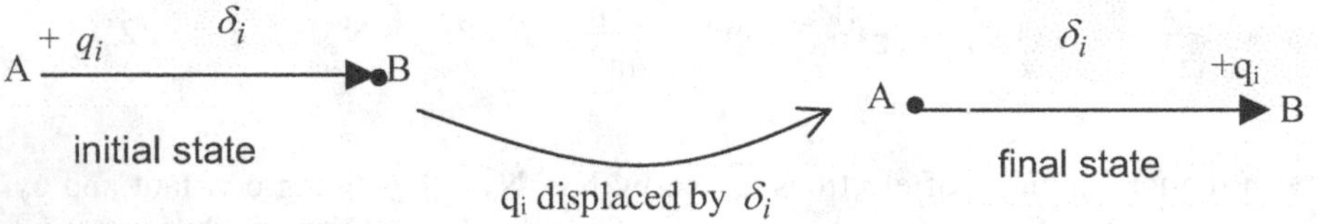

Figure 1.10. *Movement of q_i by δ_i.*

In order to obtain the final state with respect to the initial state, it suffices to superimpose on the latter the dipole $\delta\mu_i = q_i\delta_i$, as demonstrated in Figure 1.11.

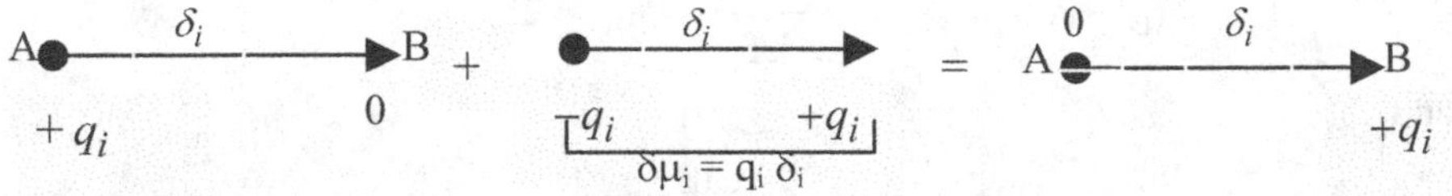

Figure 1.11. *Displacement of q_i by δ_i is the same as applying $\delta\mu_i = q_i\,\delta_i$.*

The polarization (dipole moment per unit volume) associated with the movement (δ_i) of q_i charges is the same as applying a dipole moment per unit volume equal to $P_i = n_i q_i \delta_i$, where n_i is the number of q_i charges per unit volume. If there are various charges, given by i, then the total polarization is thus:

$$P = \sum_i P_i = \sum_i n_i q_i \delta_i .$$

1.3.2.3.2. Polarization due to the displacement of N electrons per unit volume shared over a double-well potential of a given initial depth (U)

Hypothetically, it is assumed that the two wells are separated by a certain distance (a) and that they are equivalent, so that $N_A = N_B = N/2$.

Under these circumstances, the influence of a field (E) moves the electronic charges (-q) in A by +a and those in B by –a.

$$P = \sum_i n_i q_i \delta_i = N_A(-q)(+a) + N_B(-q)(-a)$$

$$= N_B qa - N_A qa = qa(N_B - N_A). \qquad (6)$$

By substituting Eq. (5) for $(N_B - N_A)$ into Eq. (6), we obtain:

$$P = \frac{N}{2}\frac{q^2a^2E}{kT}(1 - \exp[-2P_{BA}t]). \qquad (7)$$

1.3.3. Dipole rotation due to an electric field

The result of the preceding calculation also can be applied to a rotating dipole (using the dipolar polarization given in Section 1.2). The dipoles, which have moment given by $\mu_i = q_i\,\delta_i$, initially can be orientated at either of the energetically equivalent positions (E = 0) A or B. These sites are separated by the potential barrier U, as described in Figure 1.12.

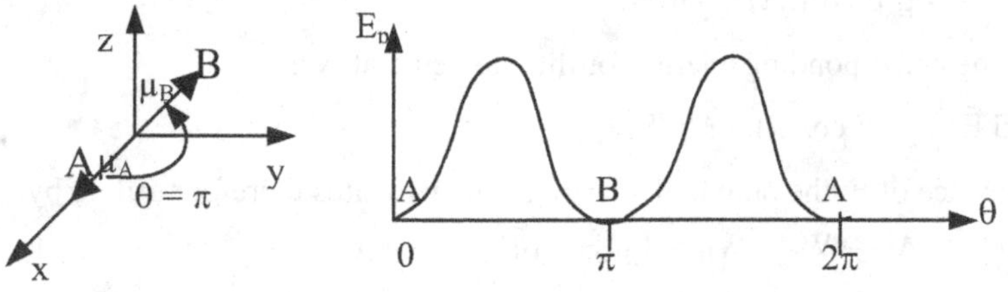

Figure 1.12. *Potential energy (Ep) as a function of dipole orientation (θ).*

This characteristic can be demonstrated in two different ways.

1.3.3.1. Using the equivalence of dipole rotation and charge displacement

As shown in Figure 1.13, turning a dipole with a moment given by $\mu = (q\,a)$ through 180° is the same as applying a dipole with a moment $p = 2\mu = 2qa$ to the system. This effectively results in the movement of a charge of 2q (see also the preceding Section 1.3.2.3.1). Equation (7) for polarization written for the movement of a charge of value 2q is thus:

$$P = \frac{N}{2}\frac{(2q)^2a^2E}{kT}(1 - \exp[-2P_{BA}t]).$$

With p = 2qa, this expression can be rewritten:

$$P = \frac{N}{2}\frac{p^2E}{kT}(1 - \exp[-2P_{BA}t]) = N\frac{2\mu^2E}{kT}(1 - \exp[-2P_{BA}t]). \qquad (7')$$

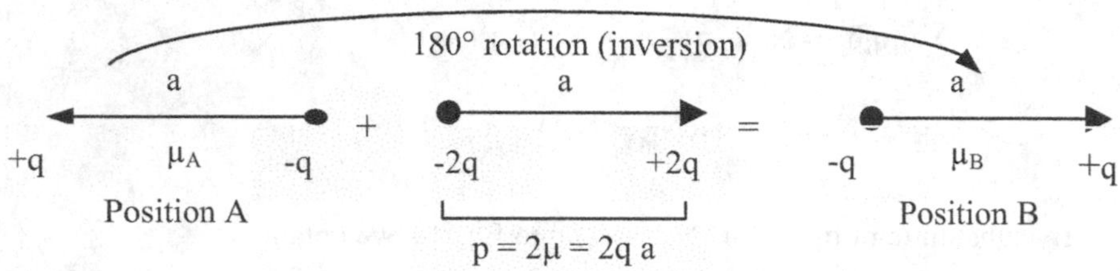

Figure 1.13. *Dipolar charge displacement following a 180° rotation.*

1.3.3.2. By direct calculation

To get some bearings we can suppose that the two positions A and B correspond to two orientations of a molecule, which differ by $\theta = \pi$ (as in Figure 1.12), and that the electric field (E) then applied is directed along the dipole (μ) associated with the molecule at B.

For the B state, the energy of the system is given by

$$W_B = -\,\vec{\mu}.\vec{E} = -\,\mu\,E\cos(2\pi) = -\,\mu\,E,$$

and therefore the corresponding energy for the molecule at A is :

$$W_A = -\,\vec{\mu}.\vec{E} = -\,\mu\,E\cos(\pi) = +\,\mu\,E.$$

In the presence of E, the potential energy of the two sites therefore differs by:

$$\Delta U = W_A - W_B = 2\,\mu\,E = pE\,,\ \text{where}\ p = 2\mu.$$

We therefore can go on to develop the calculation much as was the case in Section 1.3.2.2, with the condition that the difference in energy (qaE) between A and B is replaced by the new energy difference $\Delta U = 2\,\mu\,E = pE$ (Figure 1.9).

By denoting the number of molecules per unit volume as N, this also being the number of dipoles per unit volume, at a time t = 0 (from when E is applied) we have $N_A = N_B = N/2$, and in place of Eq. (5) we can state that

$$N_B - N_A = \frac{N}{2}\frac{pE}{kT}(1 - \exp[-2P_{BA}t]) . \qquad (5')$$

The polarization, or rather the resultant dipole moment per unit volume, is thus given by $P = (N_B - N_A)p$, and once again Eq. (7) yields Eq. (7') (see preceding paragraph).

$$P = \frac{N}{2}\frac{p^2E}{kT}(1 - \exp[-2P_{BA}t]) = N\frac{2\mu^2E}{kT}(1 - \exp[-2P_{BA}t]) \qquad (7')$$

1.3.4. Practical determination of the depth of potential wells

1.3.4.1. Fundamental formulae

Following on from Eqs. (7) and (7'), if $t \to \infty$, $P \to P_S$ (where P_S is the static polarization). P_S is therefore given by, respectively,

$$P_S = \frac{N}{2}\frac{q^2a^2E}{kT} \qquad (8)$$

or

$$P_S = 2N\frac{\mu^2E}{kT}. \qquad (8')$$

In the following Section 1.4.1, a problem directly demonstrates the use of Eq. (8) for a stationary regime (which does not go to the limits). The exercise also can be used to reproduce Eq. (8') in a close variation (the rotation of the dipoles can be substituted by their displacement as these acts are physically equivalent as detailed in Section 1.3.3.1).

The term $\exp(-2P_{BA}t)$ in Eqs. (7) and (7') represents a transition regime due to the delay (dephasing) between electrons trapped in potential wells, or dipoles, in following the applied field. As this term varies exponentially with time it is characteristic of the relaxation of the system. It is also a function of the macroscopic relaxation (Y(t)) which is defined in Chapter 3.

The double-well potential model thus represents a relaxation, much as that found through Debye's theory, and gives rise to the same types of dielectric equations (Debye equations). The term characteristic of the transition regime appears in the equation for $P_{(dipole)}$ detailed in Section 1.2.2. The form of the macroscopic relaxation function is given by $Y(t) = \exp(- t/\tau)$ and is developed in Chapter 3.

The equality of the two terms for relaxation described in the double-well model and in Debye's theory makes it possible to state that:

$$\exp(-2P_{BA}t) = \exp(-\frac{t}{\tau}), \quad \text{so that} \quad \tau = \frac{1}{2P_{BA}}.$$

With P_{BA}, given by Eq. (1'), we can deduce that:

$$\boxed{\tau = \frac{\pi}{\omega_0} e^{\frac{U}{kT}} = \tau_0 e^{\frac{U}{kT}}} . \qquad (9)$$

It is worth noting also that the term $\tau_0 = \pi / \omega_0$ represents the time required for one oscillation, as in A → B → A, for example. In addition, as U >> kT, then $\tau \gg \tau_0$.

The earlier representations of the Debye plots showed that ε'' goes through a maximum at a frequency (υ_c) which is such that $\omega_c \tau = 1$. Taking Eq. (9) for τ into account, we thus have: $1 = 2\pi\upsilon_c\tau_0 e^{U/kT}$, so that $\upsilon_c = \frac{1}{2\pi\tau_0} e^{-\frac{U}{kT}}$.

In practice, this equation is used in the form given as:

$$\boxed{\text{Log}\upsilon_c = -\frac{U}{kT} - \text{Log}2\pi\tau_0} . \qquad (10)$$

1.3.4.2. Experimental determination

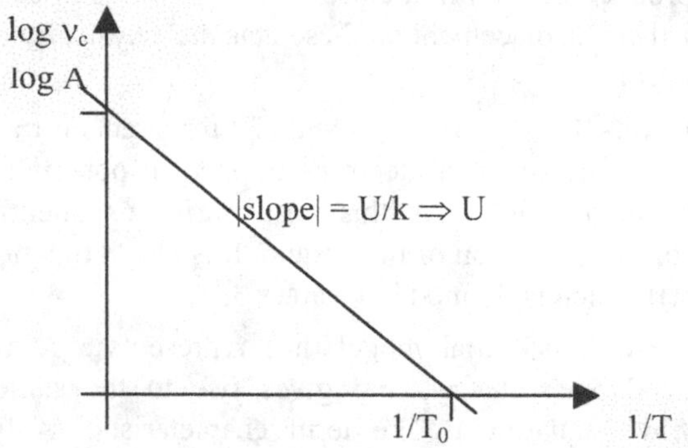

Figure 1.14. *Representation of ln ν_c =f(1/T).*

If we plot the log of the critical frequency as an inverse function of temperature, theoretically we should obtain a straight line. The slope of the line should permit a calculation of the height of the potential barrier and the ordinate at the origin the value of τ_0, as indicated in Figure 1.14. Hence, in reality the two physical magnitudes τ_0 and, most importantly, U can be determined quite facilely, as detailed below.

The law observed is this of the form $\upsilon_c = A\exp(-U/kT)$, so in turn $\ln \upsilon_c = -\frac{U}{kT} + \ln A$. When 1/T = 0, we have log ν_c = ln A, and with the ordinate at the origin giving ln A, we find that:

$$\tau_0 = 1/2\pi A.$$

Additionally, when T = T_0, then ln ν_c=0, so that $\frac{1}{T_0} = \frac{k}{U}\ln A$, which makes it possible to calculate:

$$U = kT_0 \mathrm{Ln}\, A.$$

Here, U is in eV, and k = 8.64 10^{-5} eV K^{-1} molecule^{-1}

1.4. Problems

1.4.1. Problem 1. The double-well potential at a state of equilibrium

This question concerns an electron of a given charge (-q) which, in a solid dielectric, has two equivalent equilibrium positions denoted A and B which are separated by a specified distance (a). If the probability of these transitions is of the form P_0 exp(- U/kT) where P_0 is a constant equal to the transition probability if the barrier state (U) were removed, in order to pass from A to B, and visa versa, the electron must overcome the potential barrier of height U >> kT. At equilibrium, the total number of electrons per unit volume (N) are equally shared between the two sites A and B and the average polarization is zero.

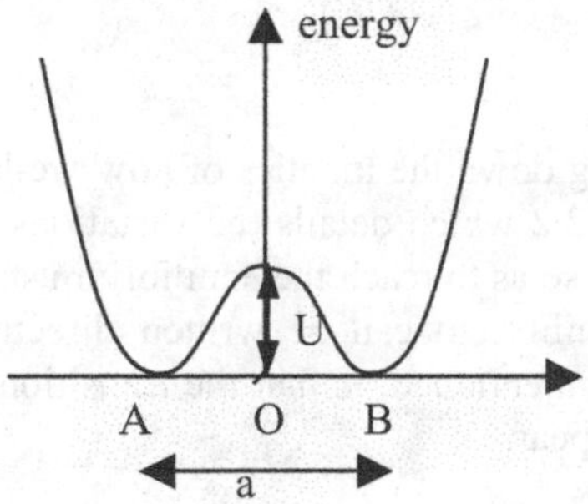

1. At a time t = 0, a field ($\vec{E}$) along AB is applied directed from B toward A.

a. Draw the new energy scheme for $\vec{E} \neq 0$. Indicate the heights U_A and U_B seen by electrons at A and B.

b. From the answers to the above, determine the new probabilities P_{AB} and P_{BA} of electrons passing from A to B and B to A, respectively.

2. N_A and N_B denote the populations of electrons per unit volume at A and B.

a. Write the condition for equilibrium for the system subject to $\vec{E}$.

b. Determine N_A and N_B at the equilibrium in the presence of $\vec{E}$; the results will be expressed as a function of N.

3. From which, determine the expression for the polarization ($\vec{P}$) in the potential well model studied above. Give an equation for the dielectric susceptibility under a weak field.

Answers to problem 1

1.

a. As indicated in Figure 1.19, the electron observes a potential barrier denoted as U_A or U_B depending on if it is at A or B, respectively. These are such that:

$$\begin{cases} U_A = U - qEa \\ U_B = U. \end{cases}$$

b. The probabilities P_{AB} and P_{BA} are given by the pair of Eqs. (1') detailed in Section 1.3.2.2, i.e.,

$$P_{AB} = \frac{\omega_0}{2\pi}\exp\left(-\frac{U - qaE}{kT}\right) \text{ and } P_{BA} = \frac{\omega_0}{2\pi}\exp\left(-\frac{U}{kT}\right).$$

2.

a. In place of writing down the kinetics of how wells A and B are filled from Eqs. (2') (see Section 1.3.2.2 which details the variations in populations with time) and then take t to infinity so as to reach the equilibrium state (Eq. (7) of Section 1.3.2.3), the equation for equilibrium can be written directly. The inconvenience of this method, it should be mentioned, is that the transition state, bound to the relaxation function, does not appear.

In effect, the concentrations at A and at B are constant; in other words, there are as many electrons moving from A to B as there are moving from B to A. If N_A and N_B represent the concentrations at A and at B, then the equilibrium is when

$$N_A P_{AB} = N_B P_{BA}.$$

b. Therefore, we have to resolve the system of two unknowns N_A and N_B with the help of two equations:

$$\begin{cases} N = N_A + N_B & (1) \text{ (Conservation of charges in the presence of } E\text{);} \\ N_A P_{AB} = N_B P_{BA}. & (2) \end{cases}$$

By substituting into Eq. (2) the values of P_{AB} and P_{BA} recalled above, and by multiplying the two parts of Eq. (2) by $\exp\left(\frac{U - qE\frac{a}{2}}{kT}\right)$, we obtain:

$$\begin{cases} N_A = N - N_B & (1') \\ N_A \exp\left(+\frac{qE\frac{a}{2}}{kT}\right) = N_B \exp\left(-\frac{qE\frac{a}{2}}{kT}\right). & (2') \end{cases}$$

By making $u = \frac{qE\frac{a}{2}}{kT}$, we then need to resolve

$$\begin{cases} N_A = N - N_B & (3) \text{ and} \\ N_A \exp(u) = N_B \exp(-u). & (4) \end{cases}$$

By carrying Eq. (3) into Eq. (4), as in:

$(N - N_B)\exp(u) = N_B \exp(-u)$,

then the multiplication of the 2 parts by exp(-u) gives

$(N - N_B) = N_B \exp(-2u)$, and hence $N_B[1 + \exp(-2u)] = N$.

Finally, $N_B = \dfrac{N}{1+\exp(-2u)} = \dfrac{N}{1+\exp(-\frac{qEa}{kT})}$.

When N_A is such that $N_A = N - N_B$, we find that

$$N_A = N - \frac{N}{1+\exp(-2u)} = N\left[1 - \frac{1}{1+\exp(-2u)}\right] = N\left[\frac{1+\exp(-2u)-1}{1+\exp(-2u)}\right].$$

Multiplying above and below by $\exp(2u)$ yields

$$N_A = N\left[\frac{1}{1+\exp(2u)}\right] = \frac{N}{1+\exp(\frac{qEa}{kT})}.$$

3.

The polarization due to the movement of electrons between A and B is given by the equation $P = qa(N_B - N_A)$. Taking the values of N_B and N_A found above on board, we have:

$$P_S = qaN(\frac{1}{1+\exp(-2u)} - \frac{1}{1+\exp(2u)}).$$

Multiplying the first bracketed term above and below by $\exp(2u)$ yields

$$P_S = qaN\left[\frac{\exp(2u)}{1+\exp(2u)} - \frac{1}{1+\exp(2u)}\right]$$

$$= qaN\frac{\exp(2u)-1}{\exp(2u)+1} = qaN\frac{\exp u - \exp(-u)}{\exp u + \exp(-u)} = qaN\frac{\sinh u}{\cosh u} = qaN\tanh u,$$

and thus

$$P_S = qaN\tanh(\frac{qaE}{2kT}).$$

Comment. Under conditions of a weak field, $\tanh(\frac{qaE}{2kT}) \cong \frac{qaE}{2kT}$, and:

$$\left.\begin{aligned} P_S &= \varepsilon_0(\varepsilon_S - 1)E = \varepsilon_0\chi_S E \\ &= qaN\frac{qaE}{2kT} \quad \text{(see Eq. (8))} \end{aligned}\right\} \Longrightarrow \left\{\begin{aligned} \chi_S &= \frac{q^2a^2N}{2\varepsilon_0 kT} \\ \varepsilon_S - 1 &= \frac{q^2a^2N}{2kT}. \end{aligned}\right.$$

This directly gives Eq. (8), found in Section 1.3.4, for the static polarization (static regime) that was brought about by taking the dynamic regime to its limit at $t \to \infty$.

1.4.2. Problem 2. The Cole–Cole diagram

Throughout this problem, the usual dielectric notation is used, i.e. $\underline{\varepsilon} = \varepsilon' - j\varepsilon''$.

1. Recall the Debye equations (for the phenomena of dielectric absorption for a single relaxation time). Show how they can be condensed to the form, $\underline{\varepsilon} = \varepsilon_\infty + \dfrac{\varepsilon_s - \varepsilon_\infty}{1 + j\omega\tau}$.

2. The Argand diagram (representation in a complex plane)

So that the image point (M) of $\underline{\varepsilon}$, defined by $\underline{\varepsilon} = \varepsilon' - j\varepsilon''$, is placed in the first quadrant the conventions of notation are detailed in the figure below.

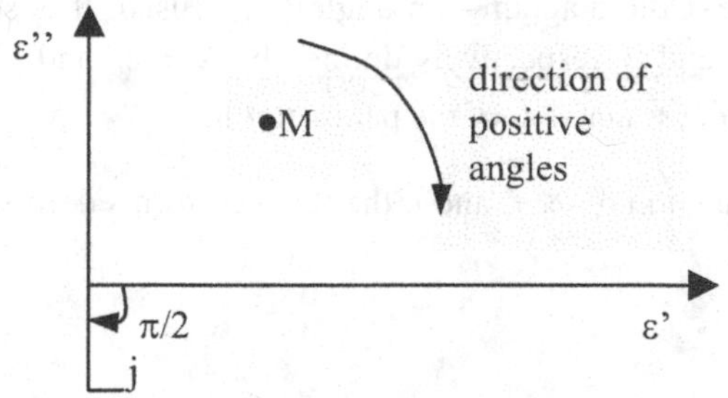

Making $v = (\varepsilon_s - \varepsilon') + j\varepsilon''$ and $u = (\varepsilon' - \varepsilon_\infty) - j\varepsilon''$, this question is set within the confines of Debye's theory.

a. Algebraically calculate the ratio of $\dfrac{v}{u}$ and specify the modulus and the argument.

b. Indicate the image of v and u on the diagram.

c. From this, determine the geometrical form in which all M points are placed.

3. Empirical observations of dielectric behavior indicate that the representation

$\varepsilon'' = f(\varepsilon')$ is rarely a perfect circle, as is otherwise predicted by Debye's theory. What is more normally found is a "flattened" half-circle, shown in the figure below, for which Cole and Cole proposed the following analytical expression:

$$\varepsilon' - j\varepsilon'' = \varepsilon_\infty + \frac{\varepsilon_S - \varepsilon_\infty}{1 + (j\omega\tau)^{(1-h)}} .$$

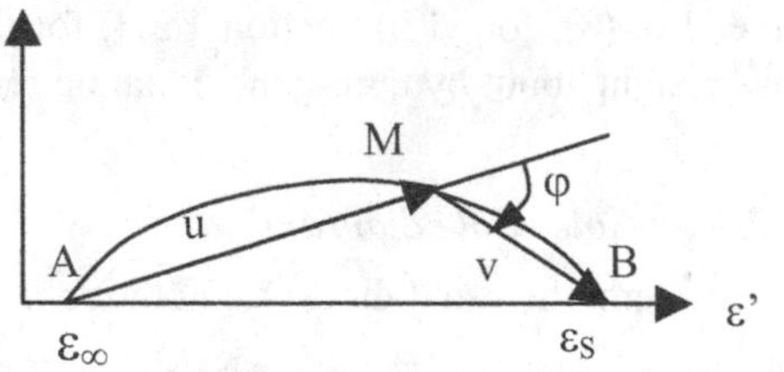

The same algebraic definitions for u and v are used as in the first question. Specify the angle (φ) between u and v that is expressed as a function of the h parameter introduced in the Cole–Cole equation. Determine the physically acceptable variation in h.

4. To designate the position of the center (G) of the circle—only the "flattened" arc appears in the Cole–Cole diagram—an angle (α) is used. It is such that $\alpha = \widehat{GAB}$ where the points A and B, respectively defined by $\varepsilon' = \varepsilon_\infty$ and $\varepsilon' = \varepsilon_S$, are both on the $O\varepsilon'$ axis. Give α as a function of the parameter h.

5. Calculate as a function of ω, τ, and h the reduced forms defined by $\dfrac{\varepsilon'}{\varepsilon_S - \varepsilon_\infty}$ and $\dfrac{\varepsilon''}{\varepsilon_S - \varepsilon_\infty}$.

Answers to problem 2

1. The Debye equations are $\varepsilon' = \varepsilon_\infty + \dfrac{\varepsilon_S - \varepsilon_\infty}{1 + \omega^2\tau^2}$ and $\varepsilon'' = \dfrac{(\varepsilon_S - \varepsilon_\infty)\omega\tau}{1 + \omega^2\tau^2}$, and make it possible to state that $\underline{\varepsilon} = \varepsilon' - j\varepsilon'' = \varepsilon_\infty + \dfrac{\varepsilon_S - \varepsilon_\infty}{1 + \omega^2\tau^2}(1 - j\omega\tau) = \varepsilon_\infty + \dfrac{\varepsilon_S - \varepsilon_\infty}{1 + j\omega\,\tau}$.

2. a. In a complex plane therefore:

$$1 + j\omega\tau = \frac{\varepsilon_S - \varepsilon_\infty}{\varepsilon' - \varepsilon_\infty - j\varepsilon''},$$

and hence $j\omega\tau = \frac{\varepsilon_s - \varepsilon_\infty - \varepsilon' + \varepsilon_\infty + j\varepsilon''}{\varepsilon' - \varepsilon_\infty - j\varepsilon''} = \frac{(\varepsilon_s - \varepsilon') + j\varepsilon''}{\varepsilon' - \varepsilon_\infty - j\varepsilon''}$.

By making $v = (\varepsilon_s - \varepsilon') + j\varepsilon''$ and $u = (\varepsilon' - \varepsilon_\infty) - j\varepsilon''$, we can write that:

$$\frac{v}{u} = j\omega\tau .$$

And as $j = \cos\left(\frac{\pi}{2}\right) + j\sin\left(\frac{\pi}{2}\right) = \exp\left(j\frac{\pi}{2}\right)$ (j is obtained from 1 with a $\pi/2$ rotation), we also have:

$$\frac{v}{u} = \omega\tau e^{j\frac{\pi}{2}} .$$

b. So for u and v we have the representation shown below in the figure.

c. As j is equivalent to a rotation of $\frac{\pi}{2}$, u and v are orthogonal. The results is that the point M, the image of ε, belongs to the circle that cuts the abscissas at $\varepsilon' = \varepsilon_\infty$ (point A) and $\varepsilon' = \varepsilon_s$ (B).

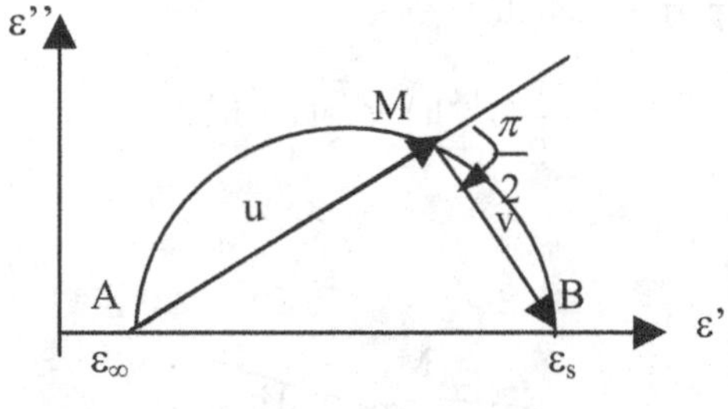

3. For the newer form, $\varepsilon' - j\varepsilon'' = \varepsilon_\infty + \frac{\varepsilon_s - \varepsilon_\infty}{1 + (j\omega\tau)^{(1-h)}}$, it can be determined that

$$1 + (j\omega\tau)^{(1-h)} = \frac{\varepsilon_s - \varepsilon_\infty}{\varepsilon' - \varepsilon_\infty - j\varepsilon''},$$

so also

$$\left(j\omega\tau\right)^{(1-h)} = \frac{\varepsilon_s - \varepsilon' + j\varepsilon''}{\varepsilon' - \varepsilon_\infty - j\varepsilon''} = \frac{v}{u}.$$

The result is that

$$\frac{v}{u} = \left(\omega\tau\right)^{(1-h)} j^{(1-h)} = \left(\omega\tau\right)^{(1-h)} \exp\left(j\frac{\pi}{2}(1-h)\right).$$

As a consequence, u and v are in effect no longer orthogonal but at an angle of $\frac{\pi}{2}(1-h)$ where h varies as a function of the "flattened" character of the circle.

In comparison with Debye's diagram, where u and v are separated by an angle of $\frac{\pi}{2}$, h now takes on a limited value, as in $h = 0$. As the semicircle flattens out, so that the limit tends toward the ε' axis, h thus tends toward 1. The range of variation in h is therefore $0 \leq h < 1$.

4. The angle $\alpha = \widehat{GAB}$ is introduced in order to define the center (G) of a circle that is no longer on the ε' axis. The angle at G given by $\beta = \widehat{AGB}$ is therefore such that its complementary angle $(2\pi - \beta)$ intercepts the same arc as the angle $\gamma = \widehat{AMB}$, which for its part is such that:

$$\gamma = \pi - \frac{\pi}{2}(1-h) = \frac{\pi}{2}(1+h).$$

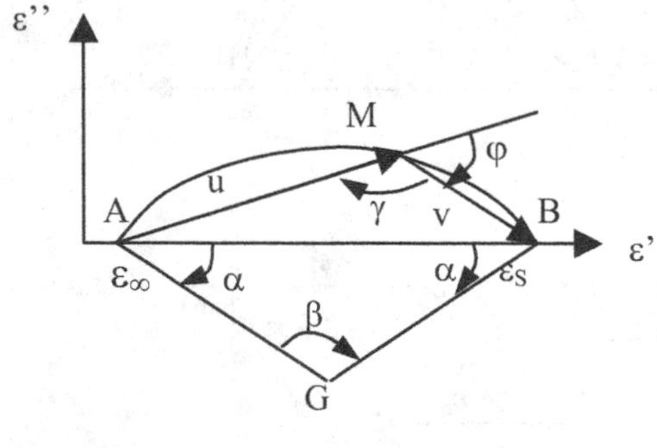

We thus have $\gamma = \frac{2\pi - \beta}{2}$, from which:

$$\beta = 2\pi - 2\gamma = \pi(1-h).$$

As $2\alpha + \beta = \pi$, it can be deduced that:

$$\alpha = \frac{\pi - \beta}{2} = \frac{\pi h}{2}.$$

5. Here we are looking to determine the reduced expressions for $\varepsilon' - j\varepsilon'' = \varepsilon_\infty + \dfrac{\varepsilon_s - \varepsilon_\infty}{1 + (j\omega\tau)^{(1-h)}}$.

As

$$j^{(1-h)} = e^{j\frac{\pi}{2}(1-h)} = e^{j\frac{\pi}{2}} e^{-j\frac{\pi h}{2}} = j\left[\cos\frac{\pi h}{2} - j\sin\frac{\pi h}{2}\right] = \sin\frac{\pi h}{2} + j\cos\frac{\pi h}{2},$$

so that by making

$$C = 1 + (\omega\tau)^{(1-h)} \sin\frac{\pi h}{2} \text{ and } D = (\omega\tau)^{(1-h)} \cos\frac{\pi h}{2},$$

we have

$$1 + (j\omega\tau)^{(1-h)} = 1 + (\omega\tau)^{(1-h)}\left[\sin\frac{\pi h}{2} + j\cos\frac{\pi h}{2}\right] = C + jD.$$

The initial equation therefore gives rise to

$$\varepsilon' - \varepsilon_\infty - j\varepsilon'' = +\frac{\varepsilon_s - \varepsilon_\infty}{C + jD}$$

$$\Rightarrow \qquad \frac{\varepsilon' - \varepsilon_\infty}{\varepsilon_s - \varepsilon_\infty} = R\left(\frac{1}{C + jD}\right) \text{ and } \frac{\varepsilon''}{\varepsilon_s - \varepsilon_\infty} = \mathrm{Im}\left(-\frac{1}{C + jD}\right).$$

With $\dfrac{1}{C + jD} = \dfrac{C - jD}{C^2 + D^2}$, these equations now become:

$$\frac{\varepsilon' - \varepsilon_\infty}{\varepsilon_s - \varepsilon_\infty} = \frac{C}{C^2 + D^2} \text{ and } \frac{\varepsilon''}{\varepsilon_s - \varepsilon_\infty} = \frac{D}{C^2 + D^2}.$$

Now having

$$C^2 + D^2 = \left[1 + (\omega\tau)^{(1-h)} \sin\frac{\pi h}{2}\right]^2 + \left[(\omega\tau)^{2(1-h)} \cos^2\frac{\pi h}{2}\right]$$

$$= 1 + (\omega\tau)^{2(1-h)} + 2(\omega\tau)^{(1-h)} \sin\frac{\pi h}{2},$$

we finally obtain:

$$\frac{\varepsilon' - \varepsilon_\infty}{\varepsilon_s - \varepsilon_\infty} = \frac{1 + (\omega\tau)^{(1-h)} \sin\frac{\pi h}{2}}{1 + (\omega\tau)^{2(1-h)} + 2(\omega\tau)^{(1-h)} \sin\frac{\pi h}{2}}$$

$$\frac{\varepsilon''}{\varepsilon_s - \varepsilon_\infty} = \frac{(\omega\tau)^{(1-h)} \cos\frac{\pi h}{2}}{1 + (\omega\tau)^{2(1-h)} + 2(\omega\tau)^{(1-h)} \sin\frac{\pi h}{2}}.$$

1.4.3. Problem 3. The Cole–Davidson diagram

For a dielectric that exhibits an asymmetric behavior with respect to frequency, that is to say that it shows "excessive" absorption at higher frequencies, the Argand diagram displays an "oblique arc" at high frequencies (see figure below).

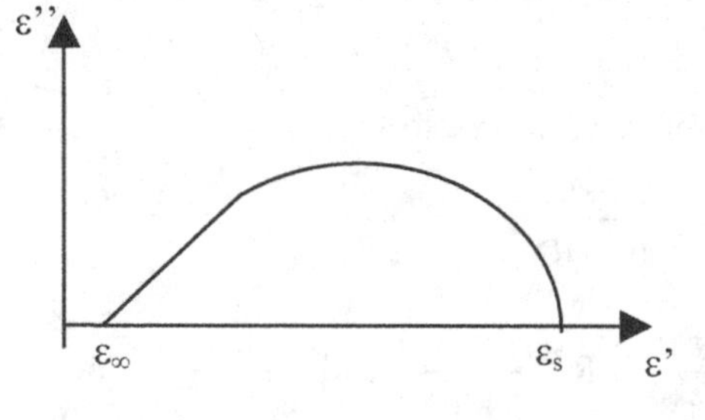

For this particular scenario, Cole and Davidson proposed an analytical model of the curves behavior, as in $\frac{\varepsilon - \varepsilon_\infty}{\varepsilon_s - \varepsilon_\infty} = \frac{1}{(1 + j\omega\tau)^\alpha}$, where $0 < \alpha \leq 1$.

In the following problem, we will look at the significance of the parameter α.

1. By making $\tan\Phi = \omega\tau$, show that the analytical equation introduced above can be written in the form: $\dfrac{\underline{\varepsilon} - \varepsilon_\infty}{\varepsilon_s - \varepsilon_\infty} = \exp(-j\alpha\Phi)\cos^\alpha\Phi$.

From this, determine the two reduced expression, as in $\dfrac{\varepsilon' - \varepsilon_\infty}{\varepsilon_s - \varepsilon_\infty}$ and $\dfrac{\varepsilon''}{\varepsilon_s - \varepsilon_\infty}$.

2. With the variable being Φ, calculate $\dfrac{d\varepsilon''}{d\varepsilon'}$.

3. In the representation given by ε'' = f(ε'), to which limit does the preceding equation ($\dfrac{d\varepsilon''}{d\varepsilon'}$) tend toward at high frequencies ($\omega \to \infty$)?

4. Under these conditions, what does α represent with respect to the plot of ε'' = f(ε')? Give a schematic representation.

Answers to problem 3

1. By multiplying the top and bottom by $(1 - j\omega\tau)^\alpha$, the analytical form proposed can be written as: $\dfrac{\underline{\varepsilon} - \varepsilon_\infty}{\varepsilon_s - \varepsilon_\infty} = \dfrac{(1 - j\omega\tau)^\alpha}{(1 + \omega^2\tau^2)^\alpha}$.

By making $\tan\Phi = \omega\tau$, we have:

$$\frac{\underline{\varepsilon} - \varepsilon_\infty}{\varepsilon_s - \varepsilon_\infty} = \frac{(1 - j\tan\Phi)^\alpha}{(1 + \tan^2\Phi)^\alpha}.$$

With $(1 + \tan^2\Phi) = \dfrac{1}{\cos^2\Phi}$, we directly obtain:

$$\frac{\underline{\varepsilon} - \varepsilon_\infty}{\varepsilon_s - \varepsilon_\infty} = \frac{\left(\dfrac{\cos\Phi - j\sin\Phi}{\cos\Phi}\right)^\alpha}{\left([1 + \tan^2\Phi]^{1/2}\right)^\alpha (1 + \tan^2\Phi)^{\alpha/2}} = \frac{\exp(-j\alpha\Phi)}{(1 + \tan^2\Phi)^{\alpha/2}} = \exp(-j\alpha\Phi)\cos^\alpha\Phi.$$

By identification of the real and imaginary parts, we obtain:

$$\frac{\varepsilon' - \varepsilon_\infty}{\varepsilon_s - \varepsilon_\infty} = \cos^\alpha\Phi\cos\alpha\Phi \quad \text{and} \quad \frac{\varepsilon''}{\varepsilon_s - \varepsilon_\infty} = \cos^\alpha\Phi\sin\alpha\Phi.$$

2. We have $\dfrac{d\varepsilon''}{d\varepsilon'} = \dfrac{d\varepsilon''}{d\Phi}\dfrac{d\Phi}{d\varepsilon'} = \dfrac{\dfrac{d\varepsilon''}{d\Phi}}{\dfrac{d\varepsilon'}{d\Phi}}$.

Making $E = \varepsilon_s - \varepsilon_\infty$, it is hence possible to state that:

$$\frac{d\varepsilon''}{d\Phi} = E\frac{d}{d\Phi}\left(\left[\cos^\alpha \Phi\right]\left[\sin \alpha\Phi\right]\right)$$
$$= E\left[\left(\cos^\alpha \Phi\right).\alpha.\cos\alpha\Phi - \left(\sin\alpha\Phi\right).\alpha\cos^{\alpha-1}\Phi\sin\Phi\right],$$

from which : $\dfrac{d\varepsilon''}{d\Phi} = E\alpha\left[\cos^{\alpha-1}\Phi\right]\left[\cos\left(\alpha+1\right)\Phi\right]$.

Similarly,

$$\frac{d\varepsilon'}{d\Phi} = E\frac{d}{d\Phi}\left(\left[\cos^\alpha \Phi\right]\left[\cos \alpha\Phi\right]\right)$$
$$= E\left[\left(-\cos^\alpha \Phi\right).\alpha.\sin\alpha\Phi - \left(\cos\alpha\Phi\right).\alpha\cos^{\alpha-1}\Phi\sin\Phi\right]$$

from which : $\dfrac{d\varepsilon'}{d\Phi} = -E\alpha\left[\cos^{\alpha-1}\Phi\right]\left[\sin\left(\alpha+1\right)\Phi\right]$.

The result is that:

$$\frac{d\varepsilon''}{d\varepsilon'} = \frac{\dfrac{d\varepsilon''}{d\Phi}}{\dfrac{d\varepsilon'}{d\Phi}} = -\operatorname{co}\tan\left(\alpha+1\right)\Phi .$$

3. On reaching the limit at higher frequencies, $\omega\tau \to \infty \Rightarrow \tan\Phi \to \infty$, which is to say that $\Phi = \dfrac{\pi}{2}$. As $\operatorname{co}\tan\left(\dfrac{\pi}{2}+\beta\right) = -\tan\beta$, we have $\operatorname{co}\tan\left(\alpha+1\right)\dfrac{\pi}{2} = \tan\dfrac{\alpha\pi}{2}$,

from which:

$$\frac{d\varepsilon''}{d\varepsilon'} = \tan\frac{\alpha\pi}{2}.$$

It is possible therefore to propose the following representation for the angle given by $\frac{\alpha\pi}{2}$:

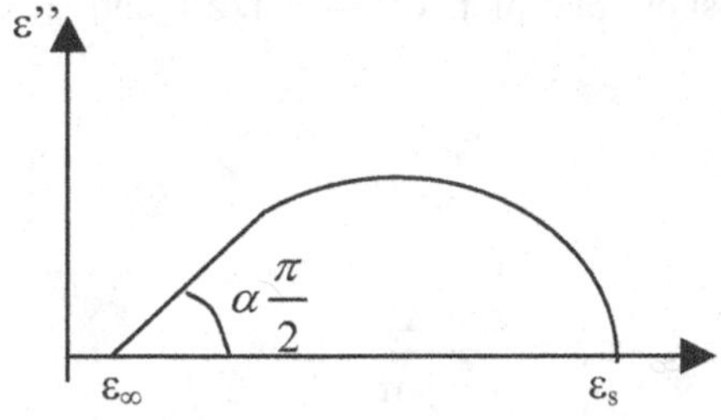

Comment. With the arc being asymmetric, the maximum value for ε" is no longer obtained when $\omega\tau = 1$, but rather when the value of $[\omega\tau]_{max}$ is such that $\left[\frac{\partial\varepsilon"}{\partial(\omega\tau)}\right]_{[\omega\tau]_{max}} = 0$.

The calculation shows that this condition is obtained when Φ is such that $\Phi = \Phi_M = \frac{1}{1+\alpha}\frac{\pi}{2}$. Hence, as $\Phi = \text{Arc}\tan\omega\tau$:

$$[\omega\tau]_{max} = \tan\left[\frac{1}{1+\alpha}\frac{\pi}{2}\right].$$

1.4.4. Problem 4. Linear relationships based on the Debye equations: the Cole–Brot equations

This question concerns a dielectric material that exhibits various Debye-type relaxation times with each mechanism characterized by a single relaxation time (τ). Therefore, we will look for various representations that make it possible to isolate a single Debye-type relaxation time that is such that $\omega_c\tau = 1$. The term ω_c is the angular frequency at which the absorption is at a maximum.

1.

a. Making $x = \omega\tau$, show that $x = \frac{\omega}{\omega_c} = \frac{\lambda_c}{\lambda}$ where λ_c designates the wavelength corresponding to ω_c.

b. From the reduced Debye equations, as in $\dfrac{\varepsilon' - \varepsilon_\infty}{\varepsilon_s - \varepsilon_\infty} = \dfrac{1}{1+\omega^2\tau^2}$ and $\dfrac{\varepsilon''}{(\varepsilon_s - \varepsilon_\infty)} = \dfrac{\omega\tau}{1+\omega^2\tau^2}$, show that plots of $\dfrac{\varepsilon''}{x} = f(\varepsilon')$ and $\varepsilon'' x = g(\varepsilon')$ are straight lines.

2.

a. Through the equation $\underline{\varepsilon} - \varepsilon_\infty = \dfrac{\varepsilon_s - \varepsilon_\infty}{1 + j\omega\tau}$, show that:

$$\varepsilon' = \varepsilon_s - (\omega\tau)\varepsilon''; \text{ and}$$

$$\varepsilon' = \varepsilon_\infty + \frac{\varepsilon''}{(\omega\tau)}.$$

If ν and ν_c denote the critical frequency, calculate $\dfrac{\nu}{\nu_c}$ as a function of $\omega\tau$. From this determine the two new expressions for ε' as a function of $x_1 = \varepsilon''\nu$ or of $x_2 = \dfrac{\varepsilon''}{\nu}$ (the Cole equations).

b. Show that it is thus possible to determine, from the graphic representations of ε' as a function of the two variables x_1 and x_2, the parameters ε_s, ε_∞ and ν_c.

3. In fact, the preceding graphic Cole representations demand an experimental determination of ε' *and* ε''. This question deals with the search for equations which make it possible to work uniquely with ε' or with ε'' (Brot's equations).

a. A graphic representation using only ε''

By making $y = \dfrac{\omega}{\varepsilon''}$ and $x = \omega^2$, show that we can obtain the linear equation

$y = ax + b$. Define a and b, and hence give the expressions for $(\varepsilon_s - \varepsilon_\infty)$ and τ as a function of these parameters a and b.

b. A graphic representation using only ε'

The values that ε' takes rest on three parameters, namely, $\varepsilon_s, \varepsilon_\infty$, and τ. Obviously, it would be impossible to determine all three simply by using a linear representation. Show that it is nevertheless possible to determine τ and ε_s if ε_∞ is known, and likewise that τ and ε_∞ can be calculated if ε_s has been found.

Answers to problem 4

1.

a. If $x = \omega\tau$, we also can write with $\omega_c \tau = 1$ that $x = \frac{\omega\tau}{\omega_c \tau} = \frac{\omega}{\omega_c} = \frac{\lambda_c}{\lambda}$ as

$\lambda = \frac{2\pi c}{\omega}$ and $\lambda_c = \frac{2\pi c}{\omega_c} = \frac{2\pi c\tau}{\omega_c \tau} = 2\pi c\tau$.

b. With $x = \omega\tau$, the reduced equations can be written as:

$$\frac{\varepsilon' - \varepsilon_\infty}{\varepsilon_s - \varepsilon_\infty} = \frac{1}{1 + \omega^2\tau^2} = \frac{1}{1 + x^2} \quad (1) \text{ and}$$

$$\frac{\varepsilon''}{(\varepsilon_s - \varepsilon_\infty)} = \frac{\omega\tau}{1 + \omega^2\tau^2} = \frac{x}{1 + x^2}. \quad (2)$$

The result of Eq. (2) is that $x = \frac{\varepsilon''}{\varepsilon_s - \varepsilon_\infty}(1 + x^2)$, and hence with Eq. (1):

$$x = \frac{\varepsilon''}{\varepsilon_s - \varepsilon_\infty} \frac{\varepsilon_s - \varepsilon_\infty}{\varepsilon' - \varepsilon_\infty} = \frac{\varepsilon''}{\varepsilon' - \varepsilon_\infty} \Rightarrow \frac{\varepsilon''}{x} = \varepsilon' - \varepsilon_\infty .$$

In addition, if we form $\varepsilon'' x$ we can obtain from Eq. (2):

$$\varepsilon'' x = (\varepsilon_s - \varepsilon_\infty)\frac{x^2}{1 + x^2} = (\varepsilon_s - \varepsilon_\infty)\frac{1 + x^2 - 1}{1 + x^2} = (\varepsilon_s - \varepsilon_\infty)\left[1 - \frac{1}{1 + x^2}\right].$$

With Eq. (1) we obtain:

$$\varepsilon'' x = (\varepsilon_s - \varepsilon_\infty)\left[1 - \frac{\varepsilon' - \varepsilon_\infty}{\varepsilon_s - \varepsilon_\infty}\right] = \varepsilon_s - \varepsilon'.$$

To summarize, we thus have $\frac{\varepsilon''}{x} = \varepsilon' - \varepsilon_\infty$ and $\varepsilon'' x = \varepsilon_s - \varepsilon'$, and hence the plots of $\frac{\varepsilon''}{x} = f(\varepsilon')$ and $\varepsilon'' x = g(\varepsilon')$ are straight lines.

2.

a. The relation $\underline{\varepsilon} - \varepsilon_\infty = \dfrac{\varepsilon_s - \varepsilon_\infty}{1 + j\omega\tau}$ gives $[\varepsilon' - \varepsilon_\infty - j\varepsilon''][1 + j\omega\tau] = \varepsilon_s - \varepsilon_\infty$,

and hence identification of the real and imaginary parts yields:

$$\begin{cases} \varepsilon' - \varepsilon_\infty + \varepsilon''\omega\tau = \varepsilon_s - \varepsilon_\infty & \Rightarrow \varepsilon' = \varepsilon_s - (\omega\tau)\varepsilon'' \\ [\varepsilon' - \varepsilon_\infty]\omega\tau - \varepsilon'' = 0 & \Rightarrow \varepsilon' = \varepsilon_\infty + \dfrac{\varepsilon''}{(\omega\tau)}. \end{cases}$$

As $\omega\tau = \dfrac{\omega\tau}{\omega_c\tau} = \dfrac{\omega}{\omega_c} = \dfrac{\nu}{\nu_c}$, we also find that:

$$\begin{cases} \varepsilon' = \varepsilon_s - \left(\dfrac{1}{\nu_c}\varepsilon''\nu\right) = f(\varepsilon''\nu) = f(x_1) & \text{(3) and} \\ \varepsilon' = \varepsilon_\infty + \left(\nu_c\dfrac{\varepsilon''}{\nu}\right) = g\left(\dfrac{\varepsilon''}{\nu}\right) = g(x_2). & \text{(4)} \end{cases}$$

b. We thus in fact have:

$$\begin{cases} \varepsilon' = \varepsilon_s - \dfrac{x_1}{\nu_c} & \text{(3') and} \\ \varepsilon' = \varepsilon_\infty + \nu_c x_2. & \text{(4')} \end{cases}$$

With $\varepsilon_s = b_1$, $\varepsilon_\infty = b_2$, and $\nu_c = a_2 = -\dfrac{1}{a_1}$, Eqs. (3') and (4') can be rewritten as:

$$\begin{cases} y_1 = a_1 x_1 + b_1 & \text{(3'')' and} \\ y_2 = a_2 x_2 + b_2. & \text{(4'')} \end{cases}$$

The ordinate at the origin gives ε_s (from Eq. (3')) and ε_∞ (from Eq. (4''). The slopes of the two lines permit a determination of ν_c.

3.

a. As $\varepsilon'' = \dfrac{(\varepsilon_s - \varepsilon_\infty)\omega\tau}{1 + \omega^2\tau^2}$, we have:

$y = \frac{\omega}{\varepsilon''} = \frac{1+\omega^2\tau^2}{(\varepsilon_s - \varepsilon_\infty)\tau} = \frac{1}{(\varepsilon_s - \varepsilon_\infty)\tau} + \frac{\tau}{(\varepsilon_s - \varepsilon_\infty)}\omega^2$. By making $x = \omega^2$, we obtain the linear relationship $y = ax + b$ with $a = \frac{\tau}{(\varepsilon_s - \varepsilon_\infty)}$ and $b = \frac{1}{(\varepsilon_s - \varepsilon_\infty)\tau}$.

The upshot is that $(\varepsilon_s - \varepsilon_\infty)^2 = \frac{\tau}{a}\frac{1}{\tau b}$, and hence

$$(\varepsilon_s - \varepsilon_\infty) = \frac{1}{\sqrt{ab}}.$$

In addition, $\tau^2 = a(\varepsilon_s - \varepsilon_\infty)\frac{1}{b(\varepsilon_s - \varepsilon_\infty)}$, from which

$$\tau = \sqrt{\frac{a}{b}}.$$

b. If ε_∞ is known, it is possible to state that: $\varepsilon' - \varepsilon_\infty = \frac{\varepsilon_s - \varepsilon_\infty}{1+\omega^2\tau^2}$, and hence

$$\frac{1}{\varepsilon' - \varepsilon_\infty} = \frac{1+\omega^2\tau^2}{\varepsilon_s - \varepsilon_\infty} = \frac{\tau^2}{\varepsilon_s - \varepsilon_\infty}\omega^2 + \frac{1}{\varepsilon_s - \varepsilon_\infty}.$$

If we plot $\frac{1}{\varepsilon' - \varepsilon_\infty} = f(\omega^2)$, the ordinate at the origin gives $(\varepsilon_s - \varepsilon_\infty)^{-1}$, from which can be calculated ε_s, while the slope gives $\frac{\tau^2}{\varepsilon_s - \varepsilon_\infty}$, from which can be determined τ.

If ε_s is known, from the equation $\varepsilon' - \varepsilon_\infty = \frac{\varepsilon_s - \varepsilon_\infty}{1+\omega^2\tau^2}$ we can draw out:

$$\varepsilon' = \varepsilon_s - \frac{(\varepsilon_s - \varepsilon_\infty)\omega^2\tau^2}{1+\omega^2\tau^2},$$

from which

$$\frac{1}{\varepsilon_s - \varepsilon'} = \frac{1}{\left(\varepsilon_s - \varepsilon_\infty\right)\omega^2\tau^2} + \frac{\omega^2\tau^2}{\left(\varepsilon_s - \varepsilon_\infty\right)\omega^2\tau^2},$$

so that :

$$\frac{1}{\varepsilon_s - \varepsilon'} = \frac{1}{\left(\varepsilon_s - \varepsilon_\infty\right)\tau^2}\frac{1}{\omega^2} + \frac{1}{\left(\varepsilon_s - \varepsilon_\infty\right)}.$$

On plotting $\frac{1}{\varepsilon_s - \varepsilon'} = f\left(\frac{1}{\omega^2}\right)$, the slope gives τ and the ordinate at the origin gives ε_∞.

Chapter 2

Characterization of Dielectrics

2.1. Introduction: representation of a dielectric with an equivalent circuit

The general form for the intensity (I) in a condenser filled with a real dielectric, detailed in Section 1.13, is given by $\underline{I} = \omega\varepsilon_r''C_0V + j\omega\varepsilon_r'C_0V = I_R + jI_C$. The corresponding impendence (Z) is such that the admittance (Y) is:

$$Y = \frac{1}{Z} = \frac{\underline{I}}{V} = \omega\varepsilon_r''C_0 + j\omega\varepsilon_r'C_0. \qquad (1)$$

Using normal notation, the intensity also can be written in the form:

$$\underline{I} = jI_C + I_R = (j\omega C + G)V\ .$$

At this level, it is often premature to conclude that a dielectric behaves like a pure capacitor (where $C = \varepsilon_r'C_0$) in parallel with a resistor. The conductance (G), where *a priori* $G = \omega\varepsilon_r''C_0$, does not relate to just one dielectric loss, but rather a number of different processes that consume energy, all of which may vary in type. In effect, G represents a number of dielectric loss mechanisms. It is for this reason that a parallel circuit based on a resistor (R_p) and a capacitor (C_p) is too simple a representation. However, a single dipole gives rise to a single type of behavior, as detailed in Section 2.2.1. Nevertheless, other types of behavior, and there are many of them, are due to other types of equivalent dipoles. Each will be singled out in this chapter, prior to studying the methods used to characterize dielectrics, which themselves occasionally necessitate equivalent dipole forms to the dielectrics.

Different analyses of dielectrics also can be carried out. As observed for the specific case of relaxation mechanisms, the tradition is to plot not only $\varepsilon' = f(\log\omega)$, $\varepsilon'' = f(\log\omega)$, but also $\varepsilon'' = f(\varepsilon')$. As is shown later on, analysis by impedance spectroscopy uses other classical representations.

By making $Z = Z' - i Z''$ and $Y = Y' - iY''$ (in simplification and by tradition, the complex impedance and admittance are written Z and Y), these representations are:

$$Y'' = f(Y') , Z'' = g(Z') , C'' = h(C').$$

This latter representation is based on the fact that the capacitance (C) can be taken in the form $C = \varepsilon\frac{S}{d}$, so that when ε is complex $\underline{C} = \underline{\varepsilon}\frac{S}{d}$.

With $\underline{C} = C' - iC''$, we immediately have:

$$\underline{C} = C'(\omega) - iC''(\omega) = \frac{S}{d}\left[\varepsilon'(\omega) - i\varepsilon''(\omega)\right]. \qquad (2)$$

To within the factor S/d, C' and C'' evolve just as ε' and ε'': In addition, Z and Y are such that by definition $Z = \frac{1}{Y}$. They are the inverse of one another. Inversions within the complex plane are such that if the representation of one of these magnitudes is a circle, then its inverse is a straight line, and visa versa.

2.2. Circuits exhibiting relaxation phenomena as possible equivalents to real dielectrics: plots of $\varepsilon' = f(\omega)$ and $\varepsilon'' = g(\omega)$

2.2.1. Parallel circuit

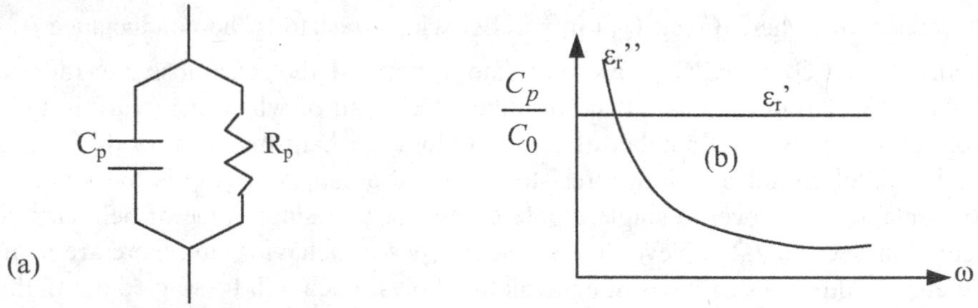

Figure 2.1. *(a) Scheme of a parallel circuit and (b) plots of ε' =(logω) and ε'' = f(logω).*

For the equivalent dipole shown in Figure 2.1a, we have:

$$\frac{1}{Z_p} = \frac{1}{R_p} + \frac{1}{Z_{Cp}}, \text{ so that with } Z_{Cp} = \frac{1}{j\omega C_p},$$

$$\frac{1}{Z_p} = \frac{1}{R_p} + j\omega C_p. \qquad (3)$$

Identification with Eq. (3) leads to:

$$\omega\varepsilon_r''C_0 + j\omega\varepsilon_r'C_0 = \frac{1}{R_p} + j\omega C_p,$$

so that identification of real and imaginary parts gives

$$\begin{cases} \varepsilon_r' = \dfrac{C_p}{C_0} \\ \varepsilon_r'' = \dfrac{1}{\omega C_0 R_p}. \end{cases} \qquad (4)$$

The use of a simple parallel circuit thus results in the representation presented in Figure 1b, which is far from the dielectric losses observed for a polar dielectric under radio frequency conditions (see also Figure 2.3 for the Debye dipolar absorption plots).

2.2.2. Circuit in series

It can appear more interesting to treat dielectric losses using a resistor in series (R_s) with a pure capacitor (C_s), as shown in Figure 2.2a. The impendence of the dipole is therefore given by:

$$Z_S = R_S + \frac{1}{j\omega C_S} = \frac{1 + j\omega C_S R_S}{j\omega C_S} \qquad (5), \text{ so that}$$

$$\frac{1}{Z_S} = \frac{j\omega C_S + \omega^2 C_S^2 R_S}{1 + \omega^2 C_S^2 R_S^2}. \text{ By identification with Eq. (1), we have:}$$

$$\begin{cases} \varepsilon_r' = \dfrac{C_S}{C_0\left[1 + (\omega C_S R_S)^2\right]} \\ \varepsilon_r'' = \dfrac{\omega C_S^2 R_S}{C_0\left[1 + (\omega C_S R_S)^2\right]}. \end{cases} \qquad (5')$$

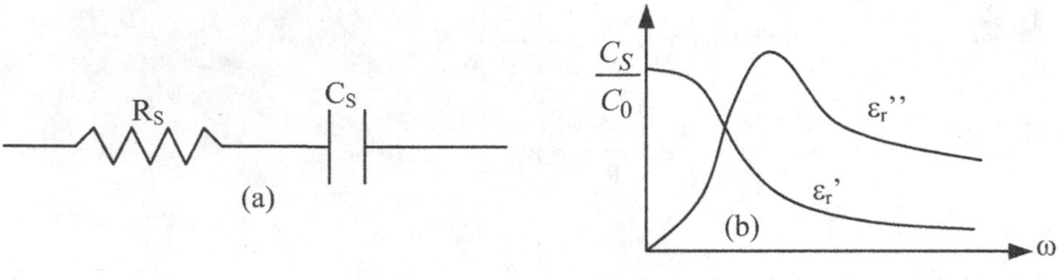

Figure 2.2. *(a) Scheme of a circuit in series and (b) plots of ε' =(logω) and ε'' = f(logω).*

Once again, the representation of a dielectric by a simple equivalent circuit does not exhibit the characteristics of $\varepsilon_r'(\omega)$ and $\varepsilon_r''(\omega)$ as found for a polar dielectric under radio frequency conditions, as indicated in Figure 2.2.

2.2.3. Association of serial and parallel circuits and relaxation plots

As discussed in Chapter 1, the plots of dipolar relaxation are characterized by a constant value at low frequencies followed by a slow decrease with respect to $\varepsilon'(\omega)$. For $\varepsilon_r''(\omega)$, the variations present a peak.

Thus, a circuit equivalent to a dielectric cannot easily consist of just a resistor (R) and a capacitor (C). Nevertheless, and as is demonstrated below, a combination of parallel and serial circuits can be used to describe most relaxation curves. In effect, from the equivalent dipole shown in Figure 2.3a it is possible to state that:

$$\frac{1}{Z} = \frac{1}{Z_S} + \frac{1}{Z_p}, \text{ where } Z_p = \frac{1}{j\omega C_p} \text{ and } Z_S = R_S + \frac{1}{j\omega C_S}.$$

From this we can determine that:

$$\frac{1}{Z} = \frac{\omega^2 C_S^2 R_S}{1 + \omega^2 C_S^2 R_S^2} + j\omega\left[C_p + \frac{C_S}{1 + \omega^2 C_S^2 R_S^2}\right]. \tag{6}$$

By making $\tau_S = R_S C_S$, identification with Eq. (1) leads to:

$$\begin{cases} \varepsilon_r' = \dfrac{C_p}{C_0} + \dfrac{C_S}{C_0}\dfrac{1}{1 + \omega^2 \tau_S^2} \\[2ex] \varepsilon_r'' = \dfrac{\omega R_S C_S^2}{C_0\left[1 + \omega^2 C_S^2 R_S^2\right]} = \dfrac{C_S}{C_0}\dfrac{\omega \tau_S}{1 + \omega^2 \tau_S^2}. \end{cases} \tag{7}$$

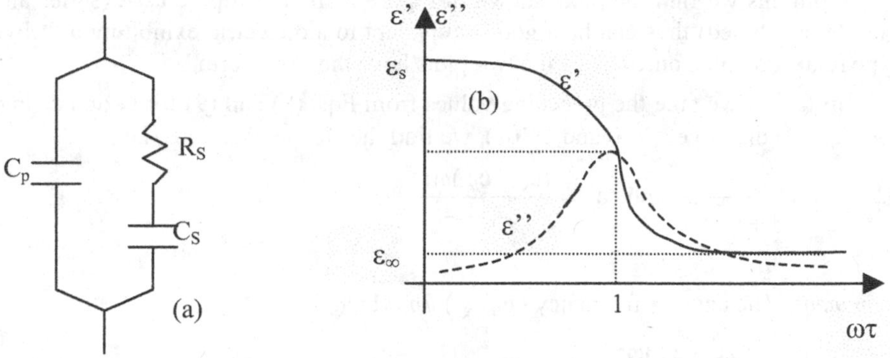

Figure 2.3. *(a) Scheme of equivalent series parallel circuit and (b) plots of ε' =(logω) and ε'' = f(logω).*

Following these equations, it is possible to say that:

- if $\omega = 0$, $\varepsilon_r'' = 0$, then ε_r' is at a maximum (the derivative $\frac{\partial \varepsilon'}{\partial \omega} = -\frac{C_S}{C_0}\frac{2\omega\tau_S^2}{\left(1+\omega^2\tau_S^2\right)^2}$ cancels out when $\omega = 0$) and is equal to

$$(\varepsilon_r')_{max} = \frac{C_p + C_S}{C_0} = \varepsilon_S \ . \qquad (8)$$

- if $\omega \to \infty$, $\varepsilon_r'' \to 0$, and ε_r' tends towards a minimal, as in

$$(\varepsilon_r')_{min} = \frac{C_p}{C_0} = \varepsilon_\infty \ . \qquad (9)$$

- the angular frequency for which ε_r'' is at a maximum is given by a solution to the equation $\frac{\partial \varepsilon_r''}{\partial \omega} = \frac{C_S}{C_0}\frac{\tau_S}{1+\omega^2\tau_S^2}\left(1 - \frac{2\omega^2\tau_S^2}{1+\omega^2\tau_S^2}\right) = 0$. From this can be determined that

$$\omega_{max} = \frac{1}{\tau_S} = \frac{1}{C_S R_S} \ . \qquad (10)$$

The maximum value for ε_r'', which is therefore when $\omega\,\tau_S = 1$, is given by $(\varepsilon_r'')_{max} = \frac{C_S}{2C_0}$, so that in addition $(\varepsilon_r'')_{max} = \frac{1}{2}(\varepsilon_S - \varepsilon_\infty)$.

From this we find the plots shown in Figure 2.3b. The dipole used (series and parallel combined) thus can be a good equivalent to a dielectric exhibiting a Debye type relaxation phenomenon, as the two plots have the same form.

In fact, if we take the preceding values from Eqs. (8) and (9) for ε_S and ε_∞ into the Eqs. (7) that give $\varepsilon_r'(\omega)$ and $\varepsilon_r''(\omega)$, we find the Debye equations, i.e.,

$$\varepsilon_r' = \varepsilon_\infty + \frac{\varepsilon_S - \varepsilon_\infty}{1+\omega^2\tau^2} \quad \text{and} \quad \varepsilon_r'' = \frac{(\varepsilon_S - \varepsilon_\infty)\omega\tau}{1+\omega^2\tau^2}.$$

Comment. The angular frequency (ω_{max}) for which

$\tan\delta = \dfrac{\varepsilon''}{\varepsilon'} = \dfrac{(\varepsilon_S - \varepsilon_\infty)\omega\tau}{\varepsilon_S + \varepsilon_\infty\omega^2\tau^2}$ is a maximum can be calculated using

$\left[\dfrac{\partial \tan\delta}{\partial\omega}\right]_{\omega_{max}} = 0$, which leads to $\omega_{max}^2 = \dfrac{1 + C_S / C_P}{\varepsilon_S^2}$, from which:

$$(\mathrm{tg}\delta)_{max} = \frac{\left(1 + C_S / C_p\right)^{1/2}}{2\left(1 + C_p / C_S\right)}.$$

Together, the above equations allow calculations of the values of Cp, R_S, and C_S for a given dielectric when plots of $\varepsilon_r'(\omega)$ and $\varepsilon_r''(\omega)$ have been obtained. In effect, Eqs. (8) and (9) can be used to determine C_p and C_S, while Eq. (10) yields R_S. These values placed in Eq. (7) give rise to the theoretical plots.

2.3. Resonating circuit

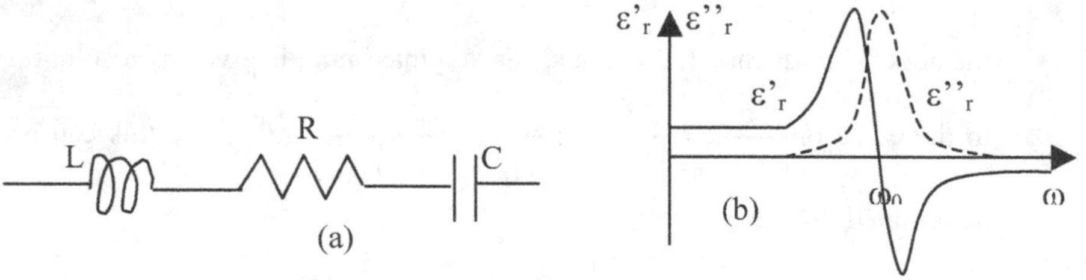

Figure 2.4. *(a) A RLC resonance circuit and (b) a resonance plot.*

Here $Z = R - \dfrac{j}{\omega C} + j\omega L = \dfrac{\omega RC - j(1-\omega^2 LC)}{\omega C}$ (11), so that:

$$\frac{1}{Z}=\frac{\omega C\left[\omega RC+j\left(1-\omega^2 LC\right)\right]}{\left[\omega^2R^2C^2+\left(1-\omega^2LC\right)^2\right]}=\frac{\omega C\left[\omega RC+j\left(1-\omega^2LC\right)\right]}{(LC)^2\left[\left(\frac{1}{LC}-\omega^2\right)^2+\omega^2\frac{R^2C^2}{L^2C^2}\right]}.$$

By making $\omega_0^2=\frac{1}{LC}$ and $\Gamma=\frac{R}{L}$, we have:

$$\frac{1}{Z}=\frac{\frac{\omega}{L^2C}\left[\omega RC+j\left(1-\omega^2LC\right)\right]}{\left[\left(\omega_0^2-\omega^2\right)^2+\omega^2\Gamma^2\right]}.$$

By identification with Eq. (1) and on making $\omega_p^2=\frac{1}{LC_0}$, we obtain:

$$\varepsilon_r'=\frac{\frac{1}{L^2CC_0}\left(1-\omega^2LC\right)}{\left[\left(\omega_0^2-\omega^2\right)^2+\omega^2\Gamma^2\right]}=\frac{\frac{1}{LC_0}\left(\frac{1}{LC}-\omega^2\right)}{\left[\left(\omega_0^2-\omega^2\right)^2+\omega^2\Gamma^2\right]}=\omega_p^2\frac{\left(\omega_0^2-\omega^2\right)}{\left[\left(\omega_0^2-\omega^2\right)^2+\omega^2\Gamma^2\right]} \qquad (12)$$

$$\varepsilon_r''=\frac{\frac{\omega R}{L^2C_0}}{\left[\left(\omega_0^2-\omega^2\right)^2+\omega^2\Gamma^2\right]}=\omega_p^2\frac{\omega\Gamma}{\left[\left(\omega_0^2-\omega^2\right)^2+\omega^2\Gamma^2\right]}. \qquad (13)$$

The plots of absorption with the resonance when $\omega=\omega_0$ are shown in Figure 2.4b. They are similar to those found in other studies on resonance phenomena, such as detailed in Chapter 8 of Volume 1 or Chapter 3 of this book.

2.4. Representation of a heterogeneous dielectric (powders) using a model of layers: two parallel circuits in series and the Maxwell–Wagner–Sillars effect

The titled representation can be carried out through the following problems:

1. C_0 denotes the capacitance of a vacuum condenser. On being filled with a real dielectric, characterized as having a complex dielectric permittivity ($\underline{\varepsilon}_r$) given by $\underline{\varepsilon}_r=\varepsilon_r'-j\varepsilon_r''$, applied to its terminals is an alternating tension (V) given by $V=V_0\,e^{j\omega t}$. Give the general expression for the admittance obtained as a function of ε_r', ε_r'', C_0, and ω.

2. When a dielectric is heterogeneous, such as is the case for a powder, then the representation can be made using a succession of electric dipoles each made up of a resistance in parallel with a capacitor. In this problem, we will limit ourselves to the two-layer model shown in Figure 2.5a.

Making:

$$C = C_1 + C_2$$

$$R = \frac{R_1 R_2}{R_1 + R_2}$$

$$\tau_1 = R_1 C_1,\ \tau_2 = R_2 C_2,\ \tau = RC,$$

determine the admittance equivalent to the model used as a function of ω, τ_1, τ_2, τ, R_1, and R_2.

3. For a model based on a bilayer dielectric, the law of the variation of the dielectric permittivity as a function of angular frequency can be found by considering the expressions for:

a. ε_r' and ε_r'' as a function of ω, τ_1, τ_2, τ, R_1, R_2, and C_0; and

b. the limiting values of ε_r' denoted ε_S and ε_∞ when ω tends toward 0 or ∞, respectively.

Give the expressions detailed in **a** and **b**.

4.Express ε_r' as a function of ε_S, ε_∞, and τ. Express ε_r'' as a function of ε_S, ε_∞, ω, τ, R_1, R_2 and C_0. Compare them to the classic Debye equations. Conclude.

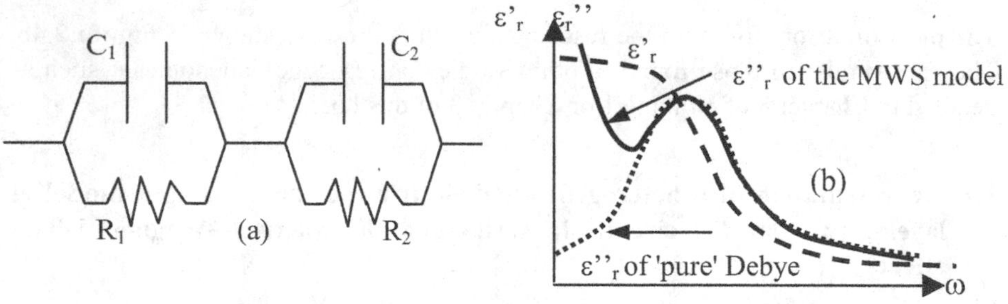

Figure 2.5. *(a) Double circuit layer and (b) characteristic plots of the Maxwell–Wagner–Sillars (MWS) effect.*

Answers

1. Again, $I = \omega\varepsilon_r''C_0V + j\omega\varepsilon_r'C_0V$, so that $Y = \dfrac{I}{V} = \omega\varepsilon_r''C_0 + j\omega\varepsilon_r'C_0$. (14)

2. The impendence of the two-layer model is given by $Z = Z_1 + Z_2$, so that

$$Z = \frac{1}{\frac{1}{R_1} + j\omega C_1} + \frac{1}{\frac{1}{R_2} + j\omega C_2} = \frac{\left[\frac{R_1 + R_2}{R_1R_2} + j\omega(C_1 + C_2)\right]R_1R_2}{(1 + j\omega R_1C_1)(1 + j\omega R_2C_2)}$$

$$= \frac{(R_1 + R_2)(1 + j\omega\tau)}{(1 + j\omega\tau_1)(1 + j\omega\tau_2)}.$$

From this can be determined that:

$$Y = \frac{1}{Z} = \frac{1}{R1 + R2}\frac{(1 + j\omega\tau_1)(1 + j\omega\tau_2)(1 - j\omega\tau\ )}{1 + \omega^2\tau^2} \qquad (15)$$

$$Y = \frac{1}{R1 + R2}\frac{1 - \omega^2\tau_1\tau_2 + \omega^2\tau(\tau_1 + \tau_2) - j\omega\tau(1 - \omega^2\tau_1\tau_2) + j\omega(\tau_1 + \tau_2)}{1 + \omega^2\tau^2}.$$

The identification with Eq. (14) gives

$$\begin{cases} \varepsilon_r' = \dfrac{1}{C_0(R_1 + R_2)}\dfrac{\tau_1 + \tau_2 - \tau + \omega^2\tau_1\tau_2\tau}{1 + \omega^2\tau^2} \\ \varepsilon_r'' = \dfrac{1}{C_0(R_1 + R_2)}\dfrac{1 - \omega^2\tau_1\tau_2 + \omega^2\tau(\tau_1 + \tau_2)}{1 + \omega^2\tau^2}. \end{cases} \qquad (16)$$

From which can be directly determined that:

- when $\omega = 0$, $\varepsilon_r' \to \varepsilon_S = \dfrac{\tau_1 + \tau_2 - \tau}{C_0(R_1 + R_2)}$
- when $\omega \to \infty$, $\varepsilon_r' \to \varepsilon_\infty = \dfrac{\tau_1\tau_2}{\tau}\dfrac{1}{C_0(R_1 + R_2)}$.

By plugging these values into Eq. (16), we have:

$$\begin{cases} \varepsilon'_r = \varepsilon_\infty + \dfrac{\varepsilon_s - \varepsilon_\infty}{1+\omega^2\tau^2} \\ \varepsilon''_r = \dfrac{1}{\omega C_0 (R_1 + R_2)} + \dfrac{(\varepsilon_s - \varepsilon_\infty)\omega\tau}{1+\omega^2\tau^2}. \end{cases} \quad (17)$$

The evolution of the plot $\varepsilon_r' = f(\omega)$ is identical to that of a Debye lot. In ε''_r the second term is identical to that of the Debye type, while the first term appears as a supplementary term that caries with $\frac{1}{\omega}$. Therefore, when $\omega \to 0$, $\varepsilon_r'' \to \infty$.

In general terms, for the layer model, or so-called Maxwell–Wagner–Sillars model, the evolution of $\varepsilon_r''(\omega)$ is that given in Figure 2.5b. This corresponds to the description of phenomenon II detailed in Section 1.1.6.

2.5. Impedance spectroscopy

Given the discussion above, it is now possible to graphically summarize the representations of ε' and ε" as a function of the angular frequency, $\varepsilon'' = f(\varepsilon')$, and also $|\text{Im}(Z)| = f[\text{Re}(Z)]$ or $\text{Im}(Y) = f[\text{Re}(Y)]$. A logarithmic scale (logω) is used so as to cover a wide range of frequencies.

2.5.1. Example using a parallel circuit

In this particular case, from Eq. (3) it is possible to write:

$$\text{Re}(Z_P) = \frac{R_P}{(1+\omega^2 R_P^2 C_P^2)}, \quad (18)$$

and

$$\text{Im}(Z_P) = \frac{-\omega R_P^2 C_P}{(1+\omega^2 R_P^2 C_P^2)}. \quad (19)$$

With the help of these equations, we can obtain a relationship between Re (Z_P) and Im (Z_P) by writing:

$$\left[\text{Re}(Z_P) - (\frac{R_P}{2})\right]^2 + \left[\text{Im}(Z_p)\right]^2 = \frac{R_P^2}{4}.$$

This equation corresponds to the equation for a circle centered on real axes $(\frac{R_P}{2}, 0)$ with a radius $\frac{R_P}{2}$. The representation of $|\text{Im}(Z_P)| = f(\text{Re}(Z_P))$ is thus given in Figure 2.6b with respect to the angular frequency (ω) which varies from 0 to $+\infty$, and:

- if $\omega = 0$, then $\mathrm{Re}(Z_p) = R_p$ and $\mathrm{Im}(Z_p) = 0$; and
- if $\omega \to \infty$, then $\mathrm{Re}(Z_p) \to 0$ and $\mathrm{Im}(Z_p) \to 0$.

The angular frequency for which $\mathrm{Im}(Z_p)$ is at a maximum is given by the condition:

$$\left[\frac{d|\mathrm{Im}(Z)|}{d\omega}\right]_{\omega_{max}} = \frac{R_P^2 C_P(1+\omega^2 R_P^2 C_P^2) - \omega R_P^2 C_P(2\omega R_P^2 C_P^2)}{(1+\omega^2 R_P^2 C_P^2)^2}$$

$$= \frac{R_P^2 C_P(1-\omega^2 R_P^2 C_P^2)}{(1+\omega^2 R_P^2 C_P^2)^2} = 0$$

from which $\omega = \omega_{max} = \dfrac{1}{R_P C_P}$.

At this angular frequency, we therefore find $|\mathrm{Im}(Z_P)|_{max} = \dfrac{R_P}{2}$ and $R_e(Z_P) = \dfrac{R_P}{2}$, hence the representations shown in Figure 2.6b.

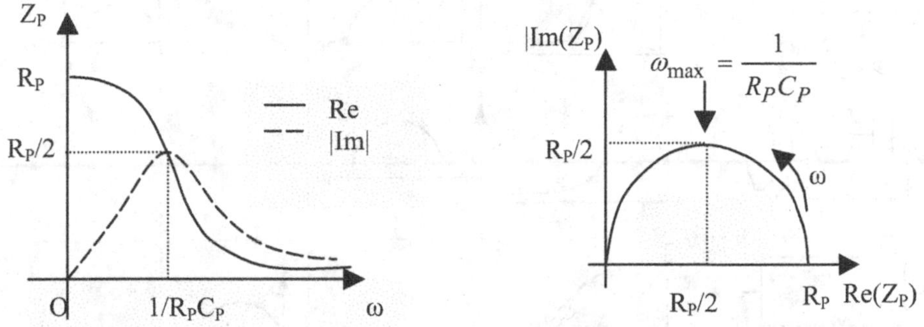

Figure 2.6. *Plots of (a) $Re(Z_p) = f(\omega)$ and $|Im(Z_p)| = f(\omega)$ and (b) $|Im(Z_p)| = f[Re(Z_p)]$.*

2.5.2. Summary

Figure 2.7 shows the assembled behaviors of dielectric behaviors, shown successively by plots of:

- ε' and ε''= f(ω) (1st column);
- ε'' = f(ε') or C'' = f(C') (2nd column);
- |Im Z| = f[Re(Z)] (3rd column) and
- |Im Y| = f[Re(Y)] (4th column).

The principal calculations not already covered and contributing to Figure 2.7 have been placed at the end of this chapter in Section 2.8.

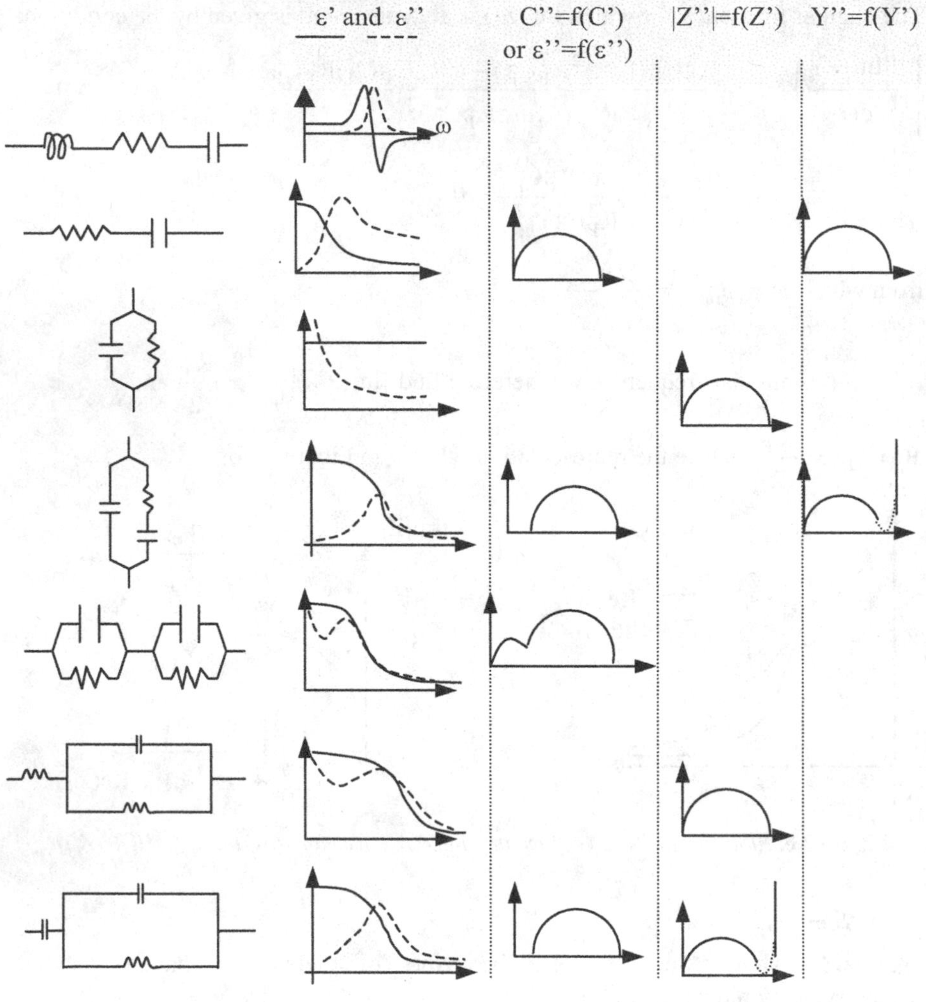

Figure 2.7. *Representations of characteristic behaviors [ε' and ε''=f(ω), C''=f(C')* or *ε''=f(ε'), | Z''|=f(Z'), Y''=f(Y')] of various circuits equivalent to real dielectrics.*

2.6. Dielectric measurements: summary of the analytical apparatus used with respect to frequency domain

The apparatus may vary with the frequency under study. The ranges involved are schematized in Figure 2.8.

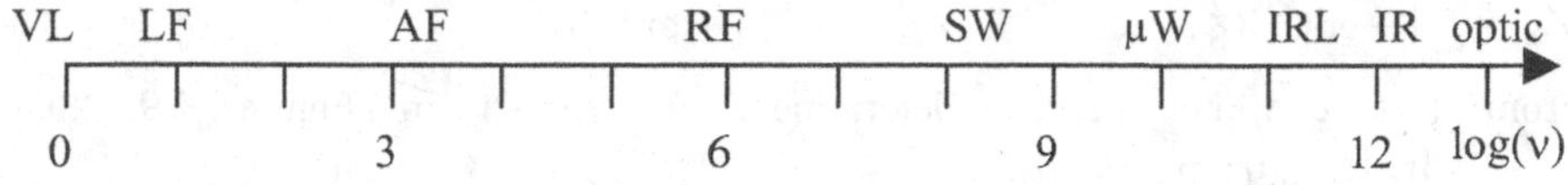

Figure 2.8. *The electromagnetic spectrum, from left to right: very low frequencies (vlf); low frequencies (VF); audio frequencies (AF); radio frequencies (RF); short wave (SW); microwave (μW); far infrared (FIR); infrared (IR); and waves in the optical domain (optic) where $\lambda = 0.8$ to $0.4\ \mu m$.*

Before venturing into the various corresponding methods, it will be useful to make a preliminary remark.

2.6.1. Opening remark: expressions for the quality factor and tangential loss for the different circuits equivalent to capacitors or wound bobbins

In general terms, the quality factor (Q) is defined in electrical technology by the relationship:

$$Q = 2\pi \frac{\text{maximum value of energy stored during time T}}{\text{energy transformed into heat during T}} = 2\pi \frac{W}{w_e}.$$

2.6.1.1. Condenser in a representation of a circuit in series (C_s-R_s)

If the intensity is in the form $i = I_0 \sin \omega t$, we thus have:

$$w_e = \int_0^T R_s I_0^2 \sin^2 \omega t \, dt = \frac{2\pi}{\omega} \frac{R_s I_0^2}{2}.$$

In addition, $W = \frac{1}{2} C_s V_{s0}^2$, where $V_{s0} = \frac{I_0}{C_s \omega}$ is the maximum tension at the condenser terminals.

Thus $W = \frac{1}{2} \frac{I_0^2}{\omega^2 C_s}$, from which $Q_s = \frac{1}{\omega C_s R_s}$.

From this can be determined that $\tan \delta_s = \frac{1}{Q_s} = \omega C_s R_s$.

The latter equation also can be obtained directly from a Fresnel diagram. In such a representation, $Z_S = R_S + \dfrac{1}{j\omega C_S}$, from which:

$$\frac{1}{Z_S} = \frac{\omega^2 C_S^2 R_S + j\omega C_S}{1+\omega^2 C_S^2 R_S^2} \quad \text{and} \quad I = \frac{V}{Z_S} = V\frac{\omega^2 C_S^2 R_S + j\omega C_S}{1+\omega^2 C_S^2 R_S^2} = I_R + I_C .$$

From this can be directly determined, as shown in Figure 2.9, that $\tan\delta = \dfrac{|I_R|}{|I_C|} = \dfrac{\omega^2 C_S^2 R_S}{\omega C_S} = \omega C_S R_S$, and which verifies that $Q = \dfrac{1}{\tan\delta}$.

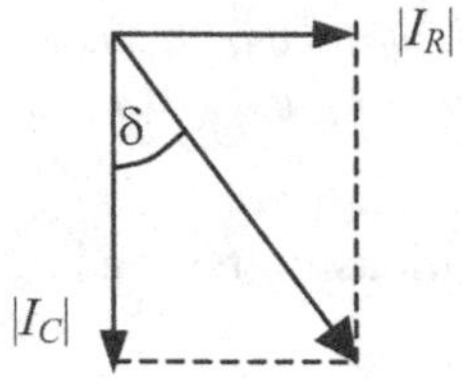

Figure 2.9. *Loss angle (δ) in a Fresnel diagram.*

2.6.1.2. Condenser in a parallel representation (parallel circuit C_p-R_p)

The tension at the terminals of the equivalent dipole are assumed to be of the form $V = V_0 \sin\omega t$, and $w_e = \dfrac{2\pi}{\omega}\dfrac{V_0^2}{2R_p}$ and $W = \dfrac{1}{2}C_p V_0^2$. The result is that

$Q_p = \omega C_p R_p$.

Thus $\tan\delta_p = \dfrac{1}{Q_p} = \dfrac{1}{\omega C_p R_p}$, and this expression also can be found through the Fresnel diagram, as $\dfrac{1}{Z_p} = \dfrac{1}{R_p} + \dfrac{1}{Z_{Cp}} = \dfrac{1}{R_p} + j\omega C_p$, so that:

$$I = \frac{V}{Z} = V\left(\frac{1}{R_p} + j\omega C_p\right) = I_R + I_C, \text{ from which:}$$

$$\tan\delta = \frac{|I_R|}{|I_C|} = \frac{1}{\omega C_p R_p}, \text{ which is what we set out to show.}$$

2.6.1.3. Quality factor for a reel represented by a circuit in series (L_p-R_p)

Just as for the condenser represented by a circuit in series, $w_e = \int_0^T R_s I_0^2 \sin^2 \omega t \, dt = \frac{2\pi}{\omega}\frac{R_s I_0^2}{2}$ as the losses are due to the Joule effect arising from the resistor in series R_s.

With $W = \frac{1}{2} L_s I_0^2$, we thus find that $Q_{L_s} = \frac{L_s \omega}{R_s}$.

2.6.1.4. Quality factor for a reel represented by a parallel circuit (L_p - R_p)

Just as for the condenser represented by a parallel circuit, we have $w_e = \frac{2\pi}{\omega}\frac{V_0^2}{2R_p}$.

With $W = \frac{1}{2} L_p I_0^2 = \frac{1}{2} L_p \frac{V_0^2}{L_p^2 \omega^2} = \frac{1}{2}\frac{V_0^2}{L_p \omega^2}$, we have $Q_{L_p} = \frac{R_p}{L_p \omega}$.

2.6.2. Very low frequencies (0 to 10 Hz)

Classically, voltmeters and ammeters are used to plot the current traversing a dielectric sample with time during the process of charging and discharging a condenser containing a dielectric. The currents can be extremely small, of the order of pA, and often a continuous current amplifier is used.

In order to obtain permittivities from continuous currents, a simple method consists of measuring the time constant for a condenser, either with or without dielectric, to discharge through a standard resister (R_0).

2.6.2.1. *First measurement*

Initially, a condenser filled with air and capacitance (C_0) is placed under a known tension (V_0). The initial charge is given by $q_0 = C_0 V_0$. At a time $t = 0$, the charged condenser is connected to the terminals of R_0 and at a time t_1 during its discharge the tension (V_1) at R_0 is recorded using a voltammeter of high internal impedance (R_v). In effect, the value of the resistor to use for the calculation is given by $R_{eq} = \frac{R_0 R_v}{R_0 + R_v}$. If $R_v >> R_0$, then $R_{eq} \approx R_0$.

We thus have $V_1 = R_0 i(t_1)$. With $i(t_1) = \frac{V_0}{R_0} \exp\left(-\frac{t_1}{R_0 C_0}\right)$, we obtain:

$$\ln\frac{V_0}{V_1}=\frac{t_1}{R_0C_0}, \qquad (20)$$

from which can be determined C_0.

2.6.2.2. *Second measurement*

A dielectric is placed between the armatures of the condenser, which now has a capacitance denoted by C, and then the unit is charged at V_0. On discharge, the time taken (τ) for the terminals at R_0 to return to V is measured. If we assume that the real condenser is represented by C in series with a resistance (R_s), then the discharge is through a total resistance given by $R_e = R_0 + R_s$. In this case,

$$V = R_0 i(\tau) = R_0\frac{V_0}{R_0+R_s}\exp\left(-\frac{\tau}{[R_0+R_s]C}\right).$$

With $\dfrac{R_0}{R_0+Rs}\approx 1$ we find

$$V \approx V_0\exp\left(-\frac{\tau}{[R_0+R_s]C}\right)=V_0\exp\left(-\frac{\tau}{R_eC}\right).$$

From this can be deduced that

$$\mathrm{Ln}\frac{V_0}{V}=\frac{\tau}{R_eC}. \qquad (21)$$

In this last equation, there are two unknown values, i.e., R_e (or more exactly, R_s) and C.

2.6.2.3. Third measurement

The aim is to determine the value of C. So, we connect in parallel with C a calibrated capacitor filled with air (and therefore with negligible losses). After charging the circuit to V_0, the time (τ') required for the same discharge as above is then recorded. This means that:

$$V = R_0 i(\tau') = R_0\frac{V_0}{R_0+R_S}\exp\left(-\frac{\tau'}{R_e[C+C_1]}\right)\approx V_0\exp\left(-\frac{\tau'}{R_e[C+C_1]}\right)\Rightarrow$$

$$\mathrm{Ln}\frac{V_0}{V}=\frac{\tau'}{R_e(C+C_1)}. \qquad (22)$$

Comparing Eqs. (21) and (22) yields $\frac{\tau}{R_e C} = \frac{\tau'}{R_e (C + C_1)}$, from which we have

$$C = \frac{\tau}{\tau'}(C + C_1), \text{ so that } C = \frac{C_1}{\tau'/\tau - 1}. \qquad (23)$$

2.6.2.4. Result

We can find R_s from Eq. (21). Simply put, we have $\varepsilon_r' = \frac{C}{C_0}$. In addition,

$\tan\delta = \frac{\varepsilon_r''}{\varepsilon_r'} = \omega C R_s$, and hence $\varepsilon_r'' = \omega R_s \frac{C^2}{C_0}$.

2.6.3. From low frequencies to radio frequencies (10 to 10^7 Hz)

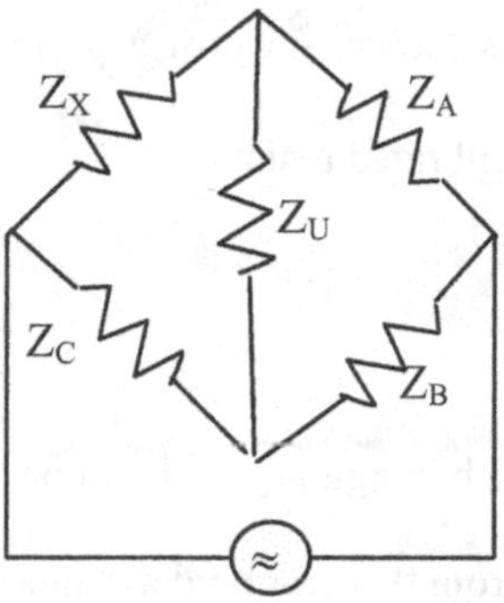

Figure 2.10. *General scheme of a bridge. At equilibrium, i = 0 through Z_U and $Z_X Z_B = Z_A Z_C$.*

Within the titled range of frequencies, a bridge as shown in Figure 2.10 is normally used. When the equilibrium point is reached, that is i = 0 in Z_U, then we have $Z_X Z_B = Z_A Z_C$. The type of condenser determines the type of bridge used, of which there are several of the more widely known described below.

2.6.3.1. The Sauty bridge for almost perfect condensers where D = tan δ =0

In this case, we simply find:

- $Z_X = \frac{1}{j\omega C_X}$ for a perfect condenser of capacitance to be determined;

- $Z_A = \dfrac{1}{j\omega C_A}$ for a perfect calibrated condenser;
- $Z_B = R_B$ for pure resistance; and
- $Z_C = R_C$ for pure resistance.

The equation for equilibrium leads to $C_X = C_A \dfrac{R_B}{R_C}$.

2.6.3.2. Wien bridge: low loss condensers where tan δ << 0.1

For such condensers, we use the series representation to give:

- $Z_X = R_X + \dfrac{1}{j\omega C_X}$, where C_X is the capacity of the condenser and R_X its resistance;
- $Z_A = R_A + \dfrac{1}{j\omega C_A}$, where C_A is the capacity of a perfect, calibrated condenser set in parallel with R_A which is a calibrated resistance;
- $Z_B = R_B$ is a pure resistance; and
- $Z_C = R_C$ is for pure resistance.

The equilibrium condition leads, with imaginary and real parts being equalized, to: $R_X R_B = R_A R_C$ and $\dfrac{R_B}{C_X} = \dfrac{R_C}{C_A}$. From this can be determined that:

$$C_X = C_A \frac{R_B}{R_C} \text{ and } \tan\delta = D_X = \omega R_X C_X = \omega C_A R_A .$$

2.6.3.3. A Nernst bridge: generally higher loss condensers where $0 \le \tan\delta \le 1$.

Here the parallel representation of a condenser is used, so that:

- $\dfrac{1}{Z_X} = \dfrac{1}{R_X} + j\omega C_X$, where the condenser under study is represented by a capacitor (CX) in parallel with R_X;
- $\dfrac{1}{Z_A} = \dfrac{1}{R_A} + j\omega C_A$, perfect, calibrated condenser C_A in parallel with the resistor R_A;

- $Z_B = R_B$, pure resistance; and
- $Z_C = R_C$, pure resistance.

By equalizing the real and imaginary parts, the equilibrium conditions gives: $R_X = R_A \frac{R_C}{R_B}$ and $C_X R_C = C_A R_B$, from which can be deduced that:

$$C_X = C_A \frac{R_B}{R_C} \quad \text{and} \quad \tan\delta = D_X = \frac{1}{\omega C_X R_X} = \frac{1}{\omega C_A R_A}.$$

2.6.3.4. *Comment*

The vectorial voltammeters with synchronous detection, which appeared on the market during the 1980s, make possible the determination of the components of a complex impedance (or its modulus and dephasing) by comparing the reference signal and the signal leaving the impedance. Such apparatuses allow an extraction of a weak signal from background noise, which can be particularly important at very low frequencies. They typically can be used within a range of frequencies from 1 Hz to 100 kHz, values that were previously outside the range of a single apparatus with a correct order of precision ($\Delta\tan\delta \approx 10^{-3}$).

2.6.4. *Radio frequencies and shortwave (10^3 to 10^8 Hz)*

At this scale of frequency, resonant (Q meter) circuits are used

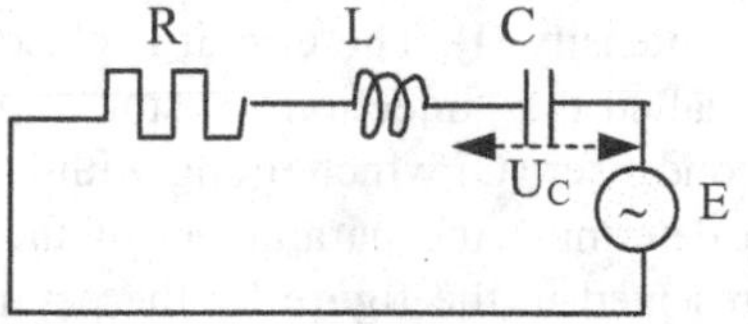

Figure 2.11. *Oscillating circuit used in Q meters.*

2.6.4.1. *General terms for a Q meter*

It is worth noting that in the oscillating circuit shown in Figure 2.11, the intensity of the current through the circuit is written in the form $I = \dfrac{E}{\sqrt{R^2 + \left(L\omega - \dfrac{1}{C\omega}\right)^2}}$.

At the point of resonance, we have $I = \dfrac{E}{R}$ and $\omega = \omega_0$ so that $\omega_0 L = \dfrac{1}{\omega_0 C}$.

The tension at the terminals of the condenser is thus given by:

$$U_{C_0} = |Z| I = \frac{1}{\omega_0 C}\frac{E}{R} = E\frac{L\omega_0}{R} = EQ,$$

where Q is the quality coefficient of the reel.

2.6.4.2. Fundamental operation

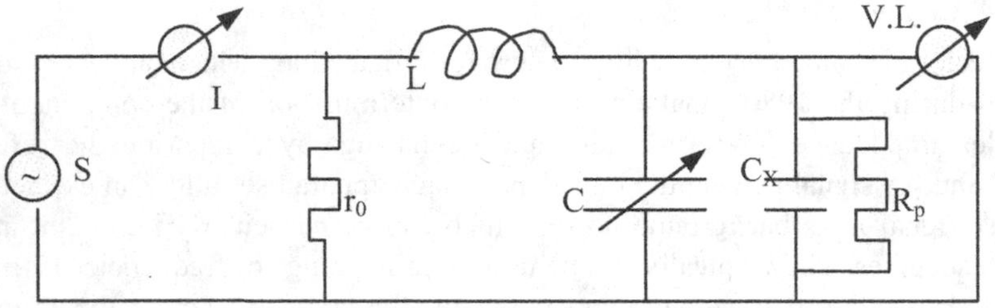

Figure 2.12. *Operational structure of a Q meter*

Figure 2.12 shows a frequency and tension variable oscillator (S) placed in a circuit that measures the current intensity (I). The circuit is closed around a constant, pure resistance (r_0) generally called the "injection resistor", and includes a high quality and calibrated variable condenser (C) which along with L makes up the basic block $r_0 - L - C$. In order to determine the parameters of the dielectric it is placed in parallel with C and represented in the figure by the parallel C_X and R_p. Then C is regulated until the maximum deviation is reached at the voltammeter (VL), which is joined to the terminals of C.

Q is determined by comparing the injected tension $r_0 I$ at the terminals of r_0 with the tension U_{C0} that appears at the terminals of C at the resonance point. We therefore find:

$$Q = \frac{U_{C_0}}{r_0 I}.$$

If I is kept constant, then for each value of U_{C0} there is a corresponding value of Q, and it is possible to directly graduate the voltammeter as a coefficient of the over voltage.

2.6.5. High frequencies

2.6.5.1. From 10^8 to 10^9 Hz: using transmission lines

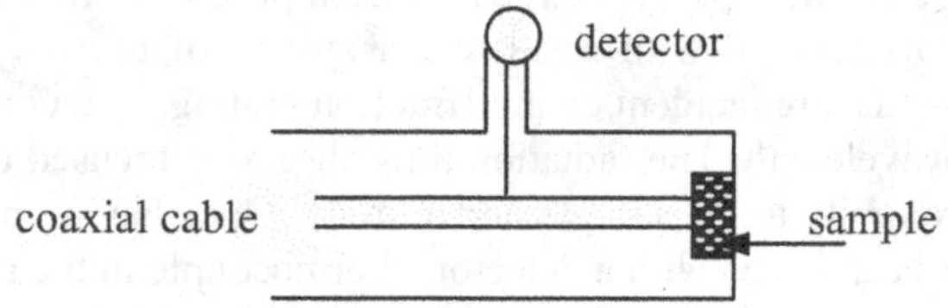

Figure 2.13. *Transmission line.*

Transmission lines work through the establishment of stationary waves by reflection of an incident wave at an end of the line. The sample under study is formed as a small disk and is placed at one end of the transmission line. A detector is used to analyze the maximum and minimum stationary waves obtained with and without the dielectric.

2.6.5.2. From 10^9 to 3×10^{11} Hz: use of microwave techniques

In this range of frequencies, waveguides and resonating cavities are used. In the waveguide, progressive waves propagate along a direction z that is normal to the sample. At the dielectric/air interface, the wave is partly reflected and partly transmitted. In the empty space that precedes the sample face, the incident and reflected waves undergo interference. It is thus possible to write that:

$$E_1 = E_0(x,y)e^{i\omega t}\left[e^{-ik_1 z} + re^{ik_1 z}\right],$$

where r denotes the reflection factor, and $k_1 = \dfrac{\omega}{c}\sqrt{\varepsilon_{r1}\mu_{r1}}$.

In the second medium, there are no parasitic reflections, and the transmitted wave is in the form $E_2 = E_0(x,y)e^{i\omega t}\, te^{-ik_2 z}$, where t is the transmission coefficient (see Chapter 4), and $k_2 = \dfrac{\omega}{c}\sqrt{\varepsilon_{r2}\mu_{r2}}$.

For nonmagnetic media $\mu_{r1} = \mu_{r2} = 1$, and k_1 and k_2 are only dependent on ε_1 and ε_2. In order to determine ε_2, the phase of the transmitted wave needs to be characterized, and the wave vector (k_2) must be taken in its complex form, as in: $k_2 = \underline{k}_2 = k_2' - ik_2''$, where k_2' relates to the real part of the permittivity and k_2'' the attenuation (by absorption) of the wave.

2.6.5.2. *Frequencies above 10^{12} Hz: spectroscopic techniques*

The classic method to analyze a spectrum is with a spectrometer that consists of a polychromatic source (white light of a large spectrum produced, for example, by an incandescent filament) coupled with a slit and a system of mirrors to generate a beam of parallel rays that are incident on a diffraction grating. The grating is used to select the required wavelength. The radiation thus chosen is focused on an exit slit, which is in close proximity to the sample under study. Once having passed through the sample, the light is analyzed with a detector (thermocouple in the infrared region and photomultiplier in the UV-visible) and recorded. The spectrum obtained represents the transmitted signal as a function of frequency.

This procedure, however, can be quite long as the grating has to turn over the required frequency range. An alternative method consists in measuring not as a function of frequency but of time. Here the wave used contains a range of frequencies that can be analyzed by a detector connected to a computer. The wave signal can then be sampled, for example, every millisecond for two seconds so as to give a high data count (2000 in this example). The application of a Fourier transformation on the stored data (which requires just a few seconds more) makes it possible to display the amplitudes and frequencies of the cosinusoidal incident waves. In effect the Fourier transformation converts a time-based spectrum to one based on a range of frequencies (further details on this can be found in most courses on the theory of data treatment).

2.7. Applied determination of dielectric parameters for frequencies below 10^8Hz (classic range for dielectric studies)

2.7.1. Condenser form

Flat or cylindrical condensers may be used, depending on whether solid (in the shape of disks) or liquid samples are used. For the latter, total influence condensers are used so that there is no distortion in the field lines separating the dielectric medium and air. A cap, with a small hole through which is passed the sample, is used, is placed so that it is part of the external electrode which surrounds the internal electrode. The field lines therefore retain the same distribution whatever the dielectric placed between the armatures, and the measurements can be carried out using a single field line distribution.

2.7.2. Connections and the effect of connecting wire capacities

2.7.2.1. *Measurement with one of the two wires grounded to earth*

This setup is normally used in combination with a Q meter or a bridge in order to measure the capacitance of shielded cables or the entry capacitance of an amplifier.

2.7.2.2. Measurement with two wires "floating" with respect to earth

In this case, the two exits of the apparatus are held independent of the bridge 'guard' which is now connected to earth. The setup, shown in Figure 2.14, which measures the connectors at the condenser, consists of three junctions around three capacitors:

- the actual capacity of the condenser between the electrodes denoted C_X;
- the auxiliary capacitors denoted C_A and C_B due to the junction wires.

The capacitors thus form a triangular system that makes up the unknown branch of the bridge in which C_X—to be determined—is included. It is now thus possible to demonstrate that if C_A and C_B are sufficiently small, their influence on the oscillator, the detector, and hence the characterization of C_X is negligible. For example, for an HP 4270 bridge (and similarly for an HP 4284 bridge), the constructor states that 50 cm of the coaxial cable (Filotex RG 58 A/U) will introduce an additional relative error of only 0.01 % while measuring a 10 nF capacitor at 10 kHz.

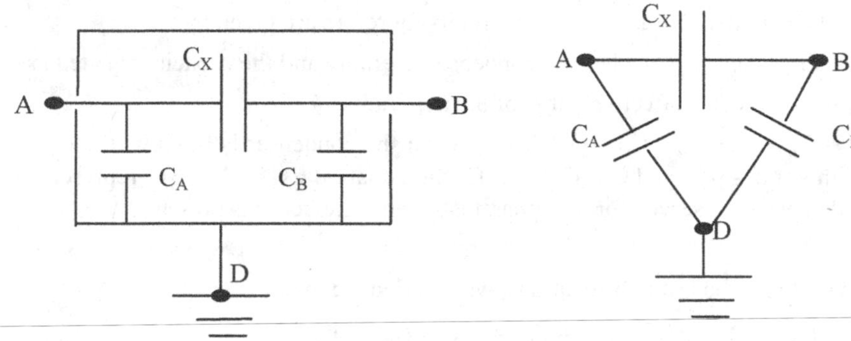

Figure 2.14. *Triangular system of capacitors (ABD) equivalent to a capacitor (C_X) setup with two connecting cables (of capacity C_A and C_B) at a bridge.*

2.7.3. Preparing equations to calculate ε' and ε''

2.7.3.1. For a dielectric represented by a parallel circuit (Nernst bridge)

As shown in Figure 2.15, G_0 and G are, respectively, the conductivities of the empty and filled condensers. The capacities due only to the active part with respect to the electrodes are, respectively, denoted C_0 and $\varepsilon' C_0$. The parasitic capacity (due to connections and electrode insulation) is represented by C_i (for a cylindrical condenser, C_i is for the most part due to the insulating ring placed between the coaxial electrodes, as indicated in Figure 2.15c).

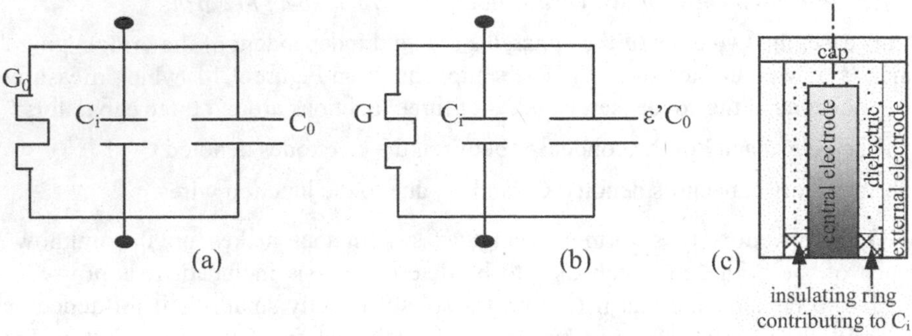

Figure 2.15. *Schemes of equivalent circuits for (a) a parallel circuit with an empty condenser; (b) with a condenser filled with dielectric or permittivity ε'; and (c) a cross section of a cylindrical condenser showing between the two coaxial electrodes the insulating ring which generates most of C_i.*

The admittances of the parallel circuits therefore are given by:

- $Y_0 = G_0 + j\omega(C_0 + C_i)$ when the condenser is empty and the capacity equated as $C_{0i} = (C_0 + C_i)$ is effectively that of the apparatus;
- $Y = G_0 + j\omega(\underline{\varepsilon}\, C_0 + C_i) = G + j\omega C_{Xi}$ when the condenser is filled with dielectric. With $\underline{\varepsilon} = \varepsilon' - j\varepsilon''$, and $G = G_0 + \omega\varepsilon'' C_0$ the equation $C_{Xi} = \varepsilon' C_0 + C_i$ represents the real capacity observed for the apparatus when a dielectric is present.

From the last two equations, we can deduce that:

- $\varepsilon' = \dfrac{C_{Xi} - C_i}{C_0} = \dfrac{C_{Xi} - (C_i + C_0) + C_0}{C_0} = \dfrac{C_{Xi} - C_{0i}}{C_0} + 1$, so that

$$\varepsilon' = 1 + \frac{\Delta C}{C_0} \quad \text{with } \Delta C = C_{Xi} - C_{0i},$$

where ΔC represents the variation in measured capacity without or with the dielectric in the condenser used to make the measurement.

- $\varepsilon'' = \dfrac{\Delta G}{\omega C_0}$, with $\Delta G = G - G_0 = \dfrac{1}{R_{Xi}} - \dfrac{1}{R_{0i}}$. As in a parallel circuit,

$\tan\delta = D_P = \dfrac{1}{\omega C_P R_P}$, we also have $\Delta G = \omega\left(D_{Xi}C_{Xi} - D_{0i}C_{0i}\right)$, where D_{Xi} and D_{0i} are the loss tangents observed for a condenser with a dielectric or without, respectively.

We thus reach:

$$\varepsilon'' = D_{Xi}\frac{C_{Xi}}{C_0} - D_{0i}\left[1+\frac{C_i}{C_0}\right].$$

Generally, C_i is small ($\approx$ 0.2 pF) with respect to C_0 ($\approx$ 20 pF), so the equation can be written more simply as:

$$\varepsilon'' = D_{Xi}\frac{C_{Xi}}{C_0} - D_{0i}\,.$$

2.7.3.2. Dielectric represented by a circuit in series (Wien bridge)

Figure 2.16 shows the scheme for a circuit equivalent to a condenser with an insulating parasitic capacitance (C_i). The impendence thus is given in the form:

$$Z_0 = R_{0s} - \frac{j}{\omega C_{0s}}, \text{ where } C_{0s} = C_0 + C_i\,.$$

When a dielectric is introduced, the impendence changes to:

$$Z = R_{0s} - \frac{j}{\omega\left(\underline{\varepsilon} C_0 + C_i\right)}.$$

A calculation analogous to that in the preceding section gives:

$$\varepsilon' = 1 + \frac{\Delta C}{C_0} \text{ et } \varepsilon'' = D_{Xs}\frac{C_{Xs}}{C_0},$$

where D_{Xs} represents the loss tangent for a cell filled with dielectric material and $C_{Xs} = \varepsilon' C_0 + C_i$ gives the corresponding capacity.

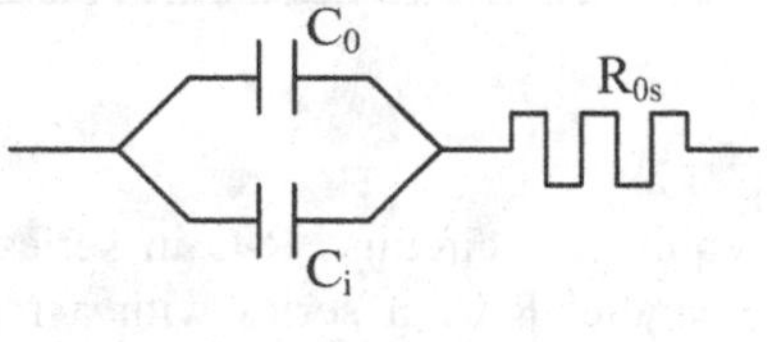

Figure 2.16. *Scheme of an equivalent circuit in a series for a vacuum-filled condenser taking the parasitic capacitor (C_i) into account.*

2.7.3.3. The Q meter

On considering losses as being due to a conductance in parallel to the condenser, the variation in admittance on introducing a dielectric into the condenser is given by (see also Section 2.7.3.1):

$$\Delta Y = Y - Y_0 = \Delta G + j\omega\Delta C$$

where $\Delta G = \omega\varepsilon'' C_0$ and $\Delta C = (\varepsilon' - 1)C_0$.

With an empty cell connected to the junction of a calibrated and variable condenser (C) identified in Figure 2.12, and with an established resonance, ΔC represents the change in the variable capacitor. This makes it possible to reestablish the resonance when the dielectric is introduced into the measuring condenser.

In this representation of dielectric losses, the coefficient for the overvoltage (of the form $Q_p = \omega C_p R_p$ from Section 2.6.1.2) is given by (where $\omega = \omega_0$):

- in a vacuum $Q_0 = \omega_0 C_0 R_0$, so that the resonance point $\left(\omega_0 C_0 = \frac{1}{\omega_0 L_0}\right)$,

$$Q_0 = \frac{R_0}{\omega_0 L} \Rightarrow G_0 = \frac{1}{R_0} = \frac{1}{\omega_0 L Q_0};$$

- with a dielectric $Q = \omega_0 C\,R$, and at resonance $\left(\omega_0 C = \frac{1}{\omega_0 L}\right)$, $Q = \frac{R}{\omega_0 L} \Rightarrow$

$$G = \frac{1}{R} = \frac{1}{\omega_0 L Q}.$$

The result is that $\Delta G = G - G_0 = \frac{1}{\omega L}\left(\frac{1}{Q} - \frac{1}{Q_0}\right) = \frac{1}{\omega L}\frac{\Delta Q}{Q Q_0}$.

The values for ε' and ε'' are thus given by:

$$\varepsilon' = 1 + \frac{\Delta C}{C_0} \text{ and } \varepsilon'' = \frac{1}{\omega^2 L C_0}\frac{\Delta Q}{Q Q_0}.$$

2.8. Problems

For the various possible equivalent circuits (R-C in series, R-C in series and in parallel with a capacitor, parallel R-C in series with a resistor Rs, parallel R-C circuit in series with a capacitor Cs, parallel R-C in series with a parallel R-C), justify by calculation the choice of one or the other of the following possible representations: Im(Z)=R(Z) or Im(Y)=R(Y). Plot |Im(Z)| = f(ω), Re(Z) = f(ω) and |Im(Z)| = f[Re(Z)] or Im(Y) = f(ω), Re(Y) = f(ω) and Im(Y) = f[Re(Y)].

2.8.1. Problem 1. R-C in series

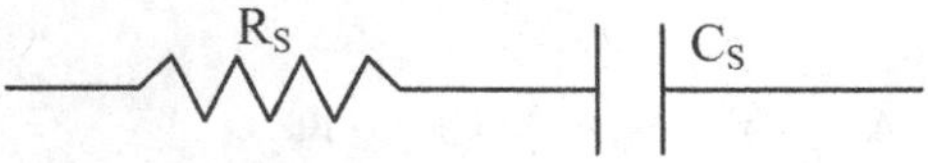

With $Zs = Rs + \frac{1}{j\omega Cs} = Rs - \frac{j}{\omega Cs}$,

we have:

$\mathrm{Re}(Z_s) = R_S$ and $|\mathrm{Im}(Z_s)| = \frac{1}{\omega C_S}$.

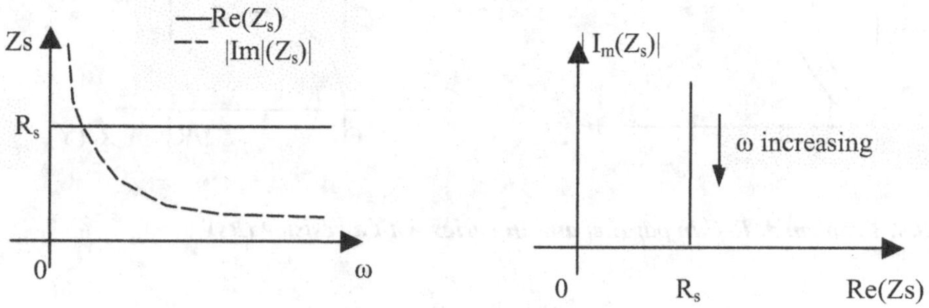

2.8.2. Problem 2. R-C in series and in parallel with a capacitor (C_p)

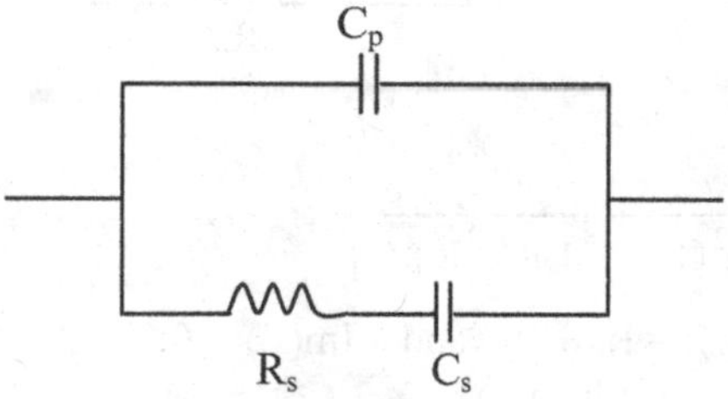

$\frac{1}{Z} = j\omega C_P + \frac{1}{Z_S}$; with $Z_S = R_S + \frac{1}{j\omega C_S} = \frac{1 + j\omega R_S C_S}{j\omega C_S}$, we have:

$$\frac{1}{Z} = j\omega C_P + \frac{j\omega C_S}{1 + j\omega R_S C_S} = j\omega C_P + \frac{j\omega C_S + \omega^2 R_S C_S^2}{1 + \omega^2 R_S^2 C_S^2},$$

$$Y = \frac{\omega^2 R_S C_S^2}{1 + \omega^2 R_S^2 C_S^2} + j\omega\left[C_P + \frac{C_S}{1 + \omega^2 R_S^2 C_S^2} \right].$$

• if $\omega = 0$, $\mathrm{Re}(Y) = 0$ and $\mathrm{Im}(Y) = 0$;

• if $\omega \to \infty$, $\mathrm{Re}(Y) \to 1/R_s$ and $\mathrm{Im}(Y) \to \infty$.

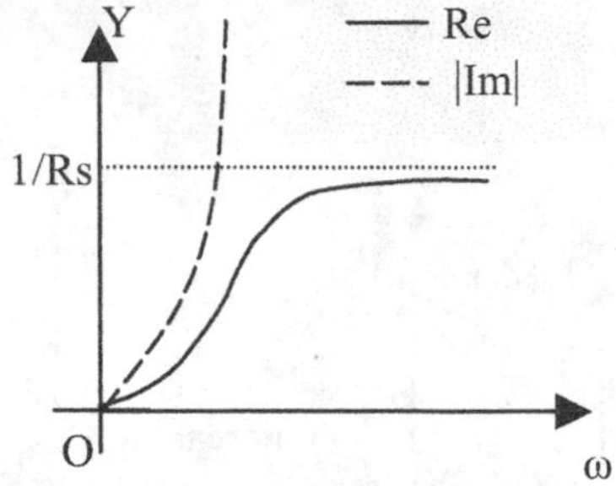

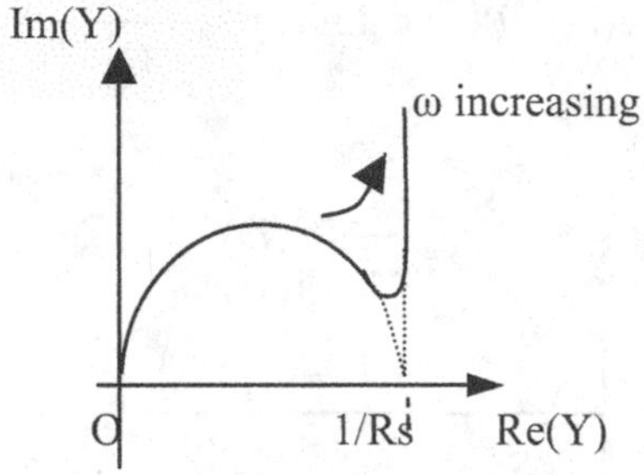

2.8.3. Problem 3. R-C in parallel and in series with a resistor (Rs)

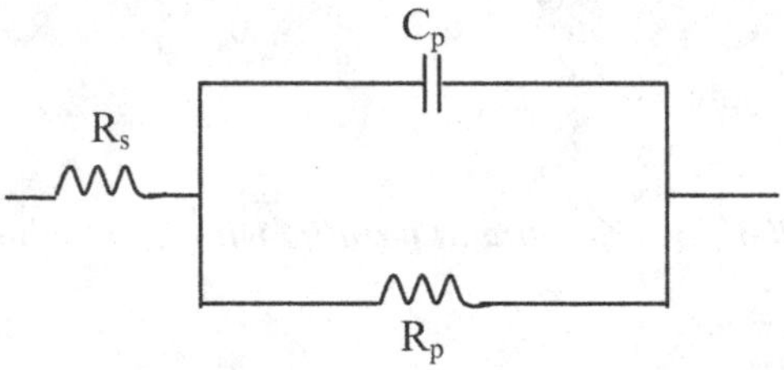

$$Z = R_S + \frac{R_P}{1+\omega^2 R_P^2 C_P^2} - j\frac{\omega R_P^2 C_P}{1+\omega^2 R_P^2 C_P^2};$$

• if $\omega = 0$, $\mathrm{Re}(Z) = R_s + R_p$ and $\mathrm{Im}(Z) = 0$;

• if $\omega \to \infty$, $\mathrm{Re}(Z) = R_s$ and $\mathrm{Im}(Z) \to 0$.

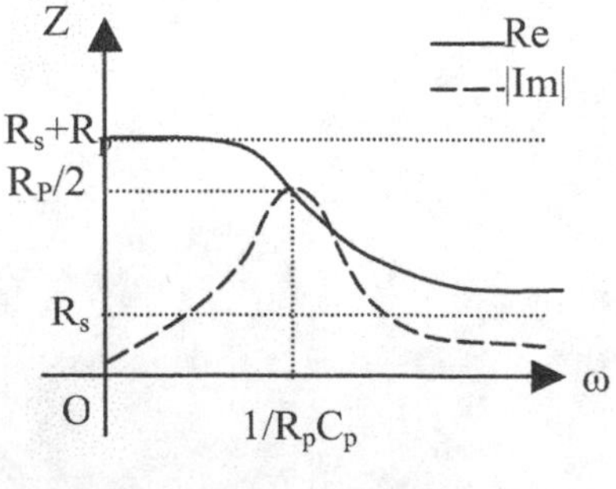

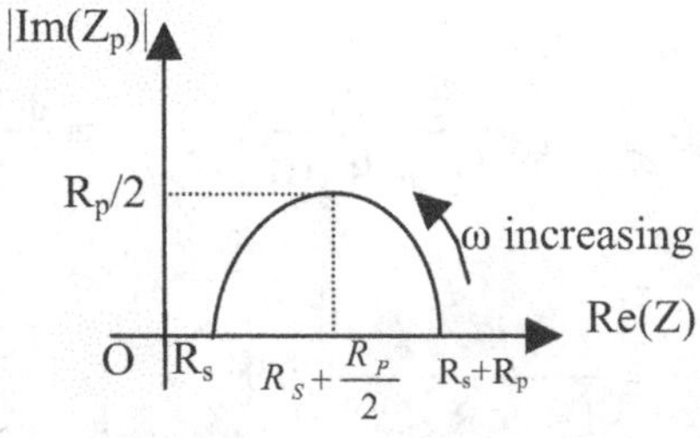

The angular frequency for which $\mathrm{Im}(Z_P)$ is at a maximum is given by:

$$\left[\frac{d\left|\mathrm{Im}(Z)\right|}{d\omega}\right]_{\omega=\omega_{max}} = \frac{R_P^2 C_P(1+\omega^2 R_P^2 C_P^2) - \omega R_P^2 C_p(2\omega R_P^2 C_P^2)}{(1+\omega^2 R_P^2 C_P^2)^2}$$

$$= \frac{R_P^2 C_P - \omega^2 R_P^4 C_P^3}{(1+\omega^2 R_P^2 C_P^2)^2} = 0,$$

from which it can be deduced that $\omega_{max} = \dfrac{1}{R_P C_P}$.

At this angular frequency, we thus have:

$$\left|\mathrm{Im}(Z)\right|_{max} = \frac{R_P}{2} \text{ and } R_e(Z) = R_S + \frac{R_P}{2}.$$

2.8.4. Problem 4. R-C in parallel and in series with a capacitor (Cs)

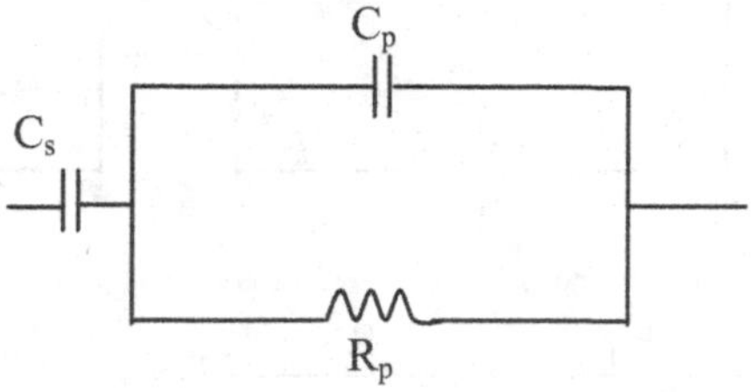

$$Z = \frac{1}{JC_S\omega} + \frac{R_P}{1+j\omega R_P C_p} = \frac{1+j\omega R_P C_p + j\omega R_P C_S}{-C_S C_p \omega^2 R_P + j\omega C_S},$$

$$Z = \frac{-R_P C_S C_P \omega^2 - jC_S\omega - jC_S C_P \omega^3 R_P^2 (C_S + C_P)}{R_P^2 C_S^2 C_P^2 \omega^4 + C_S \omega^2}$$

$$= \frac{C_S R_P \omega^2 - jC_S\omega\left[1 + R_P C_P \omega^2 (C_P + C_S)\right]}{\left[1 + R_P^2 C_P^2 \omega^2\right] C_S^2 \omega^2};$$

- if $\omega \to 0$, $Re(Z) \to Rp$ and $Im(Z) \to \infty$;
- if $\omega \to \infty$, $Re(Z) \to 0$ and $Im(Z) \to 0$.

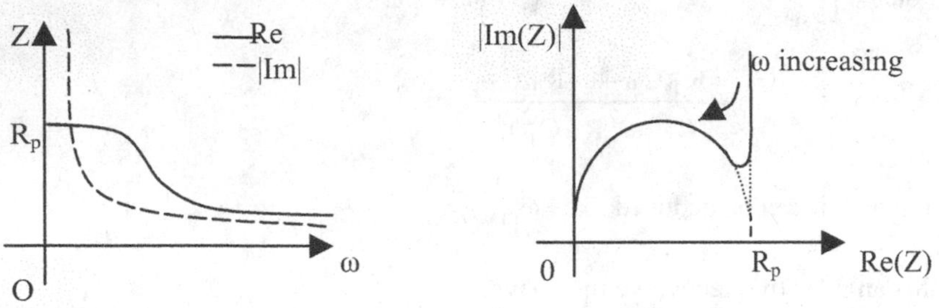

2.8.5. Problem 5. R-C in parallel and in series with a parallel R-C

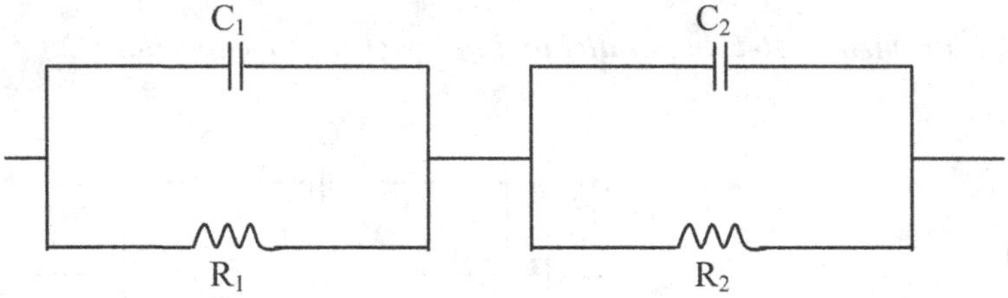

$$Z = Z_1+Z_2= \frac{R_1}{1+j\omega R_1C_1}+\frac{R_2}{1+j\omega R_2C_2}$$

$$= \frac{R_1}{(1+\omega^2R_1^2C_1^2)} - j\frac{\omega^2R_1^2C_1^2}{(1+\omega^2R_1^2C_1^2)} + \frac{R_2}{(1+\omega^2R_2^2C_2^2)} - j\frac{\omega^2R_2^2C_2^2}{(1+\omega^2R_2^2C_2^2)}$$

$$= [\frac{R_1}{(1+\omega^2R_1^2C_1^2)} + \frac{R_2}{(1+\omega^2R_2^2C_2^2)}] - j\omega[\frac{R_1^2C_1^2}{(1+\omega^2R_1^2C_1^2)} + \frac{R_2^2C_2^2}{(1+\omega^2R_2^2C_2^2)}];$$

- if $\omega \to 0$, $Re(Z) \to R_1+R_2$ and $Im(Z) \to 0$;
- if $\omega \to \infty$, $Re(Z) \to 0$ and $Im(Z) \to 0$.

The angular frequency for which the Im(Z) is at a maximum is given by the solution to the equation:

$$\frac{d|\mathrm{Im}(Z)|}{d\omega} = \frac{R_1^2C_1(1-\omega^2R_1^2C_1^2)}{(1+\omega^2R_1^2C_1^2)^2} + \frac{R_2^2C_2(1-\omega^2R_2^2C_2^2)}{(1+\omega^2R_2^2C_2^2)^2} = 0,$$

which in turn gives $R_1^2C_1(1-\omega^2R_1^2C_1^2) = 0$ and $R_2^2C_2(1-\omega^2R_2^2C_2^2) = 0$, from which can be deduced the two relaxation frequencies, as in:

$$\omega_1 = \frac{1}{R_1C_1} \quad \text{and} \quad \omega_2 = \frac{1}{R_2C_2}.$$

This is observed in the representation of a dielectric by the layer model given by Maxwell–Wagner–Sillars (MWS model).

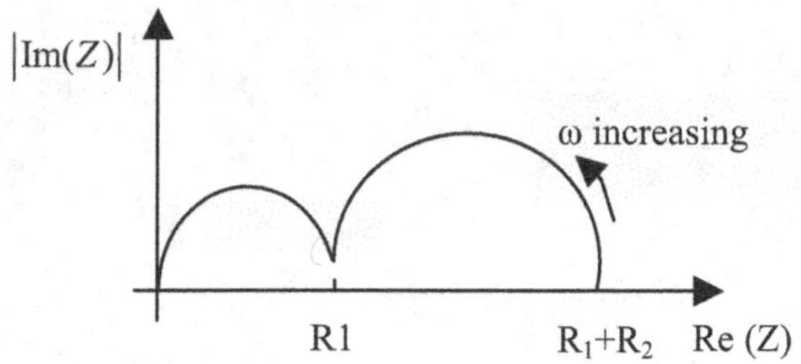

Chapter 3

Spectroscopy of Dielectrics and the Kramers–Krönig Relations

3.1. Introduction: dielectric response and direct current

3.1.1. A résumé of the components that make up the dielectric response

Chapter 1 of this volume detailed the mechanisms responsible for the phenomena of charge relaxation due to dipole moments generated by bound charges and associated with directional polarization. In solids, this mechanism gives rise to dielectric losses that generally attain a maximum in the radio frequency range. In liquids, which are less dense, the same maximum is most often displaced toward the shortwave and microwave regions.

Chapter 2 was particularly concerned with:

- recalling the form of the plots obtained by resonance absorption originally detailed in Chapter 8, Volume 1. When associated with polarization by displacement of the electric field due to bound and covalently bound electrons, the absorption is generally in the visible range. When the mechanism is due to polarization of ions, the absorption is generally in the infrared region. In both cases the quantification of the mechanisms (not detailed in this text) makes it possible to:
    - o relate electronic energy levels of various electrons to their electronic configuration and their associated quantum numbers;
    - o introduce the notion of phonons for the levels of vibrational energies that atoms and ions undergo.

• showing that the accumulated charges on either side of an interface give rise to the so-called Maxwell–Wagner–Sillars (MWS) dielectric losses associated with the interfacial polarization and at a maximum at low frequencies (*ca* 100 Hz).

In all of the above listed cases, only the displacement of bound charges (such as positive or negative charges at a permanent or induced dipole, interfacial dipoles, electronic charges bound to and as part of a covalent or ionic bond) has been reviewed. When influenced by a varying (normally sinusoidal) electric field, the dipoles change direction with movements "perturbed" by intermolecular interactions so that they follow the orientation of the electric field with a certain dephasing, and hence give rise to dielectric losses.

3.1.2. Influence of "pseudofree" charges on electric behavior

The movement of free or "pseudofree" charges can be added to the above list for certain specific cases such as, for example, electrons in a thermally generated conduction band across a large forbidden band in an insulator. Within the order of collisions against the lattice or impurities—which introduce a relaxation time (τ) equal to the average time between two successive collisions—these charges have a practically instantaneous response to an electric field.

Assuming that that τ is identical for all carriers, whatever their energy, the classic Drude model (which leads to Ohm's law) gives the following expressions:

- for direct current conductivity (see Volume 1, Section 1.3.4.2):

$$\sigma_{DC} = \sigma_\ell = \frac{nq^2}{m^*}\tau ;$$

- for alternating current conductivity (see Problem 1):

$$\sigma_{AC} = \sigma(\omega) = \frac{nq^2\tau}{m^*}\frac{1 + j\omega\tau}{1 + \omega^2\tau},$$

wherein at low frequencies $\omega\tau << 1$ always and $\sigma_{AC} \approx \sigma_{DC} = \sigma_\ell$. For example, when $\tau \approx 10^{-15}$ s, it suffices that $\omega \le 10^{14}$ Hz.

The direct current conductivity (σ_{DC}) thus is associated with the transport of charges that have enough energy to overcome the distance between the two electrodes. The current is dependent upon and extremely sensitive to changes in the parameters of electric field intensity and temperature.

Comment. For the specific case of amorphous semiconductors, there are also carriers that hop between neighboring states and limited distances, given the rapid change in the electric field. Only hops to energetic positions that are easily accessible occur. The frequency of the hops increases with the frequency of the applied field, so that the alternating conductivity associated with the charge displacement is of the form $\sigma_{\approx} = A\ \omega^s$, where A is a constant and s is a parameter often around 0.8 (Austin and Mott's law).

3.1.3. Separation of dielectric response from DC conductivity

The resultant conductivity is thus a combination of two components: one due to "pseudofree" carriers (Section 3.1.2) and the other due to dielectric losses (Section 3.1.1).

DC conductivity (Section 3.1.2) is not in fact associated with a dielectric response (Section 3.1.1), which is actually due to an excitation from an alternating field. The response can be placed in terms of:

- phenomena due to dielectric absorption (ε'') associated with a Debye dipolar relaxation, a MWS effect, or a resonance absorption; and
- a variation in the permittivity (ε') and phenomena due to dielectric dispersion.

However, on going toward very low frequencies and DC, there is a single component that appears at the level of the complex dielectric permittivity. This is because only ε'' evolves in the presence of this conduction mechanism (σ_{DC}), which otherwise does not contribute to the component of permittivity (ε').

In the following Section 3.2, there is a relatively electrotechnical discussion on the behavior of DC conductivity and how it may be isolated from the dielectric conductivity (due to bound charges). Section 3.3 details the Kramers–Krönig theory which establishes the general relation between ε' and ε''. On knowing the real component throughout the whole spectrum, this relation makes it possible to realize the imaginary component of the dielectric, and visa versa. The formula makes it possible also to characterize the CD conductivity when only ε'' changes with frequency. In effect, if the Fourier transformation of ε'' (proportional to ε') gives a constant value toward low frequencies, then ε' does not give rise to a dispersion effect and the corresponding variation in ε'' can be due only to a component of the DC conductivity.

3.2. Complex conductivity

3.2.1. General equations for real and imaginary components of conductivity

3.2.1.1. Equivalence of current density and electronic polarization current density

This section concerns only the electronic current and its equivalent, as will be seen. For the volume (or "internal part") of the material the current density is given by the equation $\vec{j}_{int} = \rho_{int}\vec{v}$.

Under the effect of an excitation provided by a sinusoidal electric field (given by $\underline{\vec{E}} = \underline{E}_0 e^{i\omega t}$), the speed of the electrons in complex notation is given in the form:

$$\underline{\vec{v}} = \frac{d\underline{\vec{r}}}{dt} = i\omega\underline{\vec{r}}\,.$$

The electronic polarization associated with the displacement ($\underline{\vec{r}}$) of electrons (e) of density n_e is given by:

$$\underline{\vec{P}} = -n_e e\underline{\vec{r}},$$

and $\rho_{int} = n_e(-e)$. From this we have

$$\underline{\vec{j}}_{int} = \rho_{int}\,\underline{\vec{v}} \overset{\underline{\vec{P}}=-n_e e\underline{\vec{r}}}{=} -i\omega n_e e\underline{\vec{r}} = i\omega\,\underline{\vec{P}} = \frac{\partial\underline{\vec{P}}}{\partial t}, \qquad (1)$$

where $\dfrac{\partial\underline{\vec{P}}}{\partial t}$ is the density of electronic polarization current which closely follows the internal current density.

3.2.1.2. Recall: total current density in a dielectric without losses ($\underline{\varepsilon}_r = \varepsilon_r$) placed between metallic electrodes

The density of the total current (j_{tot}) such that $I = \iint \vec{j}_{tot}.\vec{dS}$, so that $I = j_{tot}.S$ for a simple structure (homogeneous media between plane electrodes as detailed in Section 5.2.3, Volume 1) can be formulated with $\sigma_T = \dfrac{Q}{S}$ (total superficial charge density):

$$j_{tot} = \frac{I}{S} = \frac{dQ/dt}{S} = \frac{d}{dt}\left(\frac{Q}{S}\right) = \frac{d\sigma_T}{dt}. \qquad (2)$$

As $\sigma_T = \sigma_0 + \sigma_P$ (sum of the densities of free and polarization charges), we have:

$$j_{tot} = \frac{d\sigma_0}{dt} + \frac{d\sigma_P}{dt}, \text{ so that with } \sigma_P = P, \text{ then } j_{tot} = \frac{d\sigma_0}{dt} + \frac{dP}{dt}. \qquad (3)$$

For its part, the electric field between the electrodes that contain a dielectric of absolute permittivity (ε) can be expressed in two ways. In terms of moduli (see Volume 1, Section 5.2.3) we have either:

• $E = \frac{\sigma_T}{\varepsilon}$, which with Eq. (2) gives:

$$j_{tot} = \varepsilon\frac{dE}{dt} = \varepsilon_0\varepsilon_r\frac{dE}{dt} = j_D, \qquad (4)$$

where j_D is the displacement current; or

• $E = \frac{\sigma_0}{\varepsilon_0}$, which with Eq. (3) we have $j_{tot} = \varepsilon_0\frac{dE}{dt} + \frac{dP}{dt}$, so that with Eq. (1) it is possible to state that:

$$j_{tot} = \varepsilon_0\frac{dE}{dt} + j_{int}, \qquad (5)$$

in which the term $\varepsilon_0\frac{dE}{dt}$ is thus the displacement current in a vacuum and $j_{int} = \frac{dP}{dt}$ is the polarization current.

3.2.1.2. Density of the total current in a dielectric with losses and a relative permittivity ($\underline{\varepsilon}_r$)

In such a system, $\underline{\vec{P}} = \varepsilon_0\underline{\chi}\underline{\vec{E}} = \varepsilon_0(\underline{\varepsilon}_r - 1)\underline{\vec{E}}$, and

$$\underline{\vec{j}}_{int} = \frac{\partial\underline{\vec{P}}}{\partial t} = \varepsilon_0\underline{\varepsilon}_r\frac{\partial\underline{\vec{E}}}{\partial t} - \varepsilon_0\frac{\partial\underline{\vec{E}}}{\partial t}. \qquad (6)$$

In analogy to Eq. (5), the density of the total current is given by: $\underline{\vec{j}}_{tot} = \varepsilon_0\frac{\partial\underline{\vec{E}}}{\partial t} + \underline{\vec{j}}_{int}$, so that on taking Eq. (6) and $\underline{\vec{D}} = \varepsilon_0\underline{\varepsilon}_r\underline{\vec{E}}$ into account:

$$\underline{\vec{j}}_{tot} = \varepsilon_0\underline{\varepsilon}_r\frac{\partial\underline{\vec{E}}}{\partial t} = \frac{\partial\underline{\vec{D}}}{\partial t}. \qquad (7)$$

With $\underline{\varepsilon}_r = \varepsilon_r' - j\,\varepsilon_r''$, we have:

$$\underline{\vec{j}}_{tot} = \frac{\partial\underline{\vec{D}}}{\partial t} = \varepsilon_0\varepsilon_r'\frac{\partial\underline{\vec{E}}}{\partial t} - j\varepsilon_0\varepsilon_r''\frac{\partial\underline{\vec{E}}}{\partial t}.$$

Incidentally, it is worth noting that the displacement current given by Eq. (4) is in this case $\vec{j}_D = \varepsilon_0\varepsilon_r'\frac{\partial\underline{\vec{E}}}{\partial t}$ (and not, as might otherwise be thought possible,

$\frac{\partial \vec{\underline{D}}}{\partial t} = \varepsilon_0 \underline{\varepsilon}_r \frac{\partial \vec{\underline{E}}}{\partial t} = \vec{\underline{j}}_{tot}$).

The total current density ($\vec{\underline{j}}_{tot}$) thus can be written for a field given by $\vec{\underline{E}} = \underline{E}_0 e^{i\omega t}$, as:

$$\vec{\underline{j}}_{tot} = j\omega\varepsilon_0\varepsilon_r' \,\vec{\underline{E}} + \omega\varepsilon_0\varepsilon_r'' \,\vec{\underline{E}} . \qquad (8)$$

For its part, the complex conductivity ($\underline{\gamma}$) can be defined by the relation:

$$\vec{\underline{j}}_{tot} = \underline{\gamma}\, \vec{\underline{E}} . \qquad (9)$$

Then by writing $\underline{\gamma}$ in the form $\underline{\gamma} = \gamma' + i\gamma''$, so that $\vec{\underline{j}}_{tot} = \gamma' \,\vec{\underline{E}} + i\,\gamma'' \,\vec{\underline{E}}$, we have by identification with Eq. (8):

$$\boxed{\begin{array}{l} \gamma' = \omega\varepsilon_0\varepsilon_r'' = \omega\varepsilon'' \\ \gamma'' = \omega\varepsilon_0\varepsilon_r' = \omega\varepsilon'. \end{array}} \qquad (10)$$

The real part of the complex conductivity thus can be related to the imaginary part of the dielectric permittivity. Inversely, the imaginary part of the conductivity is equated with the real part of the dielectric permittivity.

In fact, the term ε_r'' of the dielectric permittivity can contain different components, depending on the physical origin of the real conductivity of the dielectric (residual free or pseudofree charges, dielectric losses due to bound charges).

3.2.2. Dielectric conductivity due to residual free or bound charges

3.2.2.1. Dielectric with only residual free charges

It is assumed here that the dielectric is nonpolar and therefore does not exhibit dielectric losses due to either induced or permanent dipoles. Only a few residual free charges are present, for example, electrons in a thermally generated conduction band across a large forbidden band in an insulator or "pseudofree" charges. For a purely dielectric and nonpolar contribution there is no dephasing between the excitation $\vec{\underline{E}} = \vec{\underline{E}}_0 e^{i\omega t}$ and the establishment of the response in terms of polarization of the dielectric (i.e., no dielectric absorption) so that $\vec{\underline{P}} = \varepsilon_0(\varepsilon_r - 1)\vec{\underline{E}}$ where ε_r takes on a real magnitude. In addition, $\vec{\underline{D}} = \varepsilon_0\varepsilon_r \vec{\underline{E}}$.

The resultant current for such a system is thus due to two components:

- one due to a perfect dielectric of absolute permittivity given by $\varepsilon' = \varepsilon_0\varepsilon_r$, and capacity $C = \frac{S\varepsilon'}{d}$ where S is the plane surface area and d is the distance separating the electrodes; and
- the other due to free charges that continuously generate a conductivity (γ_0) due to a conductance (G) such that $G = \frac{1}{R}$, where the resistance R for such a structure is given by $R = \rho_0 \frac{d}{S}$. With $\rho_0 = \frac{1}{\gamma_0}$, we have $G = \gamma_0 \frac{S}{d}$.

Figure 3.1 represents the resultant intensity which is such that:

$$\left.\begin{aligned} I &= (Y_C + Y_R)\,V = (i\omega C + G)V = (i\omega\varepsilon' + \gamma_0)\,S\frac{V}{d} = (i\omega\varepsilon' + \gamma_0)\,S\,E \\ &= j.S \end{aligned}\right\}$$

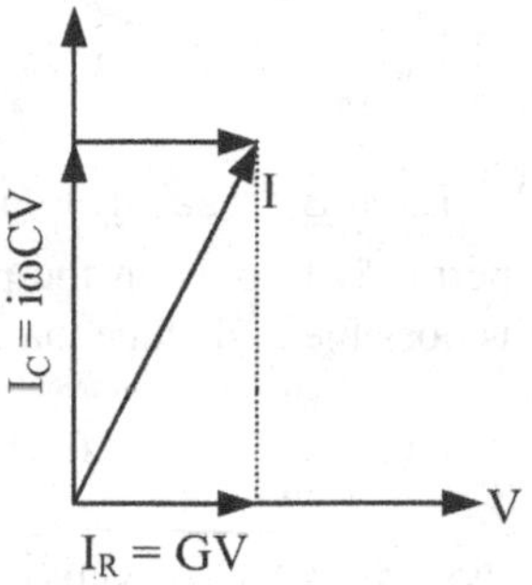

Figure 3.1. *Fresnel diagram for a dielectric with a low conductance (G).*

In terms of vectors, it is possible to state that:

$$\underline{\vec{j}} = (i\omega\varepsilon' + \gamma_0)\,\underline{\vec{E}} = i\omega\varepsilon'\,\underline{\vec{E}} + \gamma_0\,\underline{\vec{E}} \qquad (11)$$

$$= \varepsilon'\frac{\partial\underline{\vec{E}}}{\partial t} + \gamma_0\,\underline{\vec{E}} = \vec{j}_D + \vec{j}_c\,.$$

The term $\vec{j}_D = \varepsilon'\frac{\partial\underline{\vec{E}}}{\partial t}$ denotes the displacement current and the term $\vec{j}_c = \gamma_0\,\underline{\vec{E}}$ is that of the conduction current. It is advantageous to note that $\vec{j}_D$ is not underlined as its associated physical magnitude has only one component tied to the only term ε' which is real.

However, $\underline{\vec{j}}_{tot} = \underline{\varepsilon}\frac{\partial \underline{\vec{E}}}{\partial t}$, introduced in Section 3.2.1.2 and having two physical components associated with the real and imaginary part of $\underline{\varepsilon}$, has to be noted in its complex form (and thus underlined).

In global terms, we also can take into accounts these two components by:

- as above in Eq. (9), introduce a complex conductivity $\underline{\gamma} = \gamma' + i\gamma''$ such that $\underline{\vec{j}} = \underline{\gamma}\ \underline{\vec{E}}$. On identification with Eq. (11), we obtain $\underline{\gamma} = \gamma_0 + i\omega\varepsilon'$, from which:

$$\gamma' = \gamma_0$$
$$\gamma'' = \omega\varepsilon'$$

- or by introducing a complex permittivity as in $\underline{\varepsilon} = \varepsilon' - i\,\varepsilon_\ell''$ such that:

$$\underline{\vec{j}} = \underline{\varepsilon}\frac{\partial \underline{\vec{E}}}{\partial t}. \qquad (12)$$

We thus have $\underline{\vec{j}} = i\omega(\varepsilon' - i\,\varepsilon_\ell'')\ \underline{\vec{E}} = i\omega\varepsilon'\ \underline{\vec{E}} + \omega\,\varepsilon_\ell''\ \underline{\vec{E}}$ (noting that ε_ℓ'' is the imaginary component of the permittivity only in the presence of free charges). By identification with Eq. (11) it is possible to deduce that:

$$\varepsilon_\ell'' = \frac{\gamma_0}{\omega}. \qquad (13)$$

The losses simply due to free charges can be represented either by the conductivity (γ_0), or with an imaginary component of $\underline{\varepsilon}$ such that $\varepsilon_\ell'' = \gamma_0/\omega$. This latter equation represents the evolution of phenomenon I detailed in Section 1.1.6.

3.2.2.2. Dielectric with residual free and bound charges

In this case, the current vector density exhibits two components:

- one still corresponds to free charges (free electrons and ions) which generate a conductivity γ_0, and thus a conduction current density $\underline{\vec{j}}_c = \gamma_0\,\underline{\vec{E}}$, which can be written, taking into account the result from the preceding Section 3.2.2.1 as: $\underline{\vec{j}}_c = \omega\varepsilon_\ell''\underline{\vec{E}}$; and
- the other a component due to the dielectric which now, in addition, exhibits losses associated with the displacement of bound carriers (electronic polarization associated with electrons bound in covalent bonds, ionic

polarization due to ions in ionic bonds, dipole polarizations, and so on). This component is in the form $\vec{\underline{j}}_{diél} = \underline{\varepsilon}_{diel} \frac{\partial \vec{\underline{E}}}{\partial t}$, where $\underline{\varepsilon}_{diel} = \varepsilon' - i\varepsilon_d''$ and ε_d'' denotes the imaginary component of the dielectric permittivity due simply to bound charges.

We thus find:

$$\vec{\underline{j}}_{diél} = i\omega\,(\varepsilon' - i\varepsilon_d'')\,\vec{\underline{E}} = i\omega\varepsilon'\,\vec{\underline{E}} + \omega\varepsilon_d''\,\vec{\underline{E}}.$$

The total current density vector thus can be written as:

$$\vec{\underline{j}}_{tot} = \vec{j}_c + \vec{\underline{j}}_{diél} = i\omega\varepsilon'\,\vec{\underline{E}} + \omega(\varepsilon_l'' + \varepsilon_d'')\,\vec{\underline{E}}.$$

By identification with the general expression, as in $\vec{\underline{j}}_{tot} = \underline{\gamma}\,\vec{\underline{E}} = \gamma'\,\vec{\underline{E}} + i\gamma''\,\vec{\underline{E}}$ (where $\underline{\gamma} = \gamma' + i\gamma''$), we have:

$$\boxed{\begin{aligned} \gamma' &= \omega(\varepsilon_\ell'' + \varepsilon_d'') = \gamma_0 + \omega\,\varepsilon_d'' \\ \gamma'' &= \omega\,\varepsilon'. \end{aligned}} \qquad (14)$$

In the absence of free charges, $\gamma_0 = 0$ and $\gamma' = \omega\,\varepsilon_d'' = \gamma_{die}$ which is the purely dielectric conductivity. This conductivity is in fact the result of various dielectric contributions:

$$\gamma_{diel} = \gamma_{dipolar} + \gamma_{electronic\ polarization} + \gamma_{ionic\ polarization} + \ldots$$

$$= \omega\,\varepsilon_d'' = \omega\,(\varepsilon''_{dipolar} + \varepsilon''_{elec} + \varepsilon''_{ion} + \ldots). \qquad (15)$$

In the presence of free charges, the conductivity ($\gamma_0 = \omega\,\varepsilon_\ell''$) should be added. The complex part of the permittivity thus becomes

$\varepsilon'' = \varepsilon_\ell'' + \varepsilon''_{dipolar} + \varepsilon''_{elec} + \varepsilon''_{ion} + \ldots$, where $\varepsilon_\ell'' = \frac{\gamma_0}{\omega}$, and γ_0 is independent of frequency if the charges are completely free (which can follow extremely high frequencies due to the limitation in Ohm's law, as detailed in Chapter 1, Volume 1).

It thus is possible to see that ε_ℓ'' increases toward very low frequencies. This can result in the introduction of a "perturbation" term in the resulting value measured for ε" as γ_0 is no longer infinitely small. This term ε_ℓ'' is an electrical one, which takes into account all electrical measurements. However, in order to study simply the dielectric contribution, this term must be withdrawn from the characterization. This is possible by use of the theoretical relations devised by Krönig and Kramers, as indicated in Section 3.3 and in practical terms in the following Section.

3.2.2.3. Practical separation of the influences due to free charge conduction on dielectric losses caused by dipoles

For a dielectric made up of a polar medium with residual free charges, the dielectric permittivity is given by $\underline{\varepsilon} = \varepsilon' - i\,\varepsilon'' = \varepsilon' - i\,(\varepsilon''_\ell + \varepsilon''_{dip})$. It thus is possible to write:

• either $\underline{\varepsilon} = \varepsilon' - i\,(\varepsilon''_{dip} + \varepsilon''_\ell)$, from which the dielectric losses due to Debye dipolar absorptions mean that

$$\underline{\varepsilon} = (\varepsilon' - i\,\varepsilon'')_{dip} - i\varepsilon''_\ell = \varepsilon_\infty + \frac{(\varepsilon_s - \varepsilon_\infty)}{1 + i\omega\tau} - i\frac{\gamma_0}{\omega}. \qquad (16)$$

The plot of $\varepsilon'' = f(\varepsilon')$ is no longer a circle but gives rise to the form illustrated in Figure 3.2.

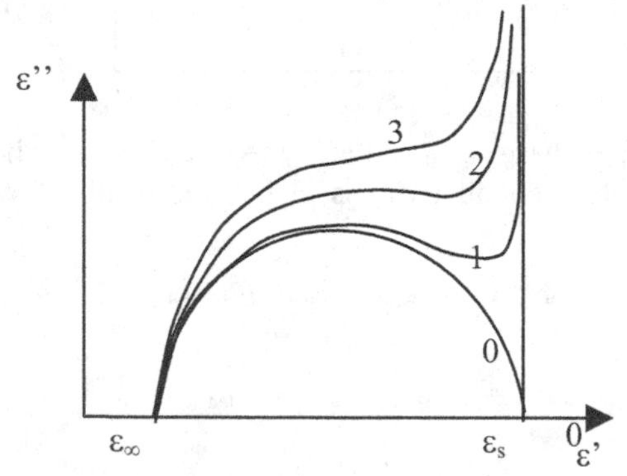

Figure 3.2. *Influence of direct current (where $\gamma_{03} > \gamma_{02} > \gamma_{01} > \gamma_{00}$) on the plot of $\varepsilon'' = f(\varepsilon')$.*

• or $\varepsilon'' = \varepsilon''_\ell + \varepsilon''_{dip}$, and if the dielectric losses are modeled on the Debye theory, then we can state that:

$$\varepsilon'' = \frac{\gamma_0}{\omega} + (\varepsilon_s - \varepsilon_\infty)\frac{\omega\tau}{1 + \omega^2\tau^2}.$$

A plot of log ε'' as a function of frequency makes it possible to easily distinguish the alternating conductivity associated with the dipolar relaxation (in general terms for bound charges) from the conductivity due to free charges.

In effect, from the preceding relation, it can be deduced that:

- when $\omega\tau << 1$, we have $\varepsilon'' = \varepsilon_I'' \approx \frac{\gamma_0}{\omega}$ (in zone I) $\Rightarrow \log \varepsilon_I'' = \log \gamma_0 - \log \omega$;
- when $\omega\tau >> 1$, we have $\varepsilon'' = \varepsilon_{II}'' \approx \frac{(\varepsilon_s - \varepsilon_\infty)}{\omega\tau}$ (in zone II) and

$$\Rightarrow \log \varepsilon_{II}'' = \log \frac{(\varepsilon_s - \varepsilon_\infty)}{\tau} - \log \omega;$$

- when $\omega\tau \approx 1$, then $\varepsilon'' \approx \frac{(\varepsilon_s - \varepsilon_\infty)}{2} + \gamma_0\tau \approx \frac{(\varepsilon_s - \varepsilon_\infty)}{2}$, and practically speaking $\gamma_0\tau \approx 10^{-18} << \frac{(\varepsilon_s - \varepsilon_\infty)}{2} \approx 10^{-11}$.

The result is that the plots of $\log \varepsilon_I'' = f(\log \omega)$ and $\log \varepsilon_{II}'' = f(\log \omega)$, which have slopes equal to –1, display a variance of $z = \log \frac{(\varepsilon_s - \varepsilon_\infty)}{\gamma_0\tau}$, which is independent of ω. This difference makes it possible to determine γ_0. From the plots of $\log \varepsilon'' = f(\log \omega)$ (shown in Figure 3.3), we can obtain $\frac{(\varepsilon_s - \varepsilon_\infty)}{2}$ and hence τ. By plugging in the values of $(\varepsilon_s - \varepsilon_\infty)$ and τ into z one determines γ_0 (which also can be obtained through extrapolation as indicated in Figure 3.3).

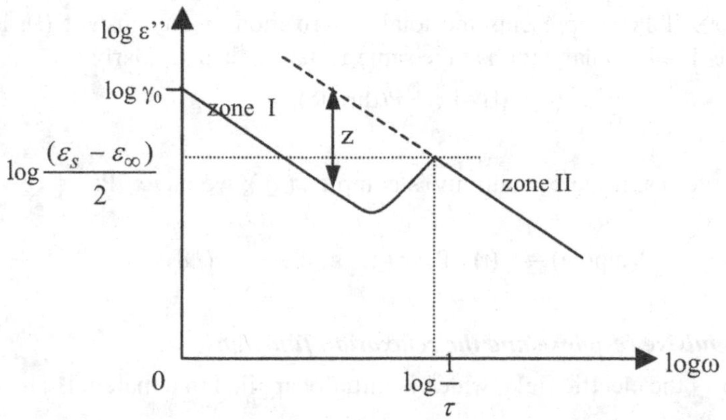

Figure 3.3. *Representation of logε'' = f(logω) permitting determination of the parameter z.*

3.3. Theoretical study of the dielectric function: the relaxation function, the Kramers–Krönig equations, and their use

3.3.1. Preliminary remarks

Here is established two fundamental equations which will then be used to formulate equations describing—in general terms—the response of a dielectric to an excitation caused by an electric field. When a field (E) is applied to a dielectric, the electrons and the nuclei (of atoms) of the medium move rapidly as they have a low inertia and their polarization is practically instantaneous. This is especially so in respect of the movements of dipolar polarizations due to the considerable inertia associated with the molecules involved. After a long period of time for which we can assume that a static regime has taken over, we can now write that $Ps(\text{total}) = P_\infty + P_s(\text{dipole})$, where:

- P_∞ denotes the polarization which is practically instantaneous, and hence follows high (infinite) frequency fields; and
- $P_s(\text{dipole})$ denotes the dipolar polarization which takes a nonnegligible time to be established.

By using the fundamental equation, as in $P = \varepsilon_0(\varepsilon_r - 1)E = (\varepsilon - \varepsilon_0)E$, we thus have:

$P_s(\text{total}) = (\varepsilon_s - \varepsilon_0)E$ and $P_\infty = (\varepsilon_\infty - \varepsilon_0)E$, from which:

$$P_s(\text{dipole}) = P_s(\text{total}) - P_\infty = (\varepsilon_s - \varepsilon_\infty)E\,. \qquad (17)$$

In addition, if P(t) represents the total polarization at any instant (t), and if P(dipole) is the dipolar polarization at the same instant t, then similarly

$$P\,(t) = P_\infty + P(\text{dipole}).$$

In general terms, if the permittivity is complex ($\underline{\varepsilon}$), we have: $P(t) = (\underline{\varepsilon} - \varepsilon_0)E$, and hence:

$$P(\text{dipole}) = P\,(t) - P_\infty = (\underline{\varepsilon} - \varepsilon_\infty)E. \qquad (18)$$

3.3.2. The impulsive response and the relaxation function

Figure 3.4 shows the electric field, which is initially applied to a material. The field can be written as:

$$E(t) = E_1\Gamma(t - t_1)\,, \qquad (19)$$

where $\Gamma(t - t_1) = 1$ when $t \leq t_1$ and $\Gamma(t - t_1) = 0$ when $t > t_1$.

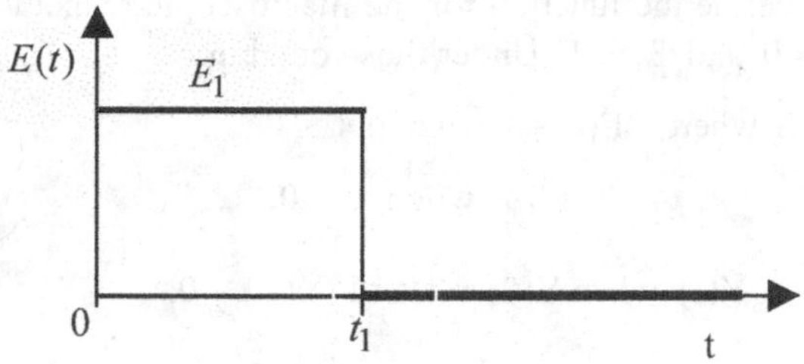

Figure 3.4. *Representation of an "exciting" electric field.*

Under a stationary regime, which corresponds to $\leq t_1$ (as $E = E_1$ = constant), the dipolar polarization (P) of the dielectric can be written, according to Eq. (17), as:

$P_s(\text{dipole}) = \chi E_1$, where by notation $\chi = (\varepsilon_s - \varepsilon_\infty)$.

By making $P_s(\text{dipole}) = P = P_1$, we also have: $\dfrac{P}{\chi} = \dfrac{P_1}{\chi} = E_1$.

Under a varying regime, we can write: $\dfrac{P(t)}{\chi} = E_1 Y(t - t_1)$, where $Y(t - t_1)$ corresponds to the function of decreasing dipolar polarization. The latter does not instantaneously cancel out at $t = t_1$ due to the dipole inertia (see Figure 3.5).

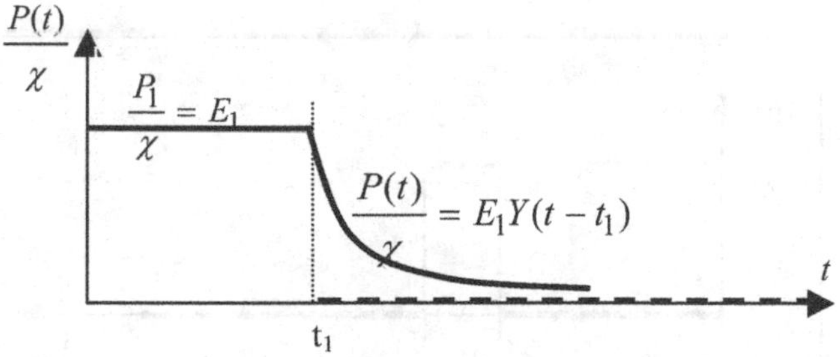

Figure 3.5. *Dielectric response [P(t)/χ] to the field [E(t)] of Figure 3.4.*

Given that when $t = t_1$ we have $\dfrac{P(t)}{\chi} = \dfrac{P(t_1)}{\chi} = E_1$, we can obtain a value for Y when $t = t_1$, as $Y(0) = 1$.

After an infinite period of time, the polarization tends toward zero (the dipoles are spread in an isotropic manner and their statistical result is zero in the absence of an orientating field after an infinite time), which means that $Y(\infty) = 0$.

We will now define the function for the macroscopic relaxation. In order to do this, we make $t_1 = 0$ and $E_1 = 1$. Under these conditions:

$$E(t) = \Gamma(t) \text{ where } \Gamma(t) = 1 \quad \text{when } t \le 0;$$
$$= 0 \quad \text{when } t > 0;$$

and $\dfrac{P(t)}{\chi} = Y(t)$ when $Y(0) = 1$ and $Y(\infty) = 0$.

$Y(t)$ thus appears as the response to the signal given by $\Gamma(t)$; and

$Y(t)$ is called the macroscopic relaxation function, as it is tied to the polarization which represents the macroscopic state of the dielectric.

3.3.3. Introducing the general expression for the response to a signal

After having considered the field $E(t) = E_1\Gamma(t - t_1)$, we now look at the "block function" for which it is assumed that:

$$\begin{cases} \text{when } (t_1 - \Delta t) < t \le t_1, \text{ the field is equal to } E_1 \\ \text{when } t \le (t_1 - \Delta t) \text{ and } t > t_1, \text{ the field is equal to } 0. \end{cases}$$

Schematically, we have the distribution presented in Figure 3.6.

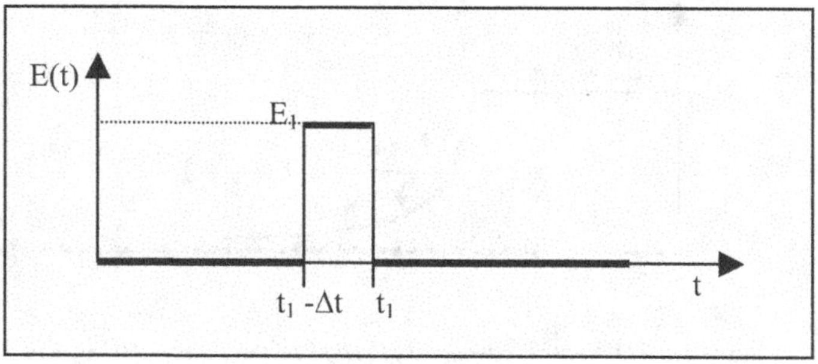

Figure 3.6. *Electric field for a "block function".*

For the field given by $E(t)$ we can write:

$$E(t) = E_1\Gamma(t - t_1) - E_1\Gamma(t - t_1 + \Delta t)$$

[where this block function is equal to (E_1 when $t \le t_1$), minus (E_1 when $t \le t_1 - \Delta t$)].

Given the preceding Section 3.3.2, with respect to the response, we can write:

$$\frac{P(t)}{\chi} = E_1 Y(t-t_1) - E_1 Y(t-t_1+\Delta t) = -E_1\left[\{Y(t-t_1+\Delta t) - Y(t-t_1)\}\right].$$

In addition, any field [E(t)] that is time dependent can be obtained by considering the result as a superposition of the block functions of an amplitude given by E_i each applied for a time t such that $t_i - \Delta t < t \le t_i$.

The individual effect of each of these block functions will give a response of the type $\frac{P(t)}{\chi}$, as indicated above.

The superposition of these block functions thus gives a field, for a linear dielectric, which gives rise to a response summing the responses of the applied signals. This can be written as:

$$\frac{P(t)}{\chi} = -\sum_i E_i(t_i)\left[\{Y(t-t_i+\Delta t) - Y(t-t_i\}\right]$$

where Δt represents the growth in t_i [or in $(t-t_i)$].

At the limit $\Delta t \to 0$, the block functions are transformed into impulse functions, and the summation becomes an integral, as in:

$$\frac{P(t)}{\chi} = -\sum_i E_i\left\{\frac{[Y(t-t_i+\Delta t) - Y(t-t_i)]}{\Delta t}\Delta t\right\} \xrightarrow[\Delta t \to 0]{} -\int E(t_i)\left\{\frac{\partial Y(t-t_i)}{\partial(t-t_i)}\right\}dt_i$$

We thus have demonstrated that it is possible to write—by changing the notation from t_i to t':

$$\frac{P(t)}{\chi} = -\int_{-\infty}^{t} E(t')\frac{\partial Y(t-t')}{\partial(t-t')}dt'$$

where the summation is overall possible instants t' from $-\infty$ up to t, the latter being the instant at which the response is characterized.

3.3.4. Relation between dielectric permittivities and the relaxation function

Under a static regime, where $E = E_s$ (the subscript s denoting "static"), we have seen in Eq. (17) that the dipolar polarization can be written as P_s (dipole) $= (\varepsilon_s - \varepsilon_\infty)E_s$. Given that $\chi = (\varepsilon_s - \varepsilon_\infty)$ we thus can state that P_s (dipole) $= \chi\ E_s$.

Under a regime that varies as a function of time, the expression for the dipolar polarization at an instant (t) is given by Eq. (18) from Section 3.3.1, so that:

$P(t) = (\underline{\varepsilon} - \varepsilon_\infty)E(t)$ where $E(t) = E_0 e^{j\omega t}$ for sinusoidal regimes.

From which can be deduced that:

$$\frac{P(t)}{\chi} = \frac{\underline{\varepsilon} - \varepsilon_\infty}{\varepsilon_s - \varepsilon_\infty} E_0 e^{j\omega t} = -\int_{-\infty}^{t} E_0 e^{j\omega t'} \frac{\partial Y(t-t')}{\partial(t-t')} dt'.$$

On changing the variable, as in $t - t' = \tau$ (where t is fixed and corresponds to the point where the polarization is calculated and t' varies, then $dt' = -d\tau$ and $t' = t \Rightarrow \tau = 0$, $t' \to -\infty \Rightarrow \tau \to +\infty$).

This gives:

$$\int_{-\infty}^{t} e^{j\omega t'} \frac{\partial Y(t-t')}{\partial(t-t')} dt' = e^{j\omega t} \int_{0}^{\infty} e^{-j\omega\tau} \frac{\partial Y(\tau)}{\partial\tau} d\tau,$$

from which

$$\frac{\underline{\varepsilon} - \varepsilon_\infty}{\varepsilon_s - \varepsilon_\infty} = -\int_{0}^{\infty} e^{-j\omega\tau} \frac{\partial Y(\tau)}{\partial\tau} d\tau.$$

By again changing the variable (by notation) to $\tau = t$, we can write in more general terms that:

$$\frac{\underline{\varepsilon} - \varepsilon_\infty}{\varepsilon_s - \varepsilon_\infty} = -\int_{0}^{\infty} e^{-j\omega t} \frac{\partial Y}{\partial t} dt = L(-Y'(t)), \qquad (20)$$

where L is the Laplace function.

3.3.5. The Kramers–Krönig relations

On integrating the preceding Eq. (20) by parts, with $u = e^{-j\omega t}$ and $dv = \frac{dY}{dt} dt$, we have:

$$\frac{\underline{\varepsilon} - \varepsilon_\infty}{\varepsilon_s - \varepsilon_\infty} = -\int_{0}^{\infty} e^{-j\omega t} \frac{dY}{dt} dt = -\left\{ \left[Y e^{-j\omega t} \right]_0^\infty + \int_{0}^{\infty} j\omega Y e^{-j\omega t} dt \right\}$$

$$= \frac{\varepsilon' - \varepsilon_\infty - j\varepsilon''}{\varepsilon_s - \varepsilon_\infty} = 1 - j\omega \int_{0}^{\infty} e^{-j\omega t} Y(t) dt$$

(as $Y(0) = 1$ and $Y(\infty) \to 0$).

Finally:

$$\frac{\varepsilon' - \varepsilon_s}{\varepsilon_s - \varepsilon_\infty} + \frac{\varepsilon_s - \varepsilon_\infty}{\varepsilon_s - \varepsilon_\infty} - \frac{j\varepsilon''}{\varepsilon_s - \varepsilon_\infty} = 1 + \left[-j\omega \int_0^\infty \cos\omega t \; Y(t)dt - \omega \int_0^\infty \sin\omega t \; Y(t)dt \right],$$

from which we have the two equations:

$$\frac{\varepsilon''(\omega)}{\varepsilon_s - \varepsilon_\infty} = \omega \int_0^\infty \cos\omega t \; Y(t)dt \qquad (21)$$

$$\frac{\varepsilon_s - \varepsilon'(\omega)}{\varepsilon_s - \varepsilon_\infty} = \omega \int_0^\infty \sin\omega t \; Y(t)dt . \qquad (22)$$

By reciprocal Fourier transformation, we directly obtain:

$$Y(t) = \frac{2}{\pi} \int_0^\infty \frac{\varepsilon''}{\varepsilon_s - \varepsilon_\infty} \cos\omega t \frac{d\omega}{\omega} \qquad (23)$$

$$Y(t) = \frac{2}{\pi} \int_0^\infty \frac{\varepsilon_s - \varepsilon'}{\varepsilon_s - \varepsilon_\infty} \sin\omega t \frac{d\omega}{\omega} . \qquad (24)$$

Plugging Eq. (24) into Eq. (21) gives

$$\frac{\varepsilon''}{\varepsilon_s - \varepsilon_\infty} = \omega \int_0^\infty \cos\omega t dt \frac{2}{\pi} \int_0^\infty \frac{\varepsilon_s - \varepsilon'(\omega')}{\varepsilon_s - \varepsilon_\infty} \sin\omega' t \frac{d\omega'}{\omega'}$$

$$= \frac{2\omega}{\pi} \lim_{R\to\infty} \int_0^R dt \int_0^\infty \frac{\varepsilon_s - \varepsilon'}{\varepsilon_s - \varepsilon_\infty} \cos\omega t \sin\omega' t \frac{d\omega'}{\omega'} ,$$

from which:

$$\varepsilon'' = \frac{2\omega}{\pi} \lim_{R\to\infty} \int_0^\infty \frac{d\omega'}{\omega'} \int_0^R (\varepsilon_s - \varepsilon') \cos\omega t \sin\omega' t dt$$

$$= \frac{2\omega}{\pi} \lim_{R\to\infty} \int_0^\infty d\omega' \left(\frac{\varepsilon_s - \varepsilon'(\omega')}{\omega'} \right) \int_0^R \cos\omega t \sin\omega' t \; dt .$$

In the sense of the distributions, the integral

$$\int_0^R \cos\omega t \sin\omega' t dt = \frac{1}{2} \int_0^R [\sin(\omega' + \omega)t + \sin(\omega' - \omega)t] dt$$

when $R \to \infty$, tends toward

$$\frac{1}{2} \left[\frac{1}{\omega + \omega'} + \frac{1}{\omega' - \omega} \right] = \frac{\omega'}{\omega'^2 - \omega^2} .$$

We thus obtain in definitive terms

$$\varepsilon''(\omega) = \frac{2\omega}{\pi}\int_0^\infty \frac{\omega'}{\omega'^2 - \omega^2}\frac{\varepsilon_s - \varepsilon'(\omega')}{\omega'}d\omega' = \frac{2\omega}{\pi}\int_0^\infty \frac{\varepsilon_s - \varepsilon'(\omega')}{\omega'^2 - \omega^2}d\omega'.$$

This relation can be expressed as a function of ε_∞:

$$\varepsilon''(\omega) = \frac{2\omega}{\pi}\int_0^\infty \frac{\varepsilon_s - \varepsilon_\infty}{\omega'^2 - \omega^2}d\omega' - \frac{2\omega}{\pi}\int_0^\infty \frac{\varepsilon'(\omega) - \varepsilon_\infty}{\omega'^2 - \omega^2}d\omega' \ ;$$

so that with $\int_0^\infty \frac{d\omega'}{\omega'^2 - \omega^2} \propto \int_0^\infty \frac{dn}{1-n^2} = \left[\frac{1}{2}\log\left|\frac{1+n}{1-n}\right|\right]_0^\infty = 0$ (in which we made $n = \frac{\omega'}{\omega}$), we can deduce that

$$\varepsilon''(\omega) = -\frac{2\omega}{\pi}\int_0^\infty \frac{\varepsilon'(\omega') - \varepsilon_\infty}{\omega'^2 - \omega^2}d\omega'. \qquad (25)$$

Similarly, by substituting Eq. (23) into Eq. (22), we obtain:

$$\varepsilon'(\omega) - \varepsilon_\infty = \frac{2}{\pi}\int_0^\infty \frac{\omega'\varepsilon''(\omega')}{\omega'^2 - \omega^2}d\omega'. \qquad (26)$$

3.3.6. Application to Debye relaxations

3.3.6.1. Form of the Y(t) function for a Debye process

Here we show that the Y(t) function therefore must be in the form $Y(t) = e^{-t/\tau}$.
Following the preceding calculations [Eq. (20)]:

$$\frac{\varepsilon - \varepsilon_\infty}{\varepsilon_s - \varepsilon_\infty} = L(-Y'(t)).$$

On calculating $L(-Y'(t))$ with $Y'(t) = -\frac{1}{\tau}e^{-t/\tau}$ we thus find:

$$L(-Y'(t) = -\int_0^\infty e^{-j\omega t} Y' dt = \frac{1}{\tau}\int_0^\infty e^{-(\frac{1}{\tau}+j\omega)t} dt = \frac{1}{\tau}\int_0^\infty e^{-\frac{1+j\omega\tau}{\tau}t} dt$$

$$= \frac{1}{\tau}\left[-\frac{\tau}{1+j\omega\tau} e^{-\frac{1+j\omega\tau}{\tau}t}\right]_0^\infty = \frac{1}{1+j\omega\tau}.$$

We clearly rediscover the equations, which describe a Debye process, which also can be obtained via a classic route (integration of the relaxation equation):

$$\frac{\underline{\varepsilon} - \varepsilon_\infty}{\varepsilon_s - \varepsilon_\infty} = \frac{1}{1+j\omega\tau}.$$

3.3.6.2. When several Debye domains are superimposed

For a discrete distribution of relaxation times (τ_i), it is possible to show that the preceding equation takes on the form:

$$\frac{\underline{\varepsilon} - \varepsilon_\infty}{\varepsilon_s - \varepsilon_\infty} = \sum_i \frac{A_i}{1+j\omega\tau_i}.$$

We can give the physical significance of A_i, then determine the relaxation function for which the limiting values can be verified.

On separating the real and imaginary values, it is possible to state:

- that $\frac{\varepsilon' - \varepsilon_\infty}{\varepsilon_s - \varepsilon_\infty} = \sum_i \frac{A_i}{1+\omega^2\tau_i^2}$, which indicates that the constants A_i are characteristic of the amplitude of various dispersion domains which verify the equation $\sum_i A_i = 1$;

- and $\frac{\varepsilon''(\omega)}{\varepsilon_s - \varepsilon_\infty} = \sum_i \frac{A_i\omega\tau_i}{1+\omega^2\tau_i^2}$. By making $m_i = \frac{t}{\tau_i}$ and $x = \omega\tau_i$, this expression

makes it possible to calculate from Eq. (23) the relaxation function, as in

$$Y(t) = \frac{2}{\pi}\sum_i A_i \int_0^\infty \frac{\cos m_i x}{1+x^2} dx.$$

Knowing that $\int_0^{\infty} \frac{\cos mx}{1+x^2} dx = \frac{\pi}{2} e^{-|m|}$ (and by noting that $m_i > 0$),

we obtain $Y(t) = \frac{2}{\pi} \sum_i A_i \frac{\pi}{2} e^{-m_i} = \sum_i A_i e^{-t/\tau_i} = \sum_i Y_i$.

(While checking that $Y(t=0) = \sum_i A_i = 1$, $Y(t \to \infty) = 0$).

Inversely, from the hypothesis that the resultant relaxation function corresponds to the sum if contributions of each relaxing domain, it is possible to state that each domain (i) is represented by a relaxation function $Y_i = A_i \exp\left(-\frac{t}{\tau_i}\right)$ so that

$Y = \sum_i Y_i = \sum_i A_i \exp\left(-\frac{t}{\tau_i}\right)$.

Performing a calculation similar to that shown in Section 3.3.6.1 and using Eq. (20) directly gives the following equation:

$$\frac{\varepsilon - \varepsilon_\infty}{\varepsilon_s - \varepsilon_\infty} = \sum_i \frac{A_i}{1 + j\omega\tau_i}$$

for which we can say Q.E.D.

3.3.7. Generalization of the Kramers–Krönig relations

The Kramers–Krönig relations, which we have just established for the relaxation phenomena induced by orientations from dielectric polarizations, are in fact valid with respect to all linear phenomena described by phenomenological processes.

3.3.7.1. Extension to induced polarizations mechanisms

So, just as the dielectric relaxation could be described with the help of a phenomenological mechanism (by bringing in the dipolar relaxation time), the mechanisms for resonance attached to induced polarization mechanisms can be studied similarly in a phenomenological manner. In effect, the absorption peaks are localized around angular frequencies (ω_k) that characterize the discrete energy levels associated with induced displacements. These absorption peaks, characterized by the function denoted $\varepsilon''(\omega)$thus can be written using delta functions, of which the frequency dependency is in the form $\varepsilon''_{(\omega)} = \sum_k A_k \delta(\omega - \omega_k)$, where ω_k is the angular frequency of the resonance under consideration.

With the help of one of the Kramers–Krönig relations, it is possible to deduce the frequency dependence of ε'(ω). The equation thus is brought from Eq. (26) and adapted to the limits of the resonance absorption. It should be noted that Eq. (26) is typical of the absorption due to relaxation mechanisms. In effect, while Eq. (26) directly yields for ε'(ω), the equation $\varepsilon'_{(\omega)} = \varepsilon_\infty + \frac{2}{\pi}\sum_k \frac{\omega_k A_k}{\omega_k^2 - \omega^2}$, we must still replace ε_∞ by ε_0. This is because the polarization given by the dipolar polarization in Eq. (18), i.e.,

$$P(\text{dipole}) = (\underline{\varepsilon} - \varepsilon_\infty)E = \varepsilon_0(\underline{\varepsilon_r} - \varepsilon_{r\infty})E,$$

needs to be replaced by the induced polarization, which is in the form

$$P_i = \varepsilon_0(\underline{\varepsilon_r} - 1)E.$$

The result is that $\varepsilon_{r\infty}$ must be replaced by 1, and that ε_∞ replaced by ε_0. This means that for resonance phenomena, we should write:

$$\varepsilon'_{(\omega)} = \varepsilon_0 + \frac{2}{\pi}\sum_k \frac{\omega_k A_k}{\omega_k^2 - \omega^2}.$$

For infinitely narrow bands (where $\omega \to \omega_k$), ε'(ω) goes to infinity at each absorption frequency, in a process known as catastrophic resonance. On making $B_k = A_k / \varepsilon_0$, we have:

$$\varepsilon'_{r(\omega)} = 1 + \frac{2}{\pi}\sum_k \frac{\omega_k B_k}{\omega_k^2 - \omega^2},$$

which closely resembles the equation found by a direct dielectric study of resonance phenomena [see Eq. (17) of Chapter 8, Volume 1].

3.3.7.2. Extension to other electromagnetic phenomena

These equations can be extended to other phenomena, notably:

- magnetic systems where Krönig–Kramer type relations exist between χ' and χ'', which are defined from the complex magnetic susceptibility, given by $\underline{\chi} = \chi' - j\chi''$; and
- conduction, where the complex conductivity given by $\underline{\gamma} = \gamma' - i\gamma''$ is such that, for example:

$$\gamma'(\omega) = \gamma_\infty + \frac{2}{\pi}\int_0^\infty \frac{\omega'\gamma''(\omega')}{\omega'^2 - \omega^2} d\omega'.$$

3.3.8. Application of the Krönig–Kramers relations

3.3.8.1. Determination of Y or ε' from ε'' for a whole spectrum

Knowing ε″ for a whole frequency range makes it possible to determine Y(t) and ε′ via Eqs. (23) and (26). This calculation is performed by numerical integration on a computer. Its interest lies in the possibility of determining ε′ with considerable precision, an otherwise difficult value to obtain from certain regions of the electromagnetic spectrum (far infrared, for example). In addition, the macroscopic relaxation function [Y(t)], which cannot be determined directly, can be characterized.

3.3.8.2. Dielectric dispersion and the identification of effects due to dielectric absorption and electric conduction

3.3.8.2.1. Estimation of the dielectric dispersion

The total dispersion of a material can be obtained by making ω tend toward zero in Eq. (26). This gives (with the following change in notation ω' → ω):

$$\varepsilon_s - \varepsilon_\infty = \frac{2}{\pi}\int_0^\infty \frac{\varepsilon''(\omega)}{\omega}d\omega = \frac{2}{\pi}\int_0^\infty \varepsilon''(\omega)d(\mathrm{Ln}\,\omega)\,. \qquad (27)$$

We can also write:

$$\int_0^\infty \varepsilon''(\omega)\frac{d\omega}{\omega} = \frac{\pi}{2}\left(\varepsilon_s - \varepsilon_\infty\right), \qquad (28)$$

where the surface delimited by a plot of ε'' as a function of ln ω is equal to $\frac{\pi}{2}\left(\varepsilon_s - \varepsilon_\infty\right)$.

In fact, to each polarization mechanism (α), we can associate an absorption with a dispersion of an amplitude ($\Delta\varepsilon_\alpha$) which is such that:

$$\Delta\varepsilon_\alpha = \frac{2}{\pi}\int_{\text{peak }\alpha} \varepsilon''(\omega)d(\mathrm{Ln}\,\omega)\,. \qquad (29)$$

3.3.8.2.2. Identification of effects due to dielectric absorption and electrical conduction by free or "pseudofree" charges

As set out in Section 3.3.7, the Krönig–Kramers relations can be established regardless of the polarization mechanism. They thus are applicable to all dispersion phenomena. Equation (29) in fact shows that for a given mechanism (α) for a polarization, the amplitude of the associated dispersion is equal to, within a factor of 2/π, to the area of the absorption peak plotted against a logarithmic scale. This therefore irrefutably demonstrates that a polarization mechanism is obligatorily accompanied by a mechanism of dielectric loss, as indicated in Figure 3.7.

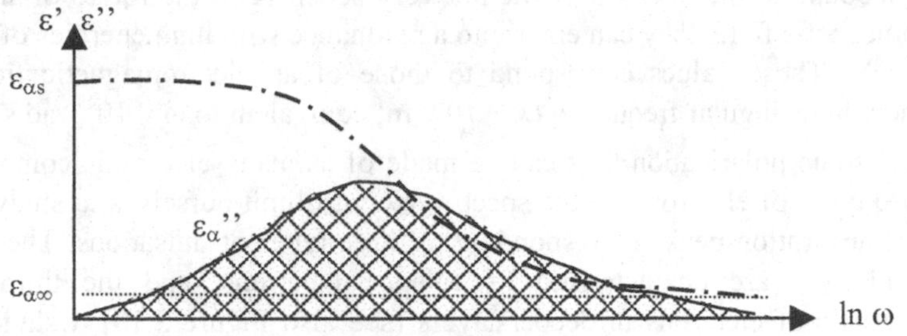

Figure 3.7. *Relaxation between an increasing dispersion for a given polarization mechanism and dielectric losses.*

In more practical terms, and by using the reverse argument, if we observe a variation in an electrical absorption which is unaccompanied by a dispersion phenomenon, then the corresponding losses cannot be the result of a polarization mechanism. In the low frequency region, and toward the continuous, such a variation in absorption (increasing ε" toward the very low frequencies) can be observed without a corresponding variation in ε'. The increase in ε" toward the very low frequencies therefore must be tied to the continuous conductivity (γ_0) such that $\varepsilon_\ell'' = \frac{\gamma_0}{\omega}$ [from Eq. (13)]. In this equation, ε_ℓ'' represents the actually measured dielectric losses resulting from dielectric effects and not a polarization mechanism. It should be remembered of course that apparatuses only measure, without discriminating between causes (such as mechanisms due to polarization and conductivity of free and pseudofree charges) which give rise to the same effects, i.e., a variation in ε". This may be contrasted though against measurements of ε', where a dispersion may be recorded for a polarization mechanism, but no change is observed when dealing with the conduction of free charge carriers.

3.4. Complete polarization of dielectrics, characteristics of spectra from dielectrics, and an introduction to spectroscopy

3.4.1. Electronic polarization and the relation between the angular frequency of an electronic resonance and the gap in an insulator

3.4.1.1. Electronic polarization associated with different types of electrons

During the study made in Chapter 8, Volume 1, on the polarization of electrons, it was only the valence electrons that were considered (as demonstrated in the values chosen for ω_{0e}). These electrons are the ones that occupy the external layers and typically participate in chemical, covalent bonding. However, we should take into account all types of electrons, notably those in the deeper, internal layers which are strongly bound to the nucleus. While not very sensitive to the forces of an external electromagnetic field, they can enter into a resonance with high energies of the order of 10^4 eV. These values correspond to those of an electromagnetic field of an extremely high angular frequency ($\lambda \approx 10^{-10}$ m, equivalent to $\omega \approx 10^{19}$ rad s^{-1}).

Electronic polarization thus can be made of as many electronic components as there are types of electrons. In the spectra, we will limit ourselves to studying three types of absorption peaks corresponding to three types of pulsations. The first two, ω_{0e1} and ω_{0e2}, are characteristic of valence electrons, and the third, ω_{0p}, is characteristic of electrons in deeper layers (see also Figure 3.10). Using quantum theory, it is possible to determine a breakdown in these levels by use of the characteristic quantum numbers for each of the electronic configurations.

3.4.1.2. Relation between static permittivity [ε_r'(0)] and the gap of an insulator (E_G)

Equation (10) detailed in Volume 1, Chapter 8, makes it possible to express the static dielectric permittivity bound to the electronic polarization as a function of the plasma angular frequency (ω_p) and the resonance pulsation frequency (ω_0), as in: $\varepsilon_r'(0) = 1 + \frac{\omega_p^2}{\omega_0^2}$. To indicate that this permittivity is attached to the contribution made by electronic (the most external of the electrons) polarization, an additional subscript e has been added giving $\varepsilon_{re}'(0)$.

In addition, if we accept as a first approximation that the gap (the size of the forbidden band) of an insulator is such that $E_G \approx \hbar\omega_0$, we can then state that:

$$\frac{\left(\hbar^2\omega_p^2\right)}{E_G^2} = \frac{\left(\hbar^2\omega_p^2\right)}{\left(\hbar^2\omega_0^2\right)} = \frac{\omega_p^2}{\omega_0^2} = \varepsilon_{re}'(0) - 1.$$

So as to set ourselves some guidelines, if we take $\hbar\omega_p = 16$ eV and $E_G \approx \hbar\omega_0 = 6$ eV, we end up with $\chi_e'(0) = \varepsilon_{re}'(0) - 1 = \frac{16^2}{6^2} \approx 7.11$, which gives $\varepsilon_{re}'(0) \approx 8.11$.

To justify the hypothesis that the gap (E_G) is such that $E_G \approx \hbar\omega_0$, we can remark that until now we have considered only the effect of the electronic polarization, along with, for these electrons, a maximum absorption (characterized by ε'') at the angular frequency $\omega = \omega_0 = \omega_{0e}$. The subscript e once again has been added to indicate that the resonance angular frequency corresponds to that for an electronic resonance. Where any steric and frictional interactions are negligible, i.e., $\omega_0 >> \Gamma = 1/\tau$, then the plot of $\varepsilon''(\omega)$ has a mid-height width which tends toward low values with respect to ω_0 (see also Volume 1, Chapter 8):

$$\Delta\omega = \omega_2 - \omega_1 = \Gamma = 1/\tau << \omega_0.$$

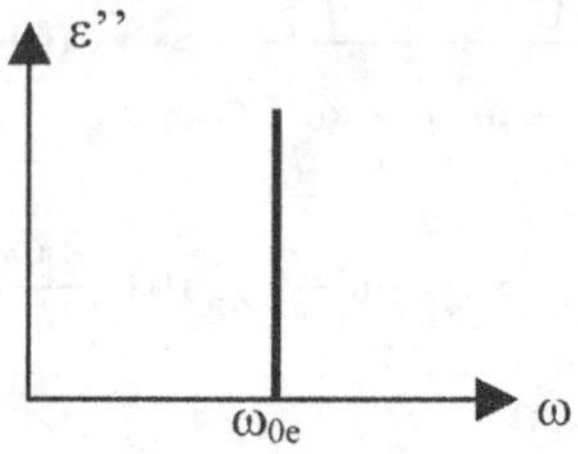

Figure 3.8. *Absorption peak in terms of ω_{0e}.*

The representation of $\varepsilon''(\omega)$ shown in Figure 3.8 is thus that of a peak situated at $\omega = \omega_0 = \omega_{0e}$. Outside of this peak, where $\omega \neq \omega_{0e}$, we find that ε'' is zero and there is no absorption. It thus is possible to state that the optical absorption (for a transition from the valence to the conduction band) corresponds to a gap with an average energy of the order of the energy of the resonance, as in $E_G \approx \hbar\omega_{0e}$.

Numerical applications:

- diamond: $E_G = 5.4$ eV, $\Rightarrow \omega_{0e} = 8 \times 10^{15}$ rad s^{-1} $\Rightarrow \lambda_0 = 2\pi c/\omega_{0e} \approx 0.25$ μm (UV);

- glass: $E_G \approx 10$ eV, $\Rightarrow \omega_{0e} = 1.5 \times 10^{16}$ rad s^{-1} $\Rightarrow \lambda_0 \approx 0.13$ μm (UV); and

- silicon: $E_G = 1.1$ eV, $\Rightarrow \omega_{0e} = 1.7 \times 10^{15}$ rad s^{-1}, and hence $\lambda_0 \approx 1.1$ μm (IR).

In dielectrics, typically, the time $\tau = \tau_e$ which corresponds to an electronic relaxation time is such that $\tau_e \approx 10^{-8}$ to 10^{-9} s. Taking the preceding values of ω_{0e} into account, where $\omega_{0e} \approx 10^{15}$ to 10^{16} rad s^{-1}, we have $\omega_{0e}\tau_e \approx 10^7 >> 1$ (which verifies the hypothesis formulated above that $\omega_0 >> \Gamma = 1/\tau$.

Thus from the equations for $\chi_e'(\omega)$ [Eqs. (6') and (9) in Volume 1, Chapter 8]:

- when $\omega << \omega_{0e}$, then $\omega\Gamma = \dfrac{\omega}{\tau_e} << \dfrac{\omega_{0e}}{\tau_e} = \dfrac{\omega_{0e}^2}{\omega_{0e}\tau_e} << \omega_{0e}^2$. The result is that (under a poorly damped regime):

$$\chi_e'(\omega) = \chi_e'(0)\frac{\omega_{0e}^2}{\left(\omega_{0e}^2 - \omega^2\right)} ; \qquad (30)$$

• when $\omega >> \omega_{0e}$, we then have

$$\chi_e'(\omega) = \chi_e'(0)\,\omega_{0e}^2 \frac{\left(\omega_{0e}^2 - \omega^2\right)}{\left(\omega_{0e}^2 - \omega^2\right)^2 + \omega^2\Gamma^2} \approx - \chi_e'(0)\,\omega_{0e}^2 \frac{-\omega^2}{\omega^4 + \omega^2\Gamma^2}$$

$$\Rightarrow \chi_e'(\omega) = - \chi_e'(0) \frac{\omega_{0e}^2}{\omega^2 + \Gamma^2} .$$

With $\omega >> \omega_{0e} \approx 10^{15}$ to 10^{16} rad s^{-1} and $\tau_e \approx 10^{-8}$ to 10^{-9} s, and hence $\omega >> 1/\tau_e = \Gamma$, we obtain the expression identical to that given as Eq. (19) in Chapter 8 of Volume 1:

$$\chi_e'(\omega) \approx -\chi_e'(0)\frac{\omega_{0e}^2}{\omega^2} . \qquad (31)$$

As in this region, $\omega >> \omega_{0e}$, the result is that $\chi_e'(\omega) \approx 0$. Equally, as $\varepsilon_{re}'' = \chi_e''(\omega) \approx 0$, we thus have $\underline{\varepsilon}_r \approx \varepsilon_{re}' \leq 1 = \varepsilon_{r0}$. The dielectric practically behaves as a vacuum because at these very high frequencies the electrons cannot follow the electric field.

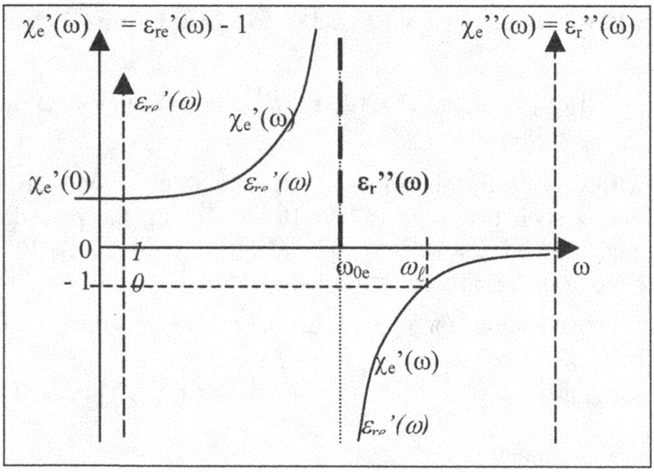

Figure 3.9. *Dispersion and absorption of dielectric resonance.*

To conclude, in dielectrics, the representations of $\chi_e'(\omega)$ [or $\varepsilon_{re}'(\omega)$] and of $\varepsilon_r''(\omega)$ have the forms shown in Figure 3.9. The plots of $\chi_e'(\omega)$ and $\varepsilon_{re}'(\omega)$ are in fact identical to the plots detailed in Figures 8.3 and 8.4 of Chapter 8, Volume 1.

3.4.2. Ionic polarization

Ionic polarization comes from the displacement of ionized atoms. Being considerably heavier than electrons by a factor of around 10^4 they cannot follow such high-field frequencies. The proper angular frequency of ions, denoted here as ω_{0i}, is mainly situated between 5 x 10^{12} and 10^{14} rad s^{-1}, and corresponds to the infrared/far infrared regions.

On using angular frequencies far enough from ω_{0i}, the system is weakly damped and the dielectric susceptibility is real ($\chi_i'' \approx 0$). The component of the ionic susceptibly thus gives rise to an equation similar to Eqs. (30) and (31). Now the subscript i denoting "ionic" replaces that of e, and:

- when $\omega \ll \omega_{0i}$ (microwave and radiofrequency region), we obtain by analogy to Eq. (30), $\chi_i(\omega) \approx \chi_i'(\omega) = \chi_i'(0)\dfrac{\omega_{0i}^2}{\left(\omega_{0i}^2 - \omega^2\right)}$. (32)

As $\omega \ll \omega_{0i}$, we have $\chi_i'(\omega) \approx \chi_i'(0) = \varepsilon_{ri}'(0) - 1$, and the susceptibility is practically constant; and

- when $\omega \gg \omega_{0i}$, we obtain through a similar analogy but with Eq. (31):

$$\chi_i(\omega) \approx \chi_i'(\omega) \approx -\chi_i'(0)\frac{\omega_{0i}^2}{\omega^2}. \qquad (33)$$

As $\omega \gg \omega_{0i}$, then $\dfrac{\omega_{0i}^2}{\omega^2} \approx 0$ and $\chi_i'(\omega \gg \omega_{0i}) \to 0$.

In effect, just as for the electronic frequencies, the ionic frequencies can be associated with different types of ions and also different types of ion displacements (harmonic and anharmonic potentials). Only quantum theory can elucidate the associated discrete energy levels.

3.4.3. Resultant polarization in an insulator

Generally speaking, the resultant polarization is the sum of all components of polarization susceptible to appear in a given medium. It is thus possible to state that:

$$\vec{P}_{total} = \vec{P}_{dipolar} + \vec{P}_{ionic} + \vec{P}_{electronic},$$

where each component can again be thought of as a resultant of a number of contributions, as discussed above for electrons and ions. For its part, the dipolar polarizations bring in various types of dipolar relaxations. These include, for example, the so-called α, β, and γ relaxations which characterize different dipolar movements in polymers (such as main chain segmental rotations).

On trying to sum up the various polarization mechanisms, it is important to bring in the polarization associated with space charges. In heterogeneous systems they generate an "interfacial" polarization (see also Section 2.5.4 of Volume 1) at interfaces (joints between grains). Given that these space charges have a low mobility, and that therefore there is a considerable time interval required for them to reach an interface, the interfacial polarization ($P_{interface}$) takes quite some time to be established. Thus, $P_{interface}$ only appears when low of very low frequencies are used (of the order of hundreds, a single or even fractions of a hertz).

The preceding expression thus should give the definitive equation:

$$\vec{P}_{total} = \vec{P}_{interface} + \vec{P}_{dipolar} + \vec{P}_{ionic} + \vec{P}_{electronic}$$

where

$$\vec{P}_{interface} = \varepsilon_0 \chi_{int} \vec{E}, \; \vec{P}_{dipolar} = \varepsilon_0 \chi_{dip} \vec{E}, \; \vec{P}_{ionic} = \varepsilon_0 \chi_{ion} \vec{E}, \; \vec{P}_{electronic} = \varepsilon_0 \chi_e \vec{E},$$

and:

$$\vec{P}_{total} = \varepsilon_0 (\chi_{int} + \chi_{dip} + \chi_i + \chi_e) \vec{E} = \varepsilon_0 \chi_{Tot} \vec{E} .$$

Hence the dielectric susceptibility also appears as a sum of the components given:

$$\chi_{Tot} = \chi_{int} + \chi_{dip} + \chi_i + \chi_e.$$

3.4.4. The resultant dielectric spectrum

The overall dielectric spectrum is presented in Figure 3.10. The figure shows both the real (plots in the upper half) and the imaginary components (lower half plots) for the susceptibilities of a material that possesses all of the types of polarization discussed so far. The material that is imagined for this figure is thus polar and contains space charges.

If the material is subject to a very high frequency electromagnetic (EM) wave (with $\omega > 10^{19}$ Hz), the oscillation is too fast for any of the polarizable elements of the material to react. The dielectric permittivity is thus close to 1, as shown in zone 1 of Figure 3.10 where the absolute permittivity is thus close to ε_0. At these frequencies the material behaves as if it were a vacuum. However, once the frequency is slightly reduced below these frequencies, the material gives rise to a relative permittivity below 1, indicating that the phase velocity propagates at a value greater than c.

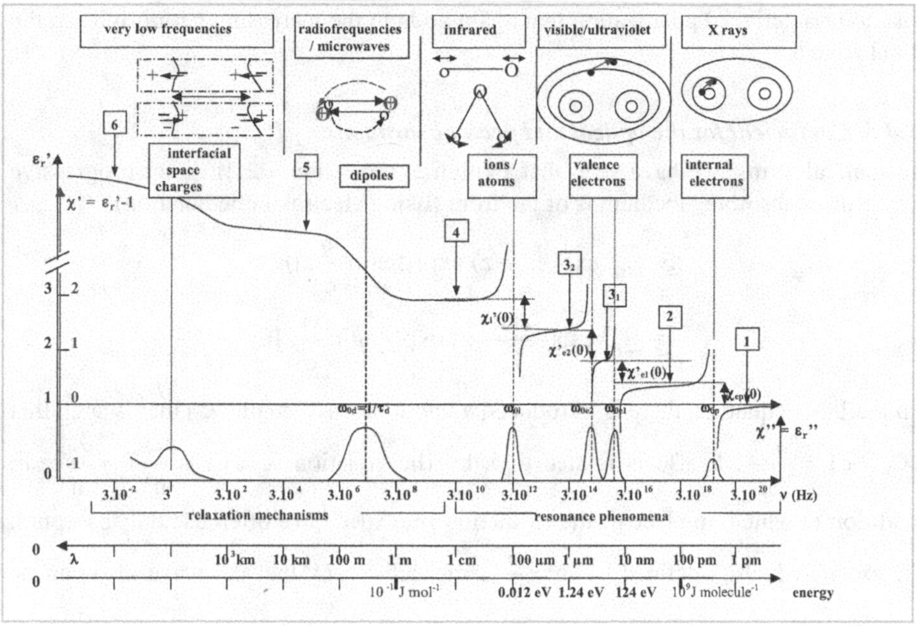

Figure 3.10. *The resultant dielectric spectrum.*

At frequencies below 10^{19} Hz, the electrons of the (deep) internal layers can be excited by the electric field. This creates an absorption at a resonance frequency ($\omega_{0p} = 10^{19}$ Hz) and a polarization of the medium that presents a higher permittivity, i.e., $\varepsilon'_r > 1$, as indicated in zone 2 of the figure.

On decreasing the frequency further, we are in a range close to that of the optical spectrum, and it is now the valence electrons that contribute to the polarization. Again here is an increase in ε'_r, as shown in zones 3_1 and 3_2, each relating to two valence electrons types.

At frequencies of the order of 3×10^{12} Hz ($\lambda \approx 100$ µm), the ions can now also add their characteristics to the polarization, so that there is once again an increase in the permittivity (see zone 4).

On reaching the Hertzian region ($\nu < 10^{10}$ to 10^{11} Hz), the dipoles can relax with the internal forces of the material. The dipolar polarization that is thus generated also contributes to the increasing value of the dielectric permittivity (zone 5).

Finally, for frequencies below around a hundred hertz, it is monocharged and slow-moving species such as positive and negative ionic impurities that now have the time to reach interfaces (grain joints if the material is heterogeneous). This gives rise to an interfacial polarization that again adds to the increasing permittivity in the final zone 6.

3.4.5. Coefficient for the optical and peak absorptions

In general terms, we have seen that (Volume 1, Section 7.2.2) direct progressive wave in an absorbing medium is of the form (using electrokinetic notation):

$$\vec{\underline{E}}=\vec{\underline{E}}_m \exp(-n''\frac{\omega}{c}z)\exp(j[\omega t-n'\frac{\omega}{c}z])$$

$$= \vec{\underline{E}}_m \exp(-\frac{2\pi n''}{\lambda_0}z)\exp(j[\omega t-n'\frac{\omega}{c}z]).$$

In the latter equation, there is introduced a vacuum wavelength (λ_0) that is such that $\lambda_0 = cT = \frac{2\pi c}{\omega}$. It also is related to λ by the equation $\lambda = vT = \frac{c}{n'}T = \frac{\lambda_0}{n'}$. In addition to which, in place of the extinction index (n''), we often use the absorption index (η) which is defined by $\eta = \frac{n''}{n'}$, in such a way that the wave also can be written as:

$$\vec{\underline{E}}=\vec{\underline{E}}_m \exp(-\frac{2\pi\eta}{\lambda}z)\exp(j[\omega t-n'\frac{\omega}{c}z]).$$

We also know that $n'^2-n''^2 = \varepsilon_r'$ and that $2\,n'\,n'' = \varepsilon_r''$ (Section 7.2.2, Volume 1), from which:

$$\varepsilon_r' = n'^2-n''^2 = n'^2\left(1-\frac{n''^2}{n'^2}\right) = \left(1-\eta^2\right)n'^2$$

$$\varepsilon_r'' = 2\,n'\,n'' = 2\,n'^2\frac{n''}{n'} = 2\,n'^2\eta.$$

In addition, the intensity (I) of the wave is proportional to the square of the amplitude of the electric field; in other words $I \propto \exp(-2\frac{2\pi\eta}{\lambda}z)$. The coefficient of the optical absorption (μ) and hence also the extinction coefficient (κ), can be introduced with the help of Lambert's law, which states that:

$$\frac{I}{I_0} = e^{-\mu z} = 10^{-\kappa z},$$

where I is the intensity of the initial wave (I_0) after having traversed a length (z) of a medium.

Hence, we need to verify that $2\frac{2\pi\eta}{\lambda} = \mu$, i.e., that $\mu = \frac{4\pi}{\lambda}\eta$.

The upshot of this is that μ is proportional to η and inversely proportional to the wavelength.

As $\varepsilon_r'' = 2\,n'^2\eta$, (so that $2\eta = \varepsilon_r''/n'^2$) hence:

$$\mu = \frac{2\pi}{\lambda n'^2}\varepsilon_r'' = \frac{\omega\varepsilon_r''}{n'c}.$$

From this the structure of μ can be directly determined from that of ε_r''.

In fact, an approach based on quantum mechanics is necessary to understand the electronic states of the bands, from which can be deduced the exact structure of the spectrum.

Comment: expression for κ

As $e^{-\mu z} = 10^{-\kappa z}$, we also can deduce that:

$$\kappa = \mu \log_{10}e = \frac{4\pi}{\lambda}\eta \log_{10}e.$$

3.5. Problems

3.5.1. Problem 1. Alternating conductivity

From an initial instant ($t_0 = 0$), we consider a number of free electrons at a concentration denoted by n_0 that undergo random collisions in a solid.

1. Using the notation n(t) for the number of electrons that have yet to undergo collisions at an instant t ($t > t_0$), and P to denote the probability that an electron will collide per unit time, the random character of the collisions makes it possible to write down the positive amount (-dn) of electrons that have collisions for the first time during an interval given by [t, t+dt] as $dn = -n(t)Pdt$.

a. Calculate the law for the evolution of n(t) as a function of time.

b. Give the law of probability that an electron at an instant t_0 will not undergo a collision during a period up to time t.

c. Indicate the law of probability for an electron that does not undergo a collision prior to instant t but has a first collision during the period between t and t + dt.

2. Using the same group, calculate the average time (τ) between two successive collisions for an electron that at an initial time $t_0 = 0$ has just undergone its last collision.

3. To the above detailed system is applied a constant electric field $\vec{E}$.

a. Give the fundamental equation for the dynamics of an electron in the system.

b. From which, determine the speed $\vec{v}(t)$ at a time t for an electron that underwent at the initial time t_0 its last collision.

c. If the average initial velocity is zero $\left(\overline{\vec{v}(t_0)} = 0\right)$, i.e., the field is insufficiently strong to enforce an orientation, calculate as a function of τ the average value of $\vec{v}(t)$ for the total number of electrons which is denoted by $\vec{v}_d$ (which is such that $\vec{v}_d = -\frac{1}{n_0}\int \frac{q\vec{E}}{m^*}(t - t_0)\,dn$). Note that in order to simplify the calculation, use $t_0 = 0$ as indicated above.

4. A time-dependent electric field is now applied to the system and it is such that $\vec{E} = \vec{E}_0 e^{j\omega t}$.

a. Determine the new equation for $\vec{v}(t)$.

b. With the help of a calculation analogous to that developed in **3c**, determine the new value for the average value of $\vec{v}(t)$ for the total number of electrons.

5.

a. Give the expression for the continuous conduction ($\sigma_=$) when $\vec{E}$ is a constant.

b. Express the real part of the alternating current ($\sigma_\approx$) obtained when the field is in the form $\vec{E} = \vec{E}_0 e^{j\omega t}$. Express the result as a function of $\sigma_=$.

Answers

1.

a. We thus have $dn = -n(t)P\,dt$, and hence $\frac{dn}{n} = -P\,dt$. Integration gives $n(t) = K\exp(-Pt)$, and the constant K can be determined using limiting conditions.

When $t = t_0$, then $n(t_0) = n_0$. The result is that $K = n_0 \exp(Pt_0)$, and finally, $n(t) = n_0 \exp(-P[t - t_0])$.

Comment. The minus sign expresses the decrease with time in the concentration of electrons that have not undergone collisions. This justifies a posteriori the insertion of the negative sign in the starting equation $dn = -n(t)Pdt$.

b. Here, n(t) represents the concentration of electrons that have not yet undergone collisions at an instant t. The number of electrons in this group (n_0) is given by $\frac{n(t)}{n_0} = \exp(-P[t - t_0])$.

c. The probability that an electron will undergo its first collision during the period between t and t + dt is given by the product of the independent probabilities, as in:

{probability of no collision up to t} x {probability of a collision between t and t+dt}

$$= \{\exp(-P[t - t_0])\}\{Pdt\} = -\frac{n(t)}{n_0}\frac{dn}{n(t)} = -\frac{dn}{n_0}.$$

2. For an electron in the group given the number i which undergoes its first collision at the instant t_i following the instant t_0, the time required for the first collision thus is given by $(t_i - t_0)$. Relative to the total number of electrons of concentration n_0, the average time required for an electron to undergo its first collision thus is $\bar{t} = \tau = \frac{1}{n_0}\sum_{i=1}^{n_0}(t_i - t_0)$. In terms of integrals, we thus have $\bar{t} = \tau = \frac{1}{n_0}\int t(-dn)$,

so that with $t_0 = 0$, and hence $dn = \exp(-Pt)Pdt$:

$$\bar{t} = \tau = \frac{1}{n_0}\int_0^\infty t\, n_0 \exp(-Pt)Pdt = \int_0^\infty t\, \exp(-Pt)Pdt .$$

By making $x = Pt$, we thus have $\tau = \frac{1}{P}\int_0^\infty x\exp(-x)dx = \frac{1}{P}$ [$\int_0^\infty x\exp(-x)dx$ is performed by integration of parts by making x = u and exp(-x) dx = dv].

We thus have:

$$\int_0^\infty x\exp(-x)dx$$

$$= \left[-x\exp(-x)\right]_0^\infty + \int_0^\infty \exp(-x)dx = \left[-x\exp(-x)\right]_0^\infty - \left[\exp(-x)\right]_0^\infty = 1$$

the result of which is that $\tau = \frac{1}{P}$.

Physically speaking, this result is reasonably obvious. If P = 10 %, the electron has a 10 % chance of undergoing a collision in 1 second. This means that on average the electron will have a collision every 10 seconds. This can be written as: $\overline{t} = \tau = 10 \text{ seconds} = \frac{1}{P}$.

3.

a. The fundamental dynamic equation is $m^* \frac{d\vec{v}}{dt} = -q\vec{E}$.

b. Integration between t and t_0 directly yields $\vec{v}(t) = \vec{v}(t_0) - \frac{q\vec{E}}{m^*}((t - t_0)$. Here,

$t_0 = 0$, $\vec{v}(t) = \vec{v}(0) - \frac{q\vec{E}}{m^*} t$.

c. On average, the initial velocity is zero ($\overline{\vec{v}(t_0)} = 0$), and with respect to the whole group of electrons, we have:

- either directly $\overline{\vec{v}(t)} = -\frac{q\vec{E}}{m^*}\overline{t}$; or
- we use the fact that $\overline{v}(t) = \frac{1}{n_0}\int v dn = \frac{1}{n_0}\int_0^\infty -\frac{qE}{m^*} t\, n_0 \exp(-Pt) P dt = -\frac{qE}{m^*}\overline{t}$.

We thus find that $\vec{v} = -\mu\vec{E}$ where $\mu = \frac{q\tau}{m^*}$, and $\sigma_= = qn\mu = \frac{nq^2}{m^*}\tau$.

4.

a. The fundamental dynamic equation remains the same (within a form close to that of the field), and the integration thus gives:

$$\vec{v}(t) = \vec{v}(t_0) - \frac{q\vec{E}_0}{m^*}\int_{t_0}^{t} \exp[j\omega t] dt = \vec{v}(t_0) - \frac{q\vec{E}_0}{m^*}\frac{1}{j\omega}\left[\exp(j\omega t) - \exp(j\omega t_0)\right].$$

b. With $\overline{\vec{v}(t_0)} = 0$, and in analogy to question 3, we have

$$\overline{v}(t) = \frac{1}{n_0}\int v dn = \frac{1}{n_0}\int_0^\infty -\frac{qE_0}{m^*}\frac{1}{j\omega}\left[e^{j\omega t} - e^{j\omega t_0}\right] n_0 \exp\left(-\frac{t - t_0}{\tau}\right)\frac{dt}{\tau}, \text{ from which:}$$

$$\overline{v}(t) = -\frac{qE_0}{m^*}\frac{1}{j\omega}\int_0^\infty \exp\left(-\frac{t - t_0}{\tau}\right)\left[e^{j\omega t} - e^{j\omega t_0}\right]\frac{dt}{\tau}.$$

A rather long calculation gives:

$\overline{\vec{v}(t)} = -\dfrac{q\vec{E}_0}{m^*}\dfrac{\tau}{1-j\omega\tau}e^{j\omega t_0} = -\dfrac{q\vec{E}_0}{m^*}\dfrac{\tau(1+j\omega\tau)}{1+\omega^2\tau^2}e^{j\omega t_0}$, so that by taking $t_0 = 0$, we have $\overline{\vec{v}(t)} = -\dfrac{q\vec{E}_0}{m^*}\dfrac{\tau(1+j\omega\tau)}{1+\omega^2\tau^2}$.

5.

a. We have seen that $\sigma_= = qn\mu = \dfrac{nq^2}{m^*}\tau$.

b. Then under alternating conditions, we have

$\mathrm{Re}\left(\overline{\vec{v}(t)}\right) = -\dfrac{q\vec{E}_0}{m^*}\dfrac{\tau}{1+\omega^2\tau^2}$, and $\mu_\approx = \dfrac{q\tau}{m^*\left(1+\omega^2\tau^2\right)}$, and hence

$\sigma_\approx = \sigma_= \dfrac{1}{1+\omega^2\tau^2}$.

For its part, the complex conductivity is given by $\underline{\sigma}_\approx = \dfrac{nq^2\tau}{m^*}\dfrac{1+j\omega\tau}{1+\omega^2\tau}$.

3.5.2. Problem 2. Optical properties of gaseous electrons

Optical properties of gaseous electrons ($\omega\tau >> 1$, in opposition to the condition $\omega\tau << 1$ for the low frequency regime given in Section 3.1.2)

As indicated in Problem 1, the collisions of gaseous electrons require that there is a parameter termed the "relaxation time". This parameter also can be introduced by considering that the collisions give rise to a heating of the system (Joule effect) which can be represented by frictional forces of a value that increases with the velocity of the electrons. We thus can state that $f_t = -m\Gamma v$ where the parameter Γ must have a dimension that is inverse to time (so that the equation is dimensionally correct). We thus can have $\Gamma = 1/\tau$, where τ has the dimension of time.

1. From the deduction of the velocity of the electrons from the fundamental dynamic equation, show how we can obtain the same expression as in Problem 1 for the conductivity. Consider the physical significance of τ.

2. In place of assuming the approximation $\omega\tau << 1$ given for low frequencies in Section 3.1.2, we now have a system based on gaseous electrons, which would mean that $\omega\tau >> 1$. Show that when $\omega > \omega_p$, where ω_p is an angular frequency to be determined, waves in the form $\underline{E} = \underline{E_0}\, exp(i\omega t)$ can propagate through the electron gas.

Answers

1. This problem is in fact the same as that in Section 8.4, Volume 1. From the fundamental dynamic equation, where $F = m\frac{dv}{dt} = \sum f = -qE - \frac{m}{\tau}v$, we can deduce the complex notations $m\frac{d\underline{v}}{dt} + \frac{m}{\tau}\underline{v} = -q\underline{E}$. In order to determine for v a solution $\underline{v} = \underline{v_0}e^{i\omega t}$, we obtain $im\omega\underline{v} + \frac{m}{\tau}\underline{v} = -q\underline{E}$ from which can be deduced that $\underline{v_0} = -\frac{q\underline{E_0}\tau}{m(1+i\omega\tau)}$.

The conduction current associated with the electrons of volume density (n_e) is given by Ohm's law: $\underline{j_\ell} = \rho_\ell\ \underline{v} = -n\,q\,\underline{v} = \frac{nq^2\tau}{m(1+i\omega\tau)}\underline{E} = \underline{\sigma_\ell}\ \underline{E}$, and hence:

$$\underline{\sigma_\ell} = \frac{nq^2\tau}{m(1+i\omega\tau)} = \frac{\sigma_=}{(1+i\omega\tau)},$$

by making $\sigma_= = \frac{nq^2}{m}\tau$.

We can see that the equation thus obtained is identical to that obtained in Problem 1, which shows that it is the same τ—and hence the same relaxation time—that intervenes in these two cases.

2. In the optical domain it is possible to use Maxwell's equations for a rapidly varying regime (where the electrons "bathe" in a vacuum): $\overrightarrow{\text{rot}}\vec{E} = -\frac{\partial\vec{B}}{\partial t}$ and $\overrightarrow{\text{rot}}\,\vec{B} = \mu_0(\vec{j}_\ell + \varepsilon_0\frac{\partial\vec{E}}{\partial t})$. The elimination of $\vec{B}$ between the two equations [which can be done by calculating $\overrightarrow{\text{rot}}(\overrightarrow{\text{rot}}\vec{E})$] gives rise to $\overrightarrow{\Delta\underline{\vec{E}}} = \mu_0(\frac{\partial\underline{\vec{j}_\ell}}{\partial t} + \varepsilon_0\frac{\partial^2\underline{\vec{E}}}{\partial t^2})$.

By using: $\underline{\vec{j}_\ell} = \underline{\sigma_\ell}\ \underline{\vec{E}}$ and $\underline{\vec{E}} = \underline{\vec{E}_0}e^{i\omega t}$, we have

$$\overrightarrow{\Delta\underline{\vec{E}_0}} = -\frac{\omega^2}{c^2}\left[1 - \frac{i\ n\ q^2\tau}{\varepsilon_0\ \ m\omega}\frac{1}{1+i\omega\tau}\right]\underline{\vec{E}_0},$$ so that with $\underline{\varepsilon}(\omega) = 1 - \frac{i\ nq^2\tau}{\varepsilon_0\ m\omega}\frac{1}{1+i\omega\tau}$,

we thus have $\overrightarrow{\Delta\underline{\vec{E}_0}} = -\frac{\omega^2}{c^2}\underline{\varepsilon}(\omega)\underline{\vec{E}_0}$.

In the optical domain, where $\omega\tau >> 1$, on setting $\omega_p{}^2 = \dfrac{ne^2}{\varepsilon_0 m}$ we obtain:

$$\underline{\varepsilon}(\omega) = \varepsilon(\omega) = 1 - \frac{\omega_p^2}{\omega^2}.$$

When $\omega > \omega_p$, we thus have $\varepsilon(\omega) > 0$, and the wave can propagate in the medium (progressive wave).

***3.5.3. Problem 3. Relation between the function of relaxation** (macroscopic magnitude) **and the autocorrelation function** (microscopic magnitude)*

This problem concerns a polar molecule with a permanent electric dipole and the possibility of occupying two equivalent equilibrium positions each separated by an angle (θ) of 180 °. By representing the problem as a pair of potential wells, show that there is an identity between the relaxation function [Y(t)] (which is of macroscopic magnitude as is bound to ε" which describes the global state of the sample) and the autocorrelation function of the dipolar moment $\vec{\mu}$, which is defined by the equation $\gamma(t) = \dfrac{\langle \vec{\mu}(t).\vec{\mu}(0)\rangle}{\langle [\vec{\mu}(0)]^2 \rangle}$; $\gamma(t)$ can be seen as being of microscopic magnitude as it is related to the orientation and moment of the dipole.

Answer

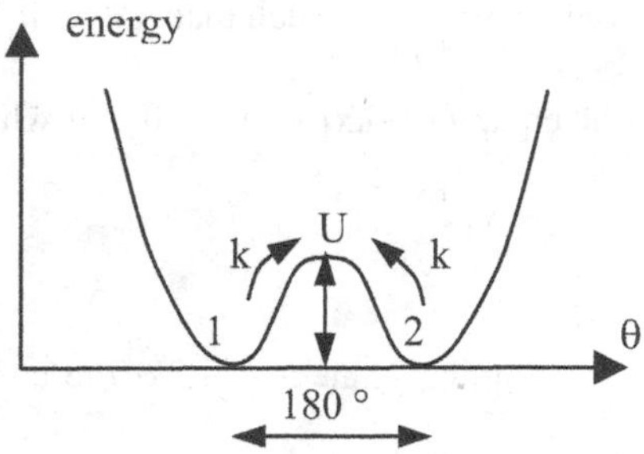

The two equivalent positions of the dipole, labeled 1 and 2 in the adjacent figure, are separated by a potential barrier of height denoted U. With their being equivalent, the dipole has an equal chance of being in 1 or in 2. The function $\gamma(t)$ thus is a sum of the two contributions γ_1 and γ_2 of the dipole initially placed in 1 or 2. Each contribution is evidently weighted by probabilities of presence at equilibrium i.e. ½.

As an example, we will use a general resolution method that will make it possible to resolve the more complex problems concerning dipole distributions in this exercise.

Denoting k for the equal transition probabilities i going from 1 to 2 or from 2 to 1, the equations that describe as a function of time the variations in the well populations are:

$$\frac{dN_1}{dt} = -kN_1 + kN_2$$

$$\frac{dN_2}{dt} = kN_1 - kN_2,$$

which can be rewritten as $\begin{pmatrix} \dfrac{dN_1}{dt} \\ \dfrac{dN_2}{dt} \end{pmatrix} = (A)\begin{pmatrix} N_1 \\ N_2 \end{pmatrix}$, where $(A) = \begin{pmatrix} -k & k \\ k & -k \end{pmatrix}$.

The proper vectors of (A) are such that:

$$\left.\begin{aligned} \det(A - \lambda I) &= \begin{vmatrix} -(k+\lambda) & k \\ k & -(k+\lambda) \end{vmatrix} = 0 \\ &= -(k+\lambda)^2 - k^2 \end{aligned}\right\} \Rightarrow \lambda^2 + 2\lambda k = 0, \quad \text{and hence:}$$

$$\begin{cases} \lambda_1 = 0 \\ \lambda_2 = -2k. \end{cases}$$

The diagonal matrix is $(D) = \begin{pmatrix} 0 & 0 \\ 0 & -2k \end{pmatrix}$. At $\lambda_1 = 0$, the corresponding proper vector $V_1\begin{cases} x_1 \\ y_1 \end{cases}$ is such that $AV_1 = \lambda_1 V_1$, so that $(A - \lambda_1 I)V_1 = 0$. We thus obtain the equation $-kx_1 + ky_1 = 0$, for which the simplest solution is $x_1 = y_1 = 1$, from which $V_1\begin{cases} x_1 = 1 \\ y_1 = 1 \end{cases}$.

Similarly, at $\lambda_2 = -2k$, the corresponding proper vector $V_2\begin{cases} x_2 \\ y_2 \end{cases}$ is such that $AV_2 = \lambda_2 V_2$, and hence $(A - \lambda_2 I)V_2 = 0$.

We thus obtain the equation $kx_2 + ky_2 = 0$ for which the simplest solution is $x_2 = -y_2 = 1$, from which $V_2\begin{cases} x_2 = 1 \\ y_2 = -1 \end{cases}$.

If Z_1 and Z_2 represent N_1 and N_2 in the new base V_1, V_2, we then can write:

$$\begin{pmatrix} \dfrac{dZ_1}{dt} \\ \dfrac{dZ_2}{dt} \end{pmatrix} = (D)\begin{pmatrix} Z_1 \\ Z_2 \end{pmatrix}, \quad \text{from which} \quad \begin{array}{l} \dfrac{dZ_1}{dt} = 0 \quad \Rightarrow Z_1 = C_1 \\ \dfrac{dZ_2}{dt} = -2kZ_2 \quad \Rightarrow Z_2 = C_2 e^{-2kt}. \end{array}$$

The "pathway matrix" (in the new base) is $(P) = \begin{pmatrix} 1 & 1 \\ 1 & -1 \end{pmatrix}$, so that:

$$\begin{pmatrix} N_1 \\ N_2 \end{pmatrix} = (P)\begin{pmatrix} Z_1 \\ Z_2 \end{pmatrix}, \text{ from which can be deduced that: } \begin{cases} N_1 = C_1 + C_2 e^{-2kt} \\ N_2 = C_1 - C_2 e^{-2kt}. \end{cases}$$

If $P_1(t)$ and $P_2(t)$ represent the probabilities of presence in wells 1 and 2 at an instant t, we then can write that $P_1(\infty) + P_1(\infty) = 1$, from which $C_1 = \dfrac{1}{2}$, so that:

$$\begin{cases} P_1(t) = \dfrac{1}{2} + C_2 e^{-2kt} \\ P_2(t) = \dfrac{1}{2} - C_2 e^{-2kt}. \end{cases}$$

Supposing that the dipole is initially in position 1:

$$\begin{cases} P_1(0) = 1 = \dfrac{1}{2} + C_2 \\ P_2(0) = 0 = \dfrac{1}{2} - C_2 \end{cases} \Rightarrow C_2 = \frac{1}{2} \quad \text{and} \quad \begin{cases} P_1(t) = \dfrac{1}{2}\left(1 + e^{-2kt}\right) \\ P_2(t) = \dfrac{1}{2}\left(1 - e^{-2kt}\right). \end{cases}$$

At an instant t, the dipole moment that follows the initial dipole position is:

$$[\mu(t)]_1 = \mu P_1(t) + (\mu\cos\theta) P_2(t) = \mu\left[P_1(t) - P_2(t)\right] = \mu e^{-2kt},$$

from which:

$$\gamma_1(t) = \frac{\left\langle \vec{\mu}(t).\vec{\mu}(0) \right\rangle_1}{\left\langle [\vec{\mu}(0)]^2 \right\rangle} = e^{-2kt}.$$

If the dipole is initially in position 2, then:

$$\begin{cases} P_2(0) = 0 = \dfrac{1}{2} + C_2 \\ P_2(0) = 1 \end{cases} \Rightarrow C_2 = -\frac{1}{2} \text{ and } \begin{cases} P_1(t) = \dfrac{1}{2}\left(1 - e^{-2kt}\right) \\ P_2(t) = \dfrac{1}{2}\left(1 + e^{-2kt}\right). \end{cases}$$

At an instant t, the dipole moment, which follows that initial position, is now:

$$[\mu(t)]_2 = \mu P_2(t) + (\mu\cos\theta) P_1(t) = \mu[P_2(t) - P_1(t)] = \mu e^{-2kt},$$

from which:

$$\gamma_2(t) = \frac{\langle \vec{\mu}(t).\vec{\mu}(0)\rangle_2}{\left\langle [\vec{\mu}(0)]^2 \right\rangle} = e^{-2kt}.$$

Taking into account the symmetry of the problem, it is normal that we find the same result as would be the case for $\gamma_1(t)$.

From this can finally be deduced that:

$$\gamma(t) = \frac{1}{2}\gamma_1(t) + \frac{1}{2}\gamma_2(t) = e^{-2kt}.$$

The relaxation time is such that $\tau = \dfrac{1}{2k}$, and the equation found for $\gamma(t)$ is identical to that for Y(t) as we end up with a Debye "relaxation" process for the double potential wells.

Chapter 4

Interactions of Electromagnetic Waves and Solid Semiconductors

After having concentrated on the electromagnetic properties of dielectrics in the last chapters, this chapter looks at semiconductors. Conduction electrons in semiconductors are more numerous and freer in their movement. We thus will look at the propagation of the de Broglie wave associated with these electrons, which will take us to a filling pattern for energy bands. This structure not only conditions the optical (at around 10^{15} Hz) and optoelectronic properties of these materials but also their electrical and magnetic properties under continuous fields (which are detailed in the following chapter).

A study of the energy levels associated with charge carriers—of a concentration to be determined—necessitates most notably a resolution of the equation for the propagation of the waves associated with these carriers. This is within a potential (potential energy) characteristic of the medium's nature, which for semiconductors is generally a solid consisting of a periodic crystalline lattice.

Without dwelling on the physical details of such systems, which are generally covered in many courses on solid, quantum, and electronic physics, this chapter details the basics needed to establish the fundamental properties of a semiconductor excited by an electromagnetic field. Just as for dielectrics, it is assumed that the excitation is sufficiently weak so as to remain in a linear regime. Nonlinear effects will be discussed in Chapter 6.

4.1. Wave equations in solids: from Maxwell's to Schrödinger's equations via de Broglie's relation

The dual particle–wave theory brought de Broglie to associate a particle of a given mass (m) with a wave of a given wavelength (λ):

$$\lambda = \frac{h}{mv}. \qquad (1)$$

For its part, the equation for the propagation of waves in a vacuum is in the form:

$$\Delta s - \frac{1}{c^2}\frac{\partial^2 s}{\partial t^2} = 0 . \qquad (2)$$

With a monochromatic wave in the form

$$s = A(x,y,z)e^{i\omega t} = A(x,y,z)e^{i2\pi\nu t}$$

we have $\Delta s = \Delta A\ e^{i\omega t}$ and $\frac{\partial^2 s}{\partial t^2} = -\omega^2 A e^{i\omega t}$.

On introducing $\lambda = 2\pi\frac{c}{\omega}$ (wavelength in a vacuum), Eq. (2) for the wave propagation can be written:

$$\Delta A + \frac{\omega^2}{c^2}A = 0 \qquad (3)$$

$$\rightleftarrows$$

$$\Delta A + \frac{4\pi^2}{\lambda^2}A = 0 . \qquad (3')$$

A particle, for example, an electron, with mass m placed in a time–independent potential energy [V(x,y,z)], has an energy (E) given by:

$$E = \frac{1}{2}mv^2 + V .$$

Its speed is thus in the form:

$$v = \sqrt{\frac{2(E-V)}{m}} . \qquad (4)$$

The de Broglie wave associated with a frequency (υ) given by $\nu = \frac{E}{h}$ can be represented by a function as in:

$$\Psi = \psi e^{i2\pi\nu t} = \psi e^{2\pi i \frac{E}{h}t} . \qquad (5)$$

In accordance with Schrödinger, the ψ function verifies a relation analogous to Eq. (3), and by taking Eqs. (1) and (4) for the wavelength into account by using

$$\lambda = \frac{h}{mv} = \frac{h}{\sqrt{2m(E-V)}} , \qquad (6)$$

we end up with

$$\Delta\psi + \frac{2m}{\hbar^2}(E - V)\psi = 0 . \qquad (7)$$

This is Schrödinger's equation. For a crystal where the potential V is periodic, there are well-defined solutions for the equation. As detailed further on, these solutions are presented in the form of permitted (valence and conduction) and forbidden ("gap" for semiconductors) bands.

4.2. Bonds within solids: weak and strong bond approximations

Conduction properties of metals have been interpreted in terms of a relatively simple theory based on free electrons. The essence of this theory was that the free conduction electrons were moving in a flat–bottomed well, of which the limits coincided with the physical limits of the solid. In three dimensions this analysis is made using a "box", which is generally treated during first-degree–level courses.

In order to take account of the electrical properties of semiconductors and insulators (where the electrons are no longer free) and to improve upon the theories for metals, it is necessary to use more elaborate potential models that bring into play finer interactions between the electrons and their environment. There are in fact two methods that may be chosen, depending on the nature of the solid.

4.2.1. Weak bonds

This approach uses a more highly developed version of the potential box model. The electrons are now supposed to interact with an internal periodic potential generated by a crystalline lattice. The potential is coulombic as it is proportional to 1/r the distance from the ions placed at nodes in the lattice.

Figure 4.1 schematizes the line of periodically arranged atoms, each separated by a given distance (a). The electrons each belong to an orbital with a radius denoted R (as in Figure 4.1a).

Figure 4.1b gives a representation of a one-dimensional potential energy of the electrons for which the condition is that of $a < 2R$. Elsewhere in this text, the common abuse of terms will be made by simply denoting the potential energy as potential.

Depending on the direction defined by a line Ox joining the nuclei of atoms, when an electron tends toward a nucleus, the potentials diverge. In fact, a study made considering only the potential Ox has little to do with reality. This is because the electrons in this study are those of conduction that reside in the external layers. With respect to a straight line (D) that does not pass through the nuclei, the electron–nuclei distance does not tend to zero and the potentials that go toward infinite values join up.

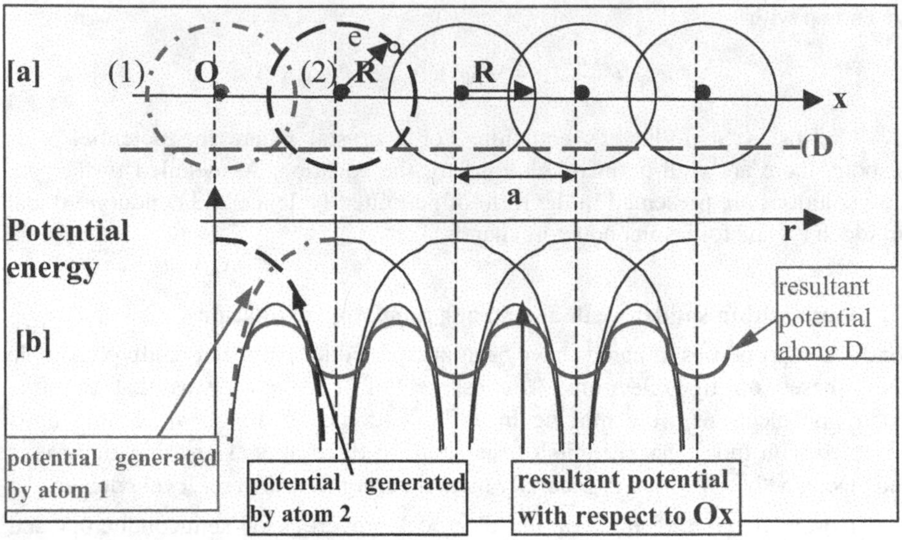

Figure 4.1. *Weak bonding and (a) atomic orbitals (s type of radius R) in a periodic lattice and (b) the resultant potential energy (thick line) as observed by electrons in a one–dimensional representation.*

In addition, the applied condition a < R decreases the barrier (due to a strong overlapping of the two potential plots) that exists midway between two adjacent nuclei. This condition makes it possible to obtain a resultant potential in a solid that exhibits weak periodic fluctuations. The initial representation of the potential as a flat-bottomed well (approximation to zero order for the free electrons) is now replaced by a container with a periodic bottom.

As a first approximation and in one dimension ($r \equiv x$), the potential can be placed in the form $V(x) = w_0 \cos \frac{2\pi}{a}x$. The term w_0, and hence the perturbation of the crystalline lattice become smaller as the condition a < 2R becomes increasingly correct. Put more practically, the smaller the term a with respect to 2R, then the smaller the perturbation, and the greater the justification for using a method by perturbation. The corresponding approximation (to the first order with a Hamiltonian for the perturbation given by $H^{(1)} = w_0 \cos \frac{2\pi}{a}x$) is that of a semifree electron which is a considerable improvement over that of the free electron, which ignores $H^{(1)}$. The resulting theory for a weak bond can be well applied to metallic bonds where electrons are easily delocalized in a lattice due to the low value of w_0.

4.2.2. Strong bonds

This approach is considerably more "chemical" in nature. It consists of deducing the properties of a solid from the actual states of the atoms of which it consists. Also considered are the chemical bonds between the atoms in terms of linear combinations of atomic orbitals. This reasoning is more acceptable than the proposition that electrons reside simply at their "resident" atom, and this approximation for strong bonds is all the more justified as the condition a $\geq 2R$ is well satisfied (see Figure 4.2a). This approximation is generally applied to covalently bonded solids where the valence electrons are localized between two atoms.

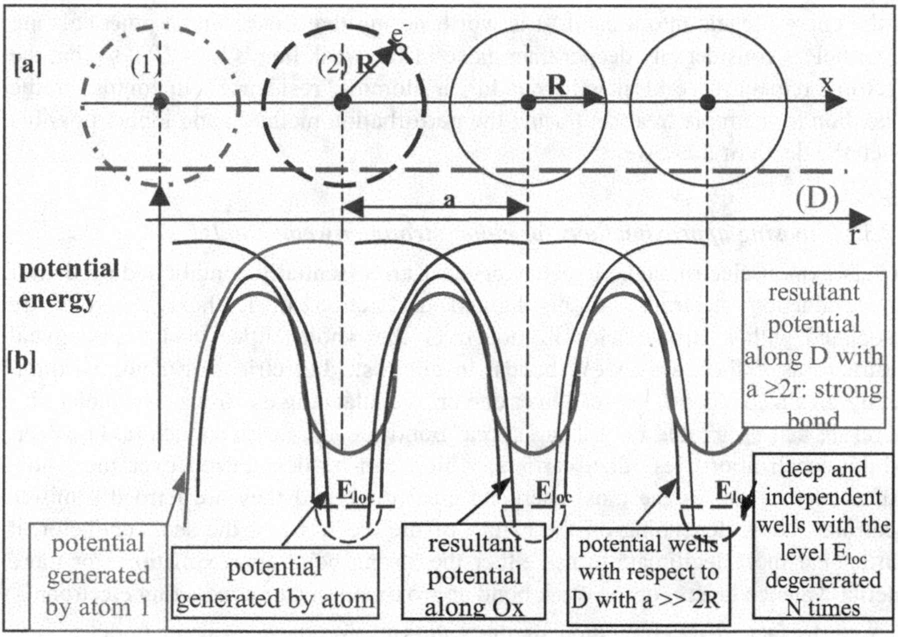

Figure 4.2. *Strong bonds: (a) atomic orbitals (s type with radius R) in a periodic lattice of period a satisfying the equation $a \geq 2R$, and (b) resultant potential energy (thick line) observed by electrons in a one-dimensional representation.*

Once again, if we study the plot of the potential with respect to Ox, we obtain a function that diverges when the electrons approach the nuclei. This discontinuity is suppressed with respect to the straight line (D) for valence electrons. There are two possible reasons for this (Figure 4.2b):

• first, if a >> 2R, then very deep potential wells appear (as there is practically no overlapping between the potentials generated by adjacent nuclei). At a limit, there is a chain of N atoms with N valence electrons, with sufficiently long bonds so as to assume that the N electrons are independent (with N very deep and independent potential wells) and the energy levels are degenerated N times (so that it is not possible to discern them from one another as they are all identical and denoted E_{loc} in the figure); and

• second, if $a \geq 2R$ and the closing up of the atoms induces a slight overlapping of the potentials generated by the nuclei, then the potential wells are no longer independent and the degeneration is increased. The electrons from one bond now can interact with the others of a neighboring bond, and we obtain a breakdown of the energy levels into a band. It is worth noting that the resultant potentials are nevertheless considerably deeper than those in the weak bonds ($a < 2R$), so that the electrons remain more localized around their atom of "residence". In contrast to the weak bonds, a simple treatment using the perturbation method is no longer possible given the depth of these wells.

4.2.3. *Choosing approximations for either strong or weak bonds*

Metals exhibit electromagnetic behaviors that are essentially conditioned by that of the conduction electrons. As highlighted in Section 4.2.1, these electrons are associated with a strong delocalization over the whole lattice, and thus demand treatment as if they were weak bonds. In contrast, dielectrics (insulators) exhibit highly localized electrons spread over one or two attaching electrons. The dielectrics therefore can be treated only using strong bond theory. Semiconductors, however, exhibit localizations less than carriers, which can be delocalized over the whole lattice (if they are of the most energetic electrons), and they are termed semifree electrons. Thus, depending on the nature of the electrons of the semiconductor, it can appear more legitimate to use either the strong bond approximation (for more internal valence layers) or the weak bond approximation (for conducting electrons).

In fact, a determination of the solutions for the energy in both cases—strong or weak bonds—yields a structure of permitted energy bands separated by a forbidden band. The strong bond approximation requires longer calculations and involves the Hückel theory for chemists or the Floquet theory for physicists. For further details, see, for example, "Optoelectronics of molecules and polymers" by A. Moliton. Weak bonds can be treated more facilely, and it is this route that is chosen here.

4.3. Evidence for the band structure in weak bonds

4.3.1. Preliminary result for the zer– order approximation

A weak bond thus corresponds to a potential in which the electrons are placed as represented in Figure 4.3, as deduced from Figure 4.1.

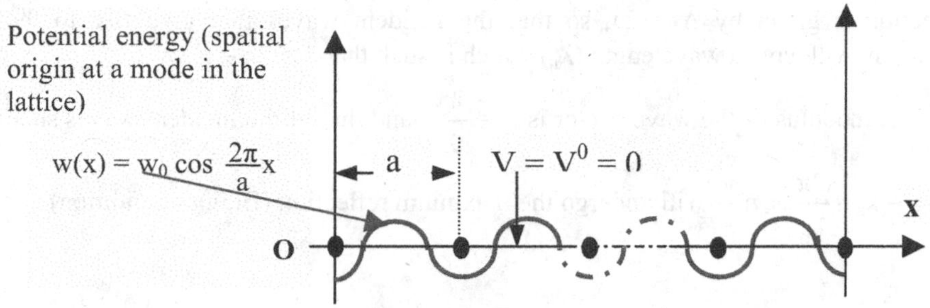

Figure 4.3. *Plot of potential energy* $w(x) = w_0 \cos \frac{2\pi}{a}x$ *demonstrating that* $w_0 < 0$.

In preliminary terms, we can recall that for a free electron (zero–order approximation) the potential follows the form of a flat-bottomed well. This is indicated in Figure 4.3 by the straight line passing through the nuclei. For this system, where $V = V^0 = 0$, Schrödinger's equation for the amplitudes is given by: $\Delta\psi^0 + \frac{2m}{\hbar^2}E\psi^0 = 0$. With $k^2 = \frac{2m}{\hbar^2}E$, by privileging the physical solutions that assure the propagation of the wave associated with the electron we have the equation $\psi^0 = Ae^{\pm ikx}$. In addition, the plot of $E = f(k)$ is obtained from $E = E^0 = \frac{\hbar^2}{2m}k^2$.

The perturbation of the potential by the lattice effect manifests itself by the generation of a periodic potential for which the first term of development in a Fourier series makes it possible to state that to a first approximation:

$V \approx V^{(1)} = w(x) = w_0 \cos \frac{2\pi}{a}x$. The wave function itself is thus perturbed. It takes on the form of a Bloch function, as in $\psi_k(x) = \psi^0 u(x) = e^{ikx}u(x)$, where u(x) is a periodic function (the lattice). This notably comes from the fact that the wave function must remain invariant with respect to a modulus transition (T_a), which would impose $u(x) = u(x+a)$.

4.3.2. Physical origin of the forbidden bands

Figure 4.4 shows a periodic chain subject to a ray of light. The ray is reflected by the lattice atoms, and following reflection a system of additive interferences occur when the difference in the step Δ of the two waves is equal to a whole number of repeated wavelengths of the incident ray.

In one dimension, the difference in step Δ between waves (1) and (2) following reflection is given by $\Delta = 2a$, so that the incident waves that give rise to the maximum will have a wavelength (λ_n) which is such that $\Delta = 2a = n\,\lambda_n$.

The modulus of the wave vector is $k = \dfrac{2\pi}{\lambda}$, and that of the incident waves such that $k = k_n = \dfrac{2\pi}{\lambda_n} = n\dfrac{\pi}{a}$ will undergo the maximum reflection (Bragg's condition).

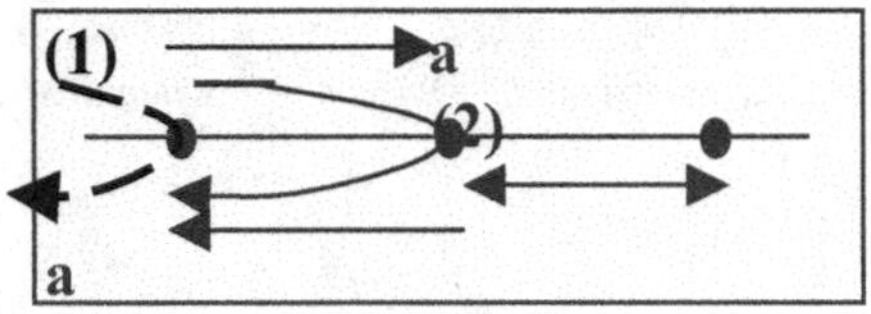

Figure 4.4. *Bragg reflection for a one-dimensional crystal.*

For a weak bond, we can assume that incident wave associated with an electron is only weakly perturbed by the linear chain and that its amplitude can be written to the zero–order approximation in the form $\psi_k^0 = Ae^{ikx}$. The time–dependent incident wave is thus $[\Psi_k^0(x,t)]_{inc.} = Ae^{i(kx-\omega t)}$. This is the expression for an incident plane progressive wave moving toward $x > 0$. When $k = k_n$ exactly, this incident wave, $[\Psi_{k_n}^0(x,t)]_{inc.} = Ae^{i(k_n x-\omega t)}$, is reflected with respect to a wave propagating toward the values $x < 0$, as in $[\Psi_{k_n}^0(x,t)]_{refl.} = Ae^{i(k_n x+\omega t)}$. The superposition of the two types of wave (incident and reflected) gives rise to the establishment of a stationary wave regime, which corresponds to two forms of the solution (symmetric and antisymmetric), i.e.,

$$\Psi^+ \propto \cos\left(\frac{n\pi}{a}x\right) e^{-i\omega t} \quad (8) \quad \text{and} \quad \Psi^- \propto \sin\left(\frac{n\pi}{a}x\right) e^{-i\omega t}. \qquad (9)$$

For each of these two solutions for the wave function for electrons satisfying

$k = k_n = n\frac{\pi}{a}$, there are two corresponding types of presence probabilities:

$$\rho^+ = \Psi^+\Psi^{+*} \propto \cos^2(\frac{n\pi}{a}x) \text{ and } \rho^- = \Psi^-\Psi^{-*} \propto \sin^2(\frac{n\pi}{a}x).$$

For the same given value of k_n, Figure 4.5 represents the presence probability densities denoted ρ^+, ρ^-, and ρ for the respective stationary, Ψ^+ and Ψ^-, and progressive waves where $[\Psi_k(x,t)] = A\, e^{i(kx-\omega t)}$. Taking its form into account, the progressive wave corresponds to ρ = constant. When $k = k_n$, this progressive wave only can exist by neglecting the reflection effect at the periodic (a) lattice and the zero–order approximation given by $V = V^0 = 0$. In addition, with $V \neq V_0$, this type of wave only can exist when $k \neq k_n$.

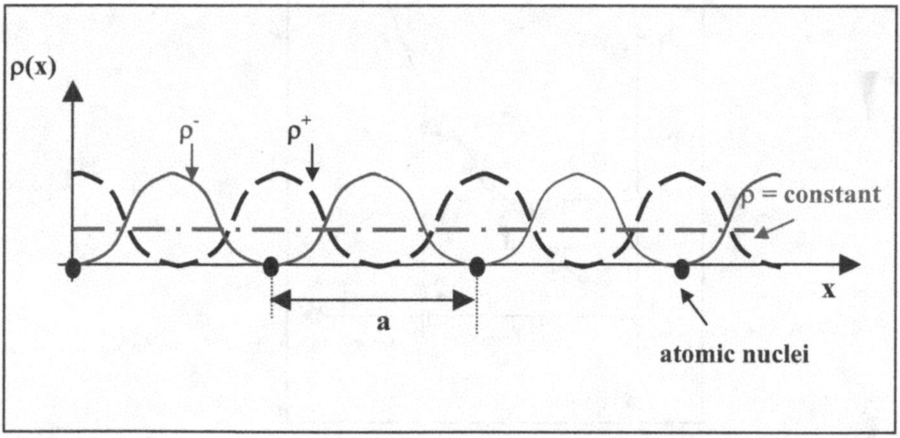

Figure 4.5. *Presence densities ρ^+, ρ^-, ρ, respectively, associated with stationary waves Ψ^+, Ψ^-, and a progressive wave.*

Figure 4.5 gives a graphic representation, which shows that:

- associated with ρ^+ is a maximum concentration of electrons in the neighborhood of the atomic nuclei. The average energy (w^+) of this configuration is the lowest (negative coulombic energy for the highest module as the electron–nuclei distance is at its smallest);
- associated with ρ^- is the maximum concentration of electrons midway between electrons. They are associated with an energy denoted w^-, which is the highest (greatest distance between electrons and nuclei and a small modulus of coulombic potential); and

- associated with ρ = constant (equal spread if electrons associated with an intermediate electron–nuclei distance) is an intermediate energy that approximates to that of a free electron.

Finally, it is the existence for the same value of k_n for k ($k = k_n = n\frac{\pi}{a}$) for the two physical solutions Ψ^+ and Ψ^-, which generate the two values for energy. Figure 4.6 shows the dispersion plot for the energy E = f(k).

The difference between the two values corresponds to a "gap" in the energy given by $E_G = w^- - w^+$. This is the forbidden band as we pass brutally from one energy w^+ to the other w^- all at the same value of $k = k_n$.

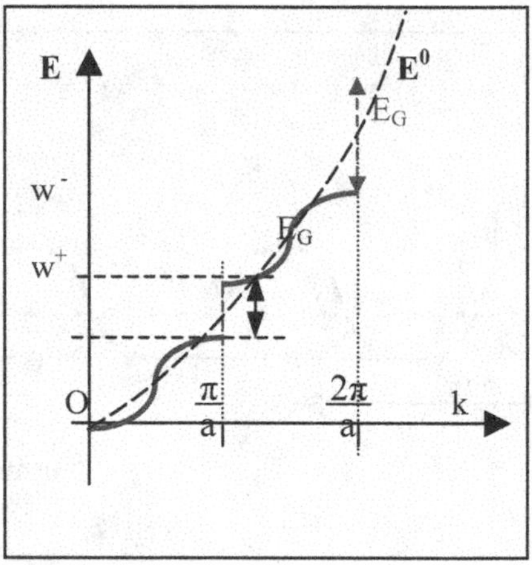

Figure 4.6. *Plot of E = f(k). The zero–order approximation, which corresponds to a perturbation potential w(x) = 0 (we ignore the interaction of electrons with the lattice as in the flat–bottomed well) so that we have an energy given by* $E^0 = \frac{\hbar^2 k^2}{2m}$. *The effect of the lattice, through the reflection of waves associated with electrons on atoms of the periodic chain, is that for each value of* $k = k_n = n\frac{\pi}{a}$ *there are two solutions for the energy, where* $w^+ < w^-$.

4.3.3. Simple estimation of the size of the forbidden band

The Ψ^+ and Ψ^- functions, normalized over a segment given by L = Na, i.e., a one–dimensional chain containing N + 1 ≈ N atoms, are such that:

$$\int_0^{L=Na} A^2\left|\Psi^+(x)\right|^2 dx = 1, \text{ and hence } \Psi^+(x) = \sqrt{\frac{2}{L}}\cos\left(\frac{\pi}{a}x\right)$$

$$\int_0^{L=Na} A^2\left|\Psi^-(x)\right|^2 dx = 1, \text{ and hence } \Psi^-(x) = \sqrt{\frac{2}{L}}\sin\left(\frac{\pi}{a}x\right).$$

The energy gap is thus equal to $E_G = w^- - w^+ = -(w^+ - w^-)$, so that in terms of quantum mechanics:

$$E_G = -\left(\left\langle\Psi^+\middle|w(x)\middle|\Psi^+\right\rangle - \left\langle\Psi^-\middle|w(x)\middle|\Psi^-\right\rangle\right) \qquad (10)$$

$$= -\frac{2w_0}{L}\int_0^L \cos\left(\frac{2\pi}{a}x\right)\left[\cos^2\frac{\pi}{a}x - \sin^2\frac{\pi}{a}x\right]dx = -\frac{2w_0}{L}\int_0^L \cos^2\left(\frac{2\pi}{a}x\right)dx = -w_0.$$

Finally, we reach $E_G = -w_0$, which is a positive value because w_0 must be negative (potential energy plot is at a minimum at the nuclei where the origin is taken, as shown in Figure 4.3).

Thus with $E_G = |w_0|$, we can conclude that the greater w_0 is, that is to say the stronger the electron–lattice interactions are, then the greater the forbidden band.

4.4. Insulator, semiconductors, and metals: charge carrier generation in the bands

4.4.1. Distinctions between an insulator, a semiconductor, and a metal

More complete calculations, with a rigorous resolution of the Schrödinger equation in the presence of a periodic potential, have been developed in solid physics books such as that by C. Kittel.

In general terms, the last completely occupied band is called the valence band and the first empty band is termed the conduction band.

The distinction between insulator, semiconductor, and metal is made from the study of the way in which these bands are filled. In a completely rigorous treatment (such as in a solid physics course) we have to determine the electronic densities relative to each band of electronic transportation. In the following paragraph, we will detail the essential results.

As a first step, and in order to simplify things, we can state that:

- When the last occupied band is completely filled (valence band) and when the following band is completely empty (conduction band), then the material is an insulator. The electronic gap (E_G), or rather the size of the forbidden band, thus is determined by the difference in energy between the lowest part of the conduction band (E_C) and the highest of the valence band (E_V). The gap $E_G = E_C - E_V$ should in principle be sufficiently large (typically > 5 eV) so as to disallow a significant passage of electrons, under a thermal agitation of kT, from the valence to the conduction band
- When we have the same scenario as in the preceding case, but the size of the gap is only of the order of 1 eV, then we have a semiconductor. With a low population of electrons in the conduction band derived from the valence band by thermal activation we have an intrinsic semiconductor. In this case, the Fermi level (average energy level for the electrons) remains localized in the middle of the gap. The introduction of doping atoms that generate levels in the neighborhood of the permitted bands (for example, n doping which forms bands of levels just under E_C) displaces the Fermi level toward the limits of the permitted bands, which thus exhibit a greater population of electrons. We have in this case a extrinsic semi-conductor.
- When the last band is but partially occupied (typically half-full) and the electrons can easily move from the inside of this band under the effect of a weak external excitation (for example, an electric field or a thermal gradient), then we have a metal.

4.4.2. Populating permitted bands

4.4.2.1. The state density functions

The state density function can be defined either within a space of energies, or within a space of wave numbers denoted k (reciprocal space).

- Definitions

In the space of energies, the state density function, denoted Z(E), is such that Z(E)dE represents the number of electronic states (each described by a wave function) of an energy in between E and E + dE, which are in a unit volume system (when one dimensional L =1, and when three dimensional V = 1).

Similarly, and still with respect to a unit volume in direct space, in the k space the state density function is denoted N(k), and is such that N(k) dk represents the number of electronic states for which the vector $\vec{k}$ is situated between $\vec{k}$ and $\vec{k} + \overrightarrow{dk}$.

The association displayed in Figure 4.7 shows that for one dimension, where k can be either positive or negative (which is due to the privileged solutions for the wave function being associated to a propagation which equally can be well toward the positive or negative values of k), we should have:

$$Z(E)\, dE = 2\, N(k)\, dk. \qquad (11)$$

The interval dE in energy space corresponds to a k space interval of 2.dk which is situated between $\vec{k}$ and $\vec{k} + \vec{dk}$ and between $-\vec{k}$ and $-(\vec{k} + \vec{dk})$.

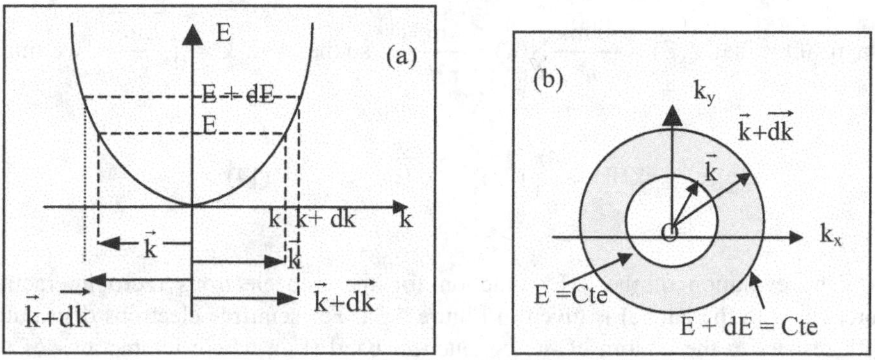

Figure 4.7. *Corresponding E and k spaces where E= ħ² k²/2m (a) in one dimension and (b) in two dimensions.*

• Determination of three dimensional state density functions

In three dimensions (3D), as illustrated in Figure 4.7 where one can imagine that the circles in the plane of the figure are cross sections of equal-energy spheres, the correspondence between the state density functions becomes $Z(E)\, dE = N(k)\, dV_k = N(k)\, 4\pi\, k^2\, dk$, from which:

$$Z(E) = 4\pi k^2 \frac{N(k)}{\frac{dE}{dk}}. \qquad (12)$$

For free electrons, we have $E = \frac{\hbar^2 k^2}{2m}$ (see Section 4.3.1) and $\frac{dE}{dk} = \frac{\hbar^2 k}{m}$, so that:

$$Z(E) = \frac{4\pi m k}{\hbar^2} N(k).$$

We now evaluate N(k). In reciprocal space, the electrons are distributed within cells of sides given by $\Delta k_x = \Delta k_y = \Delta k_z = \frac{2\pi}{L}$ and volume given by $\Delta k^3 = \frac{8\pi^3}{L^3} = \frac{8\pi^3}{V}$, where V is the volume of the material (which is taken to be equal to unity in order to determine the state density functions). As in cells of volume Δk^3 it is possible to place two electrons, of opposing spin direction, so that we have: $N(k)\,\Delta k^3 = 2$, so that:

$$N(k) = \frac{1}{4\pi^3}\,. \qquad (13)$$

The result is that $Z(E) = \frac{4\pi mk}{\hbar^2} N(k) = \frac{m}{\pi^2\hbar^2} k$, so that with $k = \sqrt{\frac{2mE}{\hbar^2}}$, we find

$$Z(E) = \frac{4\pi}{h^3}(2m)^{3/2}\sqrt{E}\,. \qquad (14)$$

The evolution of the Z(E) function for the free electrons (zero interaction potential with the lattice) is given in Figure 4.8a. For semifree electrons distributed with respect to the bottom of the conduction band (E_C), where interactions of the carriers with the lattice means replacing the mass (m) of the electrons by their effective mass (m_e*), the state density function is represented in Figure 4.8b. It can be written for the electrons as:

$$Z_C(E) = \frac{4\pi}{h^3}(2m_e^*)^{3/2}\sqrt{(E - E_C)}\,. \qquad (15)$$

Similarly for holes with effective masses denoted by m_h* distributed from the upper–most point of the valence band (E_V) downward, we have:

$$Z_V(E) = \frac{4\pi}{h^3}(2m_h^*)^{3/2}\sqrt{(E_V - E)}\,. \qquad (16)$$

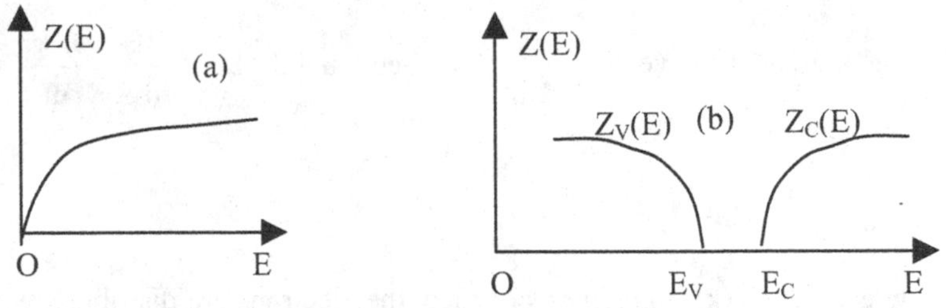

Figure 4.8. *State density functions for (a) state density function for free electrons and (b) and semifree carriers.*

4.4.2.2. Electronic densities in permitted bands

The density of electrons [n(E)] is such that if n(E) dE represents the number of electrons per unit volume which have an energy between E and E +dE, then this number is equal to the product of the number of available states between E and E + dE [in other words Z(E) dE] by the probability of state occupation. In Fermi statistics, this probability for electrons situated at a energy level denoted E is in the form:

$$F_n(E) = \frac{1}{1 + \exp\left(\frac{E - E_F}{kT}\right)}. \qquad (17)$$

The result is that the electronic density [n(E)] in the conduction band, with $A_C = \frac{4\pi}{h^3}(2m_e^*)^{3/2}$, is given by:

$$n(E) = Z_C(E).F_n(E) = A_C\left[1 + \exp\left(\frac{E - E_F}{kT}\right)\right]^{-1}\sqrt{(E - E_C}\,. \qquad (18)$$

We now will look at the equivalent case for holes in the valence band. For a given energy E, whether occupied by an electron or a hole (electron vacancy), we can write, using $F_p(E)$ to denote the probability of occupation of an energy level by a hole, that $F_n(E) + F_p(E) = 100\ \% = 1$, so that $F_p(E) = 1 - F_n(E)$.

With $A_V = \frac{4\pi}{h^3}(2m_h^*)^{3/2}$ the concentration of holes [p(E)] in the valence band is given by:

$$p(E) = Z_V(E).F_p(E) = A_V\left\{1 - \left[1 + \exp\left(\frac{E - E_F}{kT}\right)\right]^{-1}\right\}\sqrt{(E_V - E}\,. \qquad (19)$$

In practice, Eqs. (18) and (19) can be simplified for nondegenerate semiconductors. A low concentration of carriers in the bands associated with a Fermi position distanced from the band limits E_C or E_V by at least 2kT means that the Fermi function of Eq. (17) can be replaced, as we shall see, by the Boltzmann function, as in $f(E) = \exp[-(E - E_F)/kT]$.

The concentrations (population per unit volume) of electrons in the conduction band and holes in the valence band are finally obtained by integrating over the width of the permitted bands.

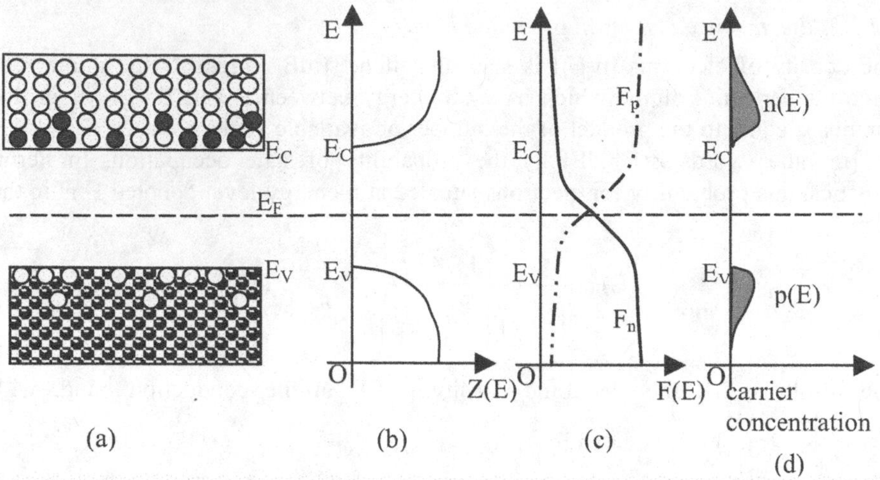

Figure 4.9. *(a) Filling bands in an intrinsic semiconductor; (b) state density functions; (c) occupation probability functions; and (d) concentration functions.*

For electrons, we need to calculate:

$$n = \int_{E_C}^{\substack{\text{highest point in}\\ \text{conduction band}}} n(E)\, dE = \int_{E_C}^{\infty} n(E)\, dE \quad . \qquad (20)$$

As elsewhere, we can extend the integration up to infinity as the probability of occupation between the highest point of the conduction band and infinity is zero (Figure 4.9). Taking Eq. (18) into account, we thus have:

$$n = A_C \int_{E_C}^{\infty} \sqrt{E - E_C}\, \frac{1}{1 + \exp\frac{E - E_F}{kT}}\, dE \, . \qquad (21)$$

With $(E - E_F) > 2\,kT$, we have $\exp\frac{E - E_F}{kT} > \exp(2) \approx 7.4$, and as 1 turns out to be negligible with respect to $\exp\frac{E - E_F}{kT}$, we thus can state that $\frac{1}{1 + \exp\frac{E - E_F}{kT}} \approx \exp\left(-\frac{E - E_F}{kT}\right)$. The Fermi function is approximated by using the Boltzmann function, and hence Eq. (21) becomes:

$$n = A_C \int_{E_C}^{\infty} \sqrt{E - E_C} \exp\left(-\frac{E - E_F}{kT}\right) dE . \qquad (22)$$

We then change the variable, $u = \dfrac{E - E_C}{kT}$, and hence $dE = (kT)\, du$. With

$$\exp\left(-\frac{E - E_F}{kT}\right) = \exp\left(-\frac{E - E_C + E_C - E_F}{kT}\right) = \exp\left(-\frac{E - E_C}{kT}\right)\exp\left(-\frac{E_C - E_F}{kT}\right),$$

we thus have $n = (kT)^{3/2} \exp\left(-\dfrac{E_C - E_F}{kT}\right) \int_0^{\infty} \sqrt{u} \exp(-u) du$.

As $\int_0^{\infty} \sqrt{u} \exp(-u) du = \dfrac{\sqrt{\pi}}{2}$, we finally obtain

$$n = N_c \exp\left(-\frac{E_C - E_F}{kT}\right), \text{ avec } N_c = 2\left(\frac{2\pi m_e^* kT}{h^2}\right)^{3/2} . \qquad (23)$$

Similarly, for holes, we have:

$$p = \int_{\text{low point of valence band}}^{E_V} p(E)\, dE = \int_{-\infty}^{E_V} p(E)\, dE . \qquad (24)$$

Between $-\infty$ and the lowest point of the valence band there is a zero probability of occupation by holes, so we can extend the integration to $-\infty$.

A similar calculation to that carried out for electrons gives:

$$p = N_v \exp\left(-\frac{E_F - E_V}{kT}\right), \text{ where } N_v = 2\left(\frac{2\pi m_p^* kT}{h^2}\right)^{3/2} . \qquad (25)$$

Notably, we can deduce from Eqs. (23) and (25) that:

$$np = N_c N_v \exp\left(-\frac{E_G}{kT}\right). \qquad (26)$$

This product is independent of the position of E_F. For a given semiconductor, the product of np is constant, in accordance with the law of mass action. For an intrinsic semiconductor, $n = p = n_i$. Combining this with the mass action law of Eq. (26), we can determine that:

$$n_i = \left(N_c N_v\right)^{1/2} \exp\left(-\frac{E_G}{2kT}\right). \qquad (27)$$

4.5. Optical properties of semiconductors: reflectivity, gap size, and the dielectric permittivity

Absorption of radiation is determined by the different types of electrons excited in the semiconductor. This response is characterized by the dielectric function. For radiative transitions, we shall see that they depend either directly or indirectly on the nature of the gap.

4.5.1. The dielectric function and reflectivity

In a semiconductor, there exist at the same time:

- valence electrons relatively well localized around their atoms; and
- throughout the lattice reasonably delocalized electrons (semi-free), which generate the transport properties specific to semiconductors.

4.5.1.1. Effect of localized electrons

The forces that attach the electrons are similar to those found in dielectrics. Thus, their response to the optical field in terms of the dielectric function is in the form of Eqs. (6) and (7) of Chapter 8, Volume 1. These electrons are characterized by an oscillation with an angular frequency (ω_0) and a frictional force given in the form $\vec{f_t} = -m\Gamma\vec{v}$. For a single type of these electrons, there are real and imaginary parts of $\underline{\varepsilon}_r$ with values respectively given by:

$$\varepsilon_r' = 1 + \omega_p^2 \frac{\left(\omega_0^2 - \omega^2\right)}{\left(\omega_0^2 - \omega^2\right)^2 + \omega^2\Gamma^2} \qquad (28)$$

$$\varepsilon_r'' = \omega_p^2 \frac{\omega\Gamma}{\left(\omega_0^2 - \omega^2\right)^2 + \omega^2\Gamma^2}. \qquad (29)$$

In these equations, ω_p is the plasma angular frequency and is such that $\omega_p^2 = \frac{n_e q^2}{\varepsilon_0 m}$, where n_e is the density of electrons (number per unit volume) of the type under consideration (with their own angular frequency ω_0 and relaxation time $\tau = 1/\Gamma$).

The function denoted ε_r'' is not equal to zero only when $\omega = \omega_0$, outside of which, ε_r'' is stuck at zero if $\Gamma = 0$ (Figure 4.10b). The dielectric function thus is real and written:

$$\varepsilon_r' = 1 + \frac{\omega_p^2}{\omega_0^2 - \omega^2}. \qquad (30)$$

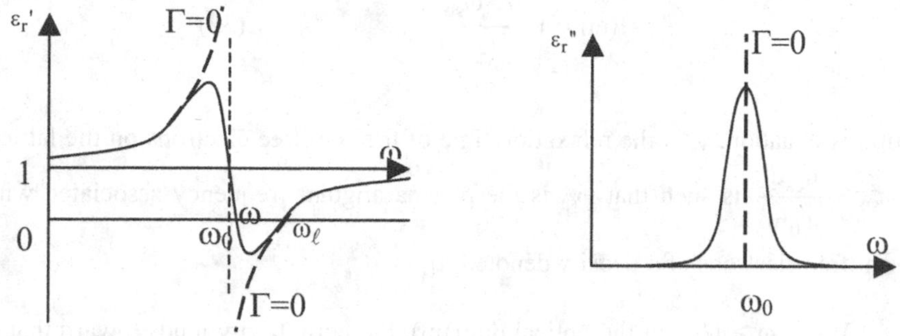

Figure 4.10. *Representative plots of the dielectric function for bound electrons: (a) $\varepsilon_r' = f(\omega)$ and (b) $\varepsilon_r'' = f(\omega)$.*

In Figure 4.10a, we can see that between angular frequencies $\omega = \omega_T$ and $\omega = \omega_\ell$, we have a negative ε_r'. As the index is defined in its most general form by $\underline{\varepsilon}_r(\omega) = \underline{n}^2$ [see Eq. (24) of Chapter 7, Volume 1], here the index must be purely imaginary so that n' = 0. Thus, an incident wave with the general form:

$$\vec{E} = \underline{\vec{E}}_m \exp(-n''\frac{\omega}{c}z)\exp(j[n'\frac{\omega}{c}z - \omega t]), \qquad (31)$$

is reduced to: $\vec{E} = \underline{\vec{E}}_m \exp\left(-n''\frac{\omega}{c}z\right)\exp(-j\omega t)$.

This is a wave that cannot propagate in the medium. The Fresnel equations (see Chapter 12, Volume 1), can be used to give the coefficients of reflection and transmission for the incident wave. For the range of frequencies given by $\omega_T < \omega < \omega_\ell$, the incident wave undergoes a reflection. In this case, the perpendicular to the incident is of the form $R = \frac{\underline{n}-1}{\underline{n}+1} = \frac{(n'-1)+in''}{(n'+1)+in''}$.

We thus can see that $|R|^2 = \frac{(n'-1)^2 + n''^2}{(n'+1)^2 + n''^2} \underset{n'\to 0}{\to} 1$.

4.5.1.2. Effect of semifree electrons (delocalized)

At ambient temperature, the presence of free or rather semifree carriers in a semiconductor brings in a specific contribution to the dielectric relaxation function. This component has a form recalling that of a metal where the returning force on the electrons is zero, as in $\omega_0 = 0$ [see also Section 8.4, Volume 1, which used the preceding Eqs. (28) and (29) for $\omega_0 = 0$]. Hence,

$$\underline{\varepsilon}_r(\omega) = 1 - \frac{\omega_{sp}^2}{\omega^2 - \dfrac{i\omega}{\tau_s}}. \qquad (32)$$

In this equation, τ_s is the relaxation time of the semifree electrons on the lattice and $\omega_{sp}^2 = \frac{n_{es} q^2}{\varepsilon_0 m}$ is such that ω_{sp} is the plasma angular frequency associated with the semifree electrons of a density denoted n_{es}.

When $\omega > 1/\tau_s$ (in the optical domain), the permittivity tends toward that given in the equation for plasmas, where $\underline{\varepsilon}_r(\omega) = 1 - \frac{\omega_{sp}^2}{\omega^2} = \varepsilon_r(\omega)$. However, when $\omega < \omega_{sp}$, then $\varepsilon_r(\omega) < 0$ and the optical index is purely imaginary. Therefore, the wave cannot propagate within the medium that thus reflects the light (see, for example, Section 11.6, Volume 1, and Figure 4.11).

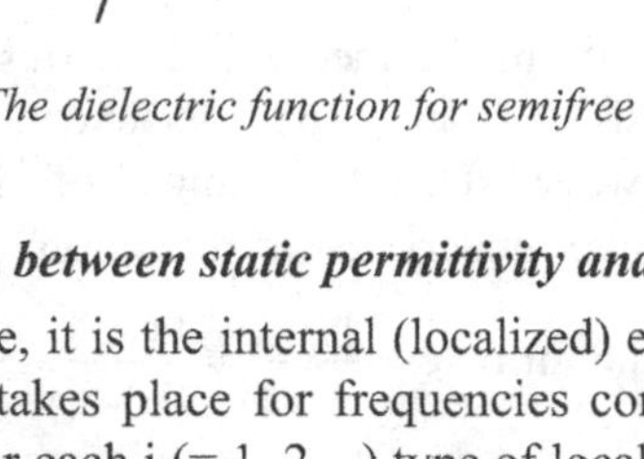

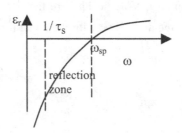

Figure 4.11. *The dielectric function for semifree electrons in the optical domain.*

4.5.2. The relation between static permittivity and the size of the gap

As discussed above, it is the internal (localized) electrons that determine the optical absorption which takes place for frequencies corresponding to the proper angular frequencies (ω_0) for each j (= 1, 2,...) type of localized electron. The term ω_{0j} is used to denote the proper angular frequency for each j type electron; hence, each of the different values of ω_{0j} locates peaks successively spread throughout the electromagnetic spectrum. Each of these is associated with a different electronic transition due to the optical absorption. For semiconductors found in column IV of the periodic table, the lowest values of ω_{0j}, denoted ω_{01}, therefore must correspond to the fundamental transition of an electron between the valence and conduction bands, as indicated in Figure 4.12. It should be noted that the reality is more complicated as the bands are not necessarily flat. This can be found, for example, in levels distributed in band tails permitted in the forbidden band.

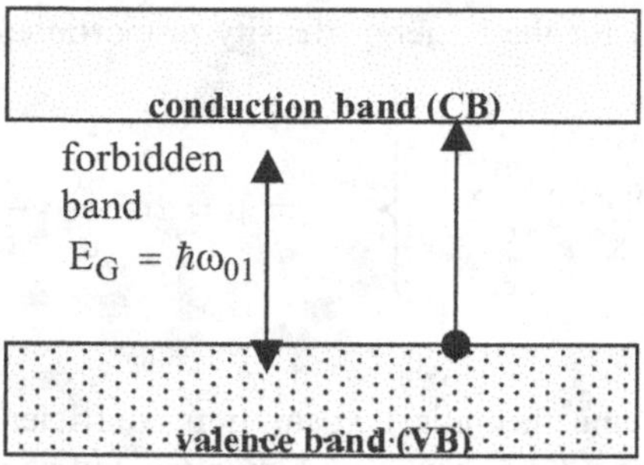

Figure 4.12. *Fundamental transition across the permitted band.*

In addition, we can relate the size of the gap (E_G) with the static permittivity [$\varepsilon'_r(0)$]. In effect, from Eq. (28), we can deduce that when $\omega = 0$, we have:

$$\varepsilon'_r(0) = 1 + \frac{\omega_p^2}{\omega_0^2} \quad \Rightarrow \quad \omega_0 = \frac{\omega_p}{\sqrt{\varepsilon'_r(0) - 1}}, \text{ so that:}$$

$$E_G \simeq \hbar\omega_0 = \frac{\hbar\omega_p}{\sqrt{\varepsilon'_r(0) - 1}}. \qquad (33)$$

We can see straight away that when the static dielectric permittivity increases, the gap decreases. This evolution well confirms the values given in the table below.

	E_G (eV)	$\varepsilon_r(0)$	$\lambda_G(\mu m)=hc/E_G$	$\hbar\omega_p$ (eV)
C	5.4	5.7	0.24	11,7
Si	1.1	12	1.1	3,6
Ge	0.7	16	1.8	2,7
Sn-α	≈ 0	24		

4.5.3. Absorption

In Eq. (31), the part that details the absorption is $\vec{E} \propto \vec{\underline{E}}_m \exp(-n''\frac{\omega}{c}z)$. As $\underline{n}^2 = \underline{\varepsilon}_r(\omega)$, and hence $(n' + i\, n'')^2 = (\varepsilon' - i\, \varepsilon'')$, we thus find by identification of the imaginary parts that $n'' = \frac{\varepsilon''}{2n'}$.

The attenuation factor μ for the I energy density proportional to the square of the amplitude follows:

$$\left.\begin{array}{l} I \propto \left[\exp\left(-n''\frac{\omega}{c}x\right)\right]^2 = \exp\left(-2n''\frac{\omega}{c}x\right) \\ \propto I_0 \exp(-\mu x) \end{array}\right\} \Rightarrow \mu = 2n''\frac{\omega}{c} = \frac{\omega\,\varepsilon''}{c\,n'}.$$

This is evidently the same theoretical absorption coefficient as that already found for dielectrics in Section 3.4.5, hence its being denoted as μ (so as to not be confused with the dielectric polarisability which is also classically denoted as α). We now will look at the form of this coefficient while taking into account the nature of the transitions specific to semiconductors.

4.6. Optoelectronic properties: electronphoton interactions and radiative transitions

4.6.1. The various absorption and emission mechanisms

4.6.1.1. Band–to–band (interband) transitions

As indicated in Figure 4.13a, these transitions are typically between the valence and conduction bands and can be divided into:

• the fundamental absorption, where the absorption of a photon generates the transition of an electron from the valence to the conduction band. The frequency of the photon (ν) must be such that $\nu > \nu_G = E_G/h$. Thus an electron–hole pair is generated, in a process used in radiation captors;

• spontaneous emission, which is the inverse of that the above, where an electron spontaneously transits from the conduction band to the balance band and gives rise to a photon; and

• a stimulated emission, which is due to a photon that irradiates the semi-conductor and induces the transition of an electron from the conduction to the valence band and the simultaneous emission of a photon of the same energy as the exciting photon. This process is central to semiconductor–based lasers.

In general terms, the band–to–band transition can involve one or more phonons. A phonon is a quasiparticle that corresponds to a quanta of a thermal vibration supplied by thermal excitation of the atoms in the material.

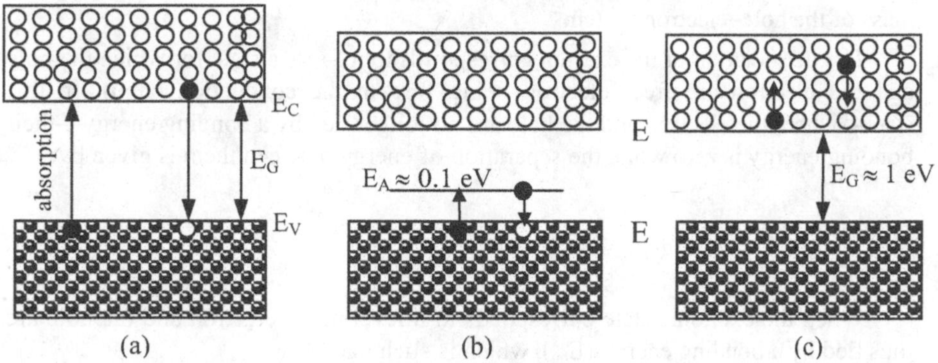

Figure 4.13. *Various transitions: (a) band to band; (b) band to impurities; and (c) intraband between free carriers.*

4.6.1.2. Transitions from the bands to impurities (Figure 4.13b)

These can result from photons that, for example, in a p-type semiconductor pass a valence band electron to the level of an acceptor where the electron is trapped. A hole thus is formed in the valence band.

4.6.1.3. Transitions due to free carriers (intraband)

A photon can transmit its energy to an electron which is thus pushed up to an empty and higher level in the band (shown for the conduction band in Figure 4.13c). This process is followed by a thermalization of the electron which relaxes back to a lower point in the conduction band while losing its surplus energy thermally to the lattice, and hence lattice vibrations.

4.6.1.4. Phonon transitions

Low–energy photons (of long wavelengths) can give up their energy to the lattice by directly exciting vibrations of the atoms, from which are generated phonons.

4.6.1.5. Excitonic transitions (Figure 4.14)

The absorption of a photon can generate an electron and a hole separated by a finite distance so that they can be bound to one another by a coulombic interaction energy. This pairing can be compared to a hydrogen atom, as here the role of the nucleus is played out by the hole (termed a Wannier exciton).

The levels of the quantified energies are defined with respect to E_G (see Figure 4.14) and are of the form $E_{Gn} = E_G - \frac{m^* e^4}{32\pi^2\varepsilon_0^2\varepsilon_r^2\hbar^2}\frac{1}{n^2}$, where m* is the reduced mass of the hole–electron system.

The dissociation of the exciton corresponds to $n \to \infty$, a state for which the hole and the electron are free (electron being free in the conduction band). In this state—where the electron and the hole are no longer tied by a bonding energy—their bonding energy is zero while the separation of energy between them is given by

$$E_G = E_{G\infty}.$$

When the excitonic state corresponds to a level n, the electron and the hole are thus tied by a bonding energy (E_{bn}), which is such that:

$$E_{bn} = E_G - E_{Gn} = \frac{m^* e^4}{32\pi^2\varepsilon_0^2\varepsilon_r^2\hbar^2}\frac{1}{n^2}. \qquad (35)$$

In other terms, this is the energy that the exciton requires to yield a separated hole and electron to the bands where they are free (noting that when $n \to \infty$, $E_{bn} \to 0$).

The energy of separation of an electron and a hole in an exciton for its part is in the form E_{Gn}. A peak due to an absorption at the lowest possible frequency corresponds to the energetic transition E_{G1}, not E_{b1} (the bonding energy).

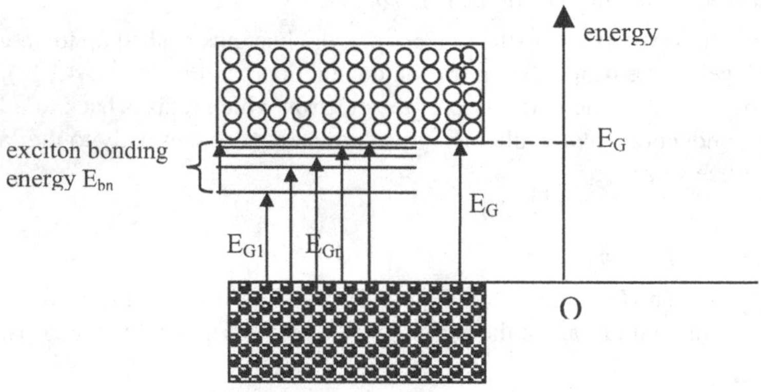

Figure 4.14. *Exciton energy level (E_{Gn}) and bonding energy level (E_{bn}).*

4.6.1.6. Result (see, for example, Figure 4.15)

All transitions contribute to the absorption coefficient. Photons with energies hν > E_G essentially induce interband transitions. The maximum absorption corresponds to the transition between zones of strong (hν > E_G) and weak absorptions. This transition is much more abrupt for direct gap semiconductors such as GaAs than for indirect gap semiconductors which include Si. The following section details the nature of these gaps and their corresponding transitions.

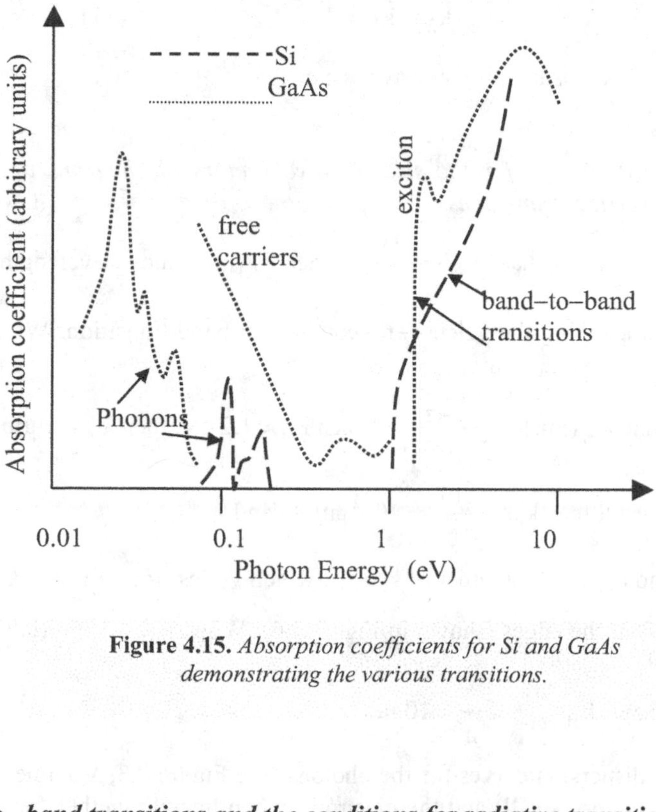

Figure 4.15. *Absorption coefficients for Si and GaAs demonstrating the various transitions.*

4.6.2. Band–to–band transitions and the conditions for radiative transitions

4.6.2.1. Equations for conservation of energy and momentum

In general terms, electronic transitions, which are accompanied by the absorption or emission of a photon, are termed radiative transitions. These processes are controlled by rules for what are elastic collisions between two particles, namely, electrons and photons. There is of course no modification of the internal energy. They state that:

- There is a conservation of energy during a transition of an electron from level E_1 in the valence band to level E_2 in the conduction band. For an absorption or emission of a photon of energy given as $E_{pt} = h\nu$, then

$$E_2 - E_1 = E_{pt} = h\nu. \qquad (36)$$

- There is a conservation of momentum given by $\vec{p} = \hbar\vec{k}$. Using the self-same evident notations, we have:

  $\vec{p}_2 - \vec{p}_1 = \vec{p}_{pt}$, from which

$$\vec{k}_2 - \vec{k}_1 = \vec{k}_{pt}, \qquad (37)$$

where $\vec{k}_{pt}$ is the photon wave vector.

4.6.2.2. Estimations for the size of wave vectors of electrons and photons, and vertical transitions

- k_{pt} is such that $k_{pt} = \dfrac{2\pi}{\lambda_{pt}}$, where λ_{pt} is the wavelength of a photon accompanying the valence–to–conduction band transition. With the gap being E_G, we have $E_G \approx h\nu_{pt}$,

  so that $\lambda_{pt}(\mu m) = \dfrac{1.24}{E_G(eV)}$. When $E_G \approx 1.2$ eV, then $\lambda_{pt} \approx 1\ \mu m = 10^3$ nm, with

  the result that $k_{pt} \approx \dfrac{10}{10^3} = 10^{-2}\,nm^{-1}$; and

- k_2 and k_1 are of the order of $k_{electron}$, which varies from 0 at the center of the band to $\dfrac{\pi}{a}$ at the edges (shown in Figure 4.6). When $a \approx 0.3$ nm (lattice repeat unit),

  we have $k_{electron} = \dfrac{\pi}{a} \approx 10\ nm^{-1}$.

The dispersion curves for the photons (see Figure 7.3, Volume 1, where $k = k_{pt}$ is in effect very small) and the electrons, when traced together for one value of m, give yield plots similar to those shown in Figure 4.16, taking into account the relative values of $k_{electron}$ and k_{pt}.

The dispersion curve for a photon practically corresponds to a vertical straight line when presented against a plot on the same scale for electron dispersion. In other terms, the wave–vector of the photon is always negligible with respect to that of an electron or a hole. As a consequence, Eq. (37) can be simplified to:

$$\vec{k}_2 \approx \vec{k}_1. \qquad (38)$$

This condition is called the selection rule for k vectors. The electronic transitions should be vertical within a graph of E = f(k). Thus, we can conclude that radiative transitions (accompanied by the absorption or emission of a photon) are vertical transitions (change in k negligible with respect to the E(k) carrier diagram).

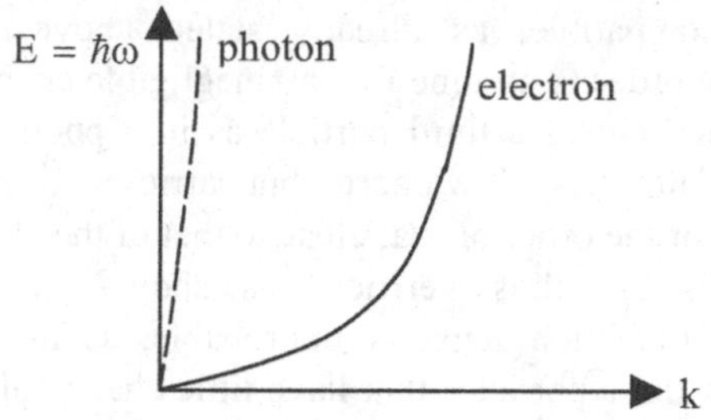

Figure 4.16. *Dispersion curves for photons and electrons.*

4.6.2.3. Direct or indirect gap semiconductors

A semiconductor has a direct gap if in the representation of E = f(k) of the energy levels of the carriers, the extremes of the permitted bands (i.e., the highest point of the valence band and the lowest of the conduction band) have the same value of k (as shown in Figure 4.17a). In contrast, an indirect gap semiconductor is so termed when these extremes are out of step and do not have the same value of k, as in Figure 4.17b. As a consequence, and taking Eq. (38) into account for k vectors, in the neighborhood of the fundamental gap, the direct gap materials assist relatively large radiative transitions. However, indirect gap semiconductors necessitate the intervention of a third particle (phonon) of large momentum in order to undergo oblique transitions between the extremes of the permitted bands.

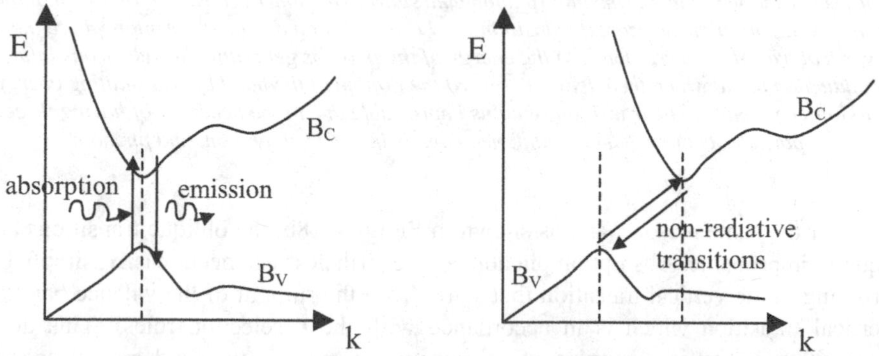

Figure 4.17 *Semiconductors: (a) direct gap and (b) indirect gap.*

In more precise terms, the radiative electron–hole recombination is unlikely in an indirect gap semiconductor. In effect, the conservation of energy is respected (electronic transition emitting a photon) and the conservation of momentum is not due to the oblique nature of the conduction to valence band transition. The transition brings into play a considerable variation in the wave vector (and the quantity of momentum), which is incompatible, as already stated above, with the low momentum of the photon. In order to assume this nonnegligible change in the wave vector, the system needs the help of a third particle as in a phonon (indicated in Figure 4.18a). In effect, the latter has a low energy but carries a considerable degree of movement (with k values or the order of π/a, close to that of the electrons) and the two conservation equations are thus verified (as shown in Figure 4.18a). Nevertheless, the resulting transition remains improbable, as it necessitates the simultaneous intervention of three particles that have little chance of doing so. This is why silicon, which has an indirect gap, has a level of recombination considerably below that of GaAs, which has a direct gap.

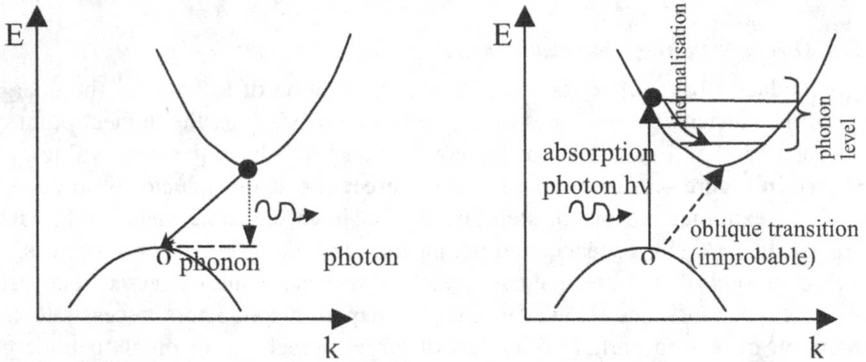

Figure 4.18. *For an indirect gap: (a) oblique transition during emission, with conservation of energy in the vertical transition (photon emission) and moment conservation (by phonon intervention) in the horizontal transition; and (b) transition during absorption where the vertical transition corresponds to the energy of the photons generating the electrons in the conduction band, which then deexcite toward the bottom of the band by transmitting energy to lattice phonons. The transition remains improbable due to the necessity of having three particles come together simultaneously, as in electron, photon, and phonon.*

For the same reasons, and as shown in Figure 4.18b, the oblique transitions are equally improbable. Absorption phenomena nevertheless can occur easily, simply by bringing in the vertical transition that starts from the summit of the valence band (a vertical transition which is in accordance with the k selection rules). This does necessitate, however, a surplus of energy with respect to the fundamental energy gap. The excess energy with respect to the bottom of the conduction band is then loss by an electron that is thermalized at the bottom of the conduction band through a loss of energy through the lattice in the form of phonons.

4.7. Level of absorption and emission

We now turn to the probability densities for emission or absorption of a photon during direct band-to-band transitions. There are three factors that determine the result. They are the densities of the states with which the photons interact, the probability of occupation in those states, and the probabilities of transitions between the states brought into play.

4.7.1. Optical function of the state density

4.7.1.1. Relation for the carrier energy levels and the effective masses

In Eq. (36), the energy level—denoted E_2—for a free electron in the conduction band with respect to the energy origin (E_C) at the bottom of the conduction band is given in the form $\frac{\hbar^2 k^2}{2m}$ (see Section 4.3.1). Being semifree given its interactions with the lattice, the electron mass needs to be replaced by the effective mass m_e* so that the energy E_2 is in the form:

$$E_2 = E_C + \frac{\hbar^2 k^2}{2m_e^*}. \qquad (39)$$

For their part, the holes (with the same wave number k as for the electrons–direct gap conductor) in the valence band are situated under the summit (E_V) of this band. They have an energy (E_1) given by (where m_h* is the effective mass of a hole)

$$E_1 = E_V - \frac{\hbar^2 k^2}{2m_h^*}. \qquad (40)$$

Taking the difference between Eqs. (39) and (40) we obtain with $E_G = E_C - E_V$:

$$E_2 - E_1 = E_G + \frac{\hbar^2 k^2}{2m_e^*} + \frac{\hbar^2 k^2}{2m_h^*} = h\nu. \quad (41)$$

By making $\frac{1}{m_r^*} = \frac{1}{m_e^*} + \frac{1}{m_v^*}$, we deduce that:

$$k^2 = \frac{2m_r^*}{\hbar^2}(h\nu - E_G). \qquad (42)$$

The ratio of Eq. (42) in Eqs. (39) and (40) finally gives:

$$E_2 = E_C + \frac{m_r^*}{m_e^*}\left(h\nu - E_G\right) \qquad (43)$$

$$E_1 = E_V - \frac{m_r^*}{m_v^*}\left(h\nu - E_G\right). \qquad (44)$$

4.7.1.2. State densities [ρ(ν)] for which photons (hν) interact in a direct gap semiconductor

The calculation of the number [ρ(ν)] of states implicated per unit time (for a unit volume of material) in the direct optical transitions a priori should bring in the densities of states in the valence and conduction bands.

For example, in a direct gap semiconductor, from each of the E_2 states situated within a range of energies (dE_2), a radiation of frequency ν can be emitted, which is within a zone of frequencies given by dν. The E_2 and the energy of the radiation hν also are related by Eq. (43) for a direct gap semiconductor. In addition, in such a conductor, there is a relation between the number of energy levels in the range dE2 from which the emission can occur and the number of emissions within the range dν. It can be written as:

$$Z_C(E_2)dE_2 = \rho(\nu)\, d\nu,$$

where $\frac{dE_2}{d\nu}$ is determined from Eq. (43).

From this, $\rho(\nu) = \frac{dE_2}{d\nu} Z_C(E_2) = \frac{hm_r^*}{m_e^*} Z_C(E_2)$, from which we can deduce with Eq. (15) that:

$$\rho(\nu) = \frac{\left(2m_r^*\right)^{3/2}}{\pi\hbar^2}\left(h\nu - E_G\right)^{1/2}, \text{ with } h\nu \geq E_G. \qquad (45)$$

The state density function with which the photons (of energy hν) can interact, otherwise termed the optical density-of-states function [ρ(ν)], is represented in Figure 4.19. It increases in accordance with the square root of (hν - E_G).

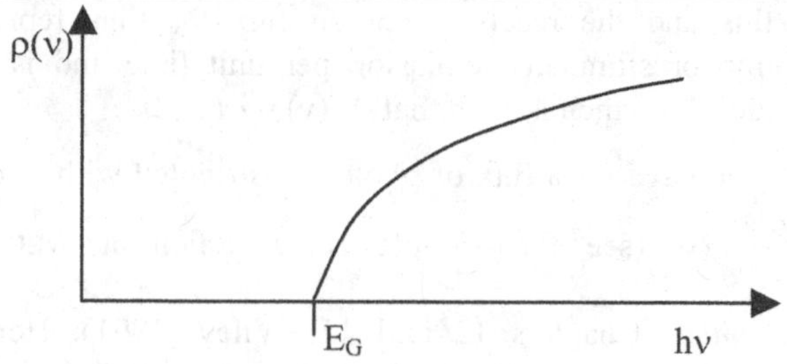

Figure 4.19. *Relation between ρ(ν) and hν.*

4.7.2. Probabilities of occupation

The probability [$F_e(\nu)$] for each level that photon emission will occur [between E_2 and E_1 according to Eq. (41)], is obtained by multiplying the probability of the function "the E_2 level in the conduction band is full" by the probability of the function "that the level E_1 of the valence band is empty", as in:

$$F_e(\nu) = F_c(E_2)[1 - f_v(E_1)], \qquad (46)$$

where $F_c(E_2)$ is the Fermi function for electrons in the E_2 level of the conduction band and $F_v(E_1)$ is the Fermi function for electrons in the E_1 level of the valence band.

The probability that the required occupation of the different levels for the absorption of a photon, denoted as $f_a(\nu)$, can be calculated in a similar manner, as in:

$$F_a(\nu) = F_v(E_1)[1 - F_c(E_2)]. \qquad (47)$$

4.7.3. Probabilities for radiative transitions

The radiative lifetime can be denoted as τ_r. The probability that a carrier (which conforms to the conditions of recombination) spontaneously produces radiation with a frequency ν is $P(\nu) = 1/\tau_r$.

In the presence of a photon flux of density Φ_ν, of received frequency ν (photons per unit time and unit surface), the system can absorb a part of the flux or produce a stimulated emission. The probability per unit time (probability density) for one carrier to make the transition by absorption or by emission (by stimulation) of a photon (frequency ν) depends on the efficient section of the incident flux with the medium. In atomic physics, we introduce a term for the efficient section for the transition for the material denoted $\sigma(\nu)$.

The product of this and the received photon flux Φ_ν thus represents the probability of absorption or stimulated emission per unit time and is otherwise termed the probability density which is such that $W(\nu) = \sigma(\nu)\,\Phi_\nu$.

In atomic physics, we have for a flux of photons distributed with a form factor $[g(\nu)]$ that $\sigma(\nu) = \dfrac{\lambda^2}{8\pi\tau_r} g(\nu)$ (see, for example, B.E.A. Saleh and M.C. Teich in *Fundamentals of Photonics*, Chapters 12 and 15, Wiley, 1991). Here we are concerned with a single frequency centered about the principal frequency (ν) for absorption or emission, so that the absorption or stimulated emission probability density is given by

$$W(\nu) = \frac{\lambda^2}{8\pi\tau_r}\Phi_\nu .$$

4.7.4. Overall level of emission or absorption transitions

Finally, the levels of spontaneous or stimulated emission (for a volume V = 1) and absorption are in the form, respectively:

$$r_{sp}(\nu) = \frac{1}{\tau_r}\rho(\nu)F_e(\nu) \tag{48}$$

$$r_{st}(\nu) = \frac{\lambda^2}{8\pi\tau_r}\Phi_\nu\,\rho(\nu)\,F_e(\nu) \tag{49}$$

$$r_{ab}(\nu) = \frac{\lambda^2}{8\pi\tau_r}\Phi_\nu\,\rho(\nu)\,F_a(\nu). \tag{50}$$

4.7.5. Absorption coefficient

If $\Phi_\nu(z)$ and $\Phi_\nu(z) + d\Phi_\nu(z)$ represent the density of flux entering and leaving a cylinder of length dz and section S = 1, then the flux density of lost photons is given by $d\Phi_\nu(z)$. This is equal to the difference between the density of the absorbed photon flux and the flux density of photons emitted by stimulation along the length (dz) of the cylinder, so that:

$$d\Phi_\nu(z) = \left[r_{ab}(\nu) - r_{st}(\nu)\right]dz .$$

With the help of Eqs. (49) and (50), we deduce:

$$\frac{d\Phi_\nu(z)}{dz} = \frac{\lambda^2}{8\pi\tau_r}\rho(\nu)\left[F_a(\nu) - F_e(\nu)\right]\Phi_\nu(z) = \mu(\nu)\Phi_\nu(z). \tag{51}$$

This relation defines the absorption coefficient μ(ν), which is thus in the form:

$$\mu(\nu) = \frac{\lambda^2}{8\pi\tau_r}\rho(\nu)\left[F_a(\nu) - F_e(\nu)\right] = \frac{\lambda^2}{8\pi\tau_r}\rho(\nu)F_g(\nu)\,. \qquad (52)$$

From Eqs. (46) and (47), the factor introduced $[\,F_g(\nu)\,]$ is such that:

$$F_g(\nu) = \left[F_a(\nu) - F_e(\nu)\right] = F_v(E_1) - F_c(E_2)\,. \qquad (53)$$

At the thermodynamic equilibrium, where the Fermi's pseudolevel coincides with the Fermi level (E_F), the functions $F_v(E)$ and $F_c(E)$ coincide with the Fermi function, as in: $F(E) = \dfrac{1}{1+\exp\left[(E - E_F)/kT\right]}$ and $F_g(\nu) = F(E_1) - F(E_2)$. We can thus remark that in this case of thermodynamic equilibrium, when $E_2 > E_1$, then the term $F_g(\nu) > 0$ and there is an absorption of radiation (the gain in radiation can only be obtained outside of equilibrium under electrical or optical injection regimes.

The introduction of Eq. (45) into Eq. (52) gives rise to:

$$\mu(\nu) = D\left(h\nu - E_G\right)^{1/2}\left[F(E_1) - F(E_2)\right], \qquad (54)$$

where $D = \dfrac{\sqrt{2}\,\lambda^2 m_r^{*3/2}}{h^2\tau_r}$.

With E_F placed in the gap of the material, far enough from the limits of the permitted bands, then $F(E_1) \approx 1$ and $F(E_2) \approx 0$, from which $F(E_1) - F(E_2) \approx 1$. In this case, we can write that:

$$\mu(\nu) = D\left(h\nu - E_G\right)^{1/2} = \frac{\sqrt{2}\,\lambda^2 m_r^{*3/2}}{h^2\tau_r}\left(h\nu - E_G\right)^{1/2}$$
$$= \frac{\sqrt{2}\,c^2 m_r^{*3/2}}{n^2\nu^2 h^2\tau_r}\left(h\nu - E_G\right)^{1/2}. \qquad (55)$$

A plot of $(\mu h^2\nu^2)^2$ as a function of $(h\nu$ - $E_G)$ makes it possible to obtain E_G by extrapolation.

The coefficient $\sigma(\nu)$ for the efficient section introduces the λ^2 (or $1/\nu^2$) term into the expression for D, and hence for μ. This term, obtained from results of atomic physics, is introduced in different ways for semiconductors, depending on the author.

So, Pankove (*Optical Processes in Semiconductors*, Dober Publishing, 1975) indicates that the absorption coefficient is described for:

- permitted direct transitions as

$$\mu(h\nu) = A^*\left(h\nu - E_G\right)^{1/2} \text{ where } A^* \approx \frac{q^2\left(\dfrac{2m_h^* m_e^*}{m_h^* + m_e^*}\right)^{3/2}}{nch^2 m_e^*};$$

- forbidden direct transitions as

$$\mu(h\nu) = A'\left(h\nu - E_G\right)^{1/2} \text{ where } A' = \frac{4}{3}\frac{q^2\left(\dfrac{2m_h^* m_e^*}{m_h^* + m_e^*}\right)^{5/2}}{nch^2 m_e^* m_h^* h\nu};$$

- indirect transitions between valleys: transitions with equally possible emission or absorption of phonons of energy E_{pn} can occur and

$$\mu_e(h\nu) = \frac{A\left(h\nu - E_G - E_{pn}\right)^2}{1 - \exp\left(-\dfrac{E_{pn}}{kT}\right)} \quad \text{when } h\nu > E_G + E_{pn}$$

$$\mu_a(h\nu) = \frac{A\left(h\nu - E_G + E_{pn}\right)^2}{\exp\left(\dfrac{E_{pn}}{kT}\right) - 1} \quad \text{when } h\nu > E_G - E_{pn}.$$

Note that the form of the denominator of the absorption coefficients comes from the statistical distribution of phonons, which is itself derived from the Bose–Einstein distribution.

4.8. Problem

For indirect intervalley transitions, show schematically how the radiative transitions (hν) with either emission or absorption of phonons of energy E_{pn} can intervene. Determine the condition on the energy of these phonons. How can the gap E_G be determined?

Answer

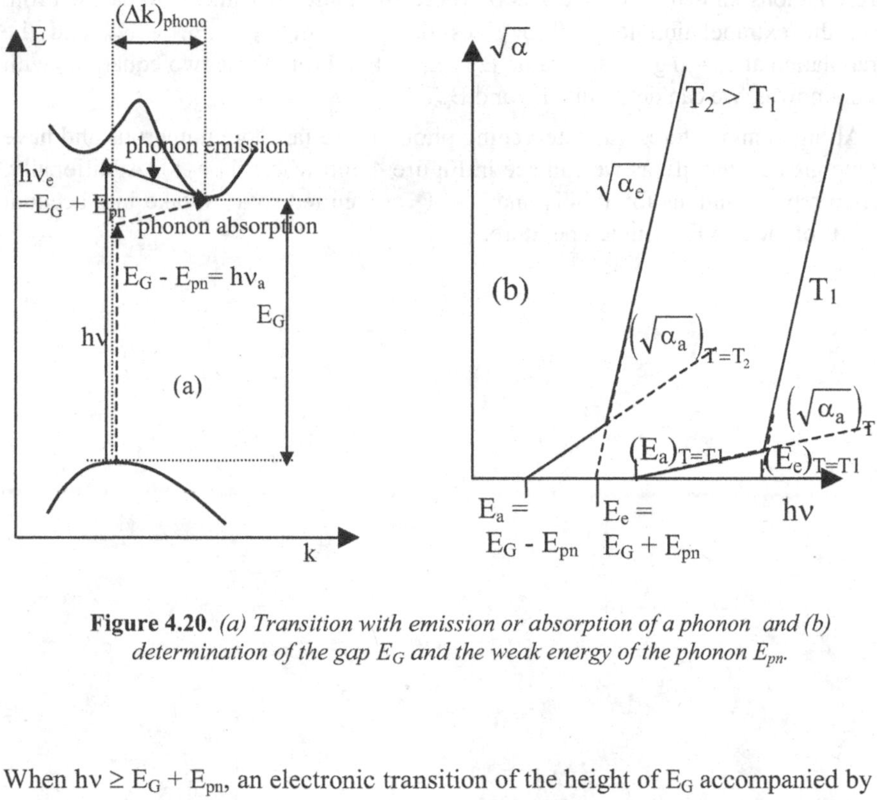

Figure 4.20. *(a) Transition with emission or absorption of a phonon and (b) determination of the gap E_G and the weak energy of the phonon E_{pn}.*

When $h\nu \geq E_G + E_{pn}$, an electronic transition of the height of E_G accompanied by the emission of a phonon with energy E_{pn} is possible (Figure 4.20). For this condition on hν, an electronic transition by E_G accompanied by the absorption of a phonon also is possible, as this mechanism only requires the energy given by $E_G - E_{pn}$. In addition, under this condition, we have:

$$\mu(h\nu) = \mu_e(h\nu) + \mu_a(h\nu).$$

(Inversely, if we "only" have $h\nu > E_G - E_{pn}$, then only the transition with phonon absorption is possible).

Also when $h\nu > E_G + E_{pn}$, and if the temperature is not very high, then the density of phonons is relatively low (due to the low degree of thermally induced quantic vibrations). The denominator μ_a is very high for a low temperature and the coefficient μ_a, associated with the condition $h\nu > E_G - E_{pn}$, is very low. At the same time, the phonon emission mechanism (associated with an absorption factor μ_e considerably greater than μ_a) is preponderant and corresponds to the condition $h\nu > E_G + E_{pn}$.

In addition if we trace the plots of $\sqrt{\mu(h\nu)} = f(h\nu)$, then we have the representations shown in Figure 4.20b where there are two linear domains. From these, the extrapolation to $\mu_a = 0$ gives the value of $E_a = E_G - E_{pn}$ and the extrapolation at $\mu_e = 0$ gives the value $E_e = E_G + E_{pn}$. From these two equations with two unknowns, we can determine E_G and E_{pn}.

At high temperatures, the intervening phonons are far more numerous and have a value greater than μ_a, as we can see in Figure 4.20b where $T_1 > T_2$. Additionally, the two representations for $T = T_1$ and $T = T_2$ are shifted so as to take into account the shift of the gap E_G with temperature.

Chapter 5

Electrical and Magnetic Properties of Semiconductors

5.1. Introduction

This chapter looks at the electronic and magnetic properties of semiconductors in a similar way to the study made on insulators in the volume, *Basic Electromagnetism and Materials*. While the applied fields will be essentially static, there will be a chance to follow, for example, microwave emissions due to the Gunn effect in semiconductors, which act as solid sources.

Initially, the properties can be studied by using a small perturbation, notably a weak electric field, so that with respect to the thermodynamic equilibrium, the n and p concentrations of the carriers are not modified in the bands of the applied field. In effect, only the position of the carriers with respect to the internal levels can change (see Figure 5.1a). This essentially concerns electrons.

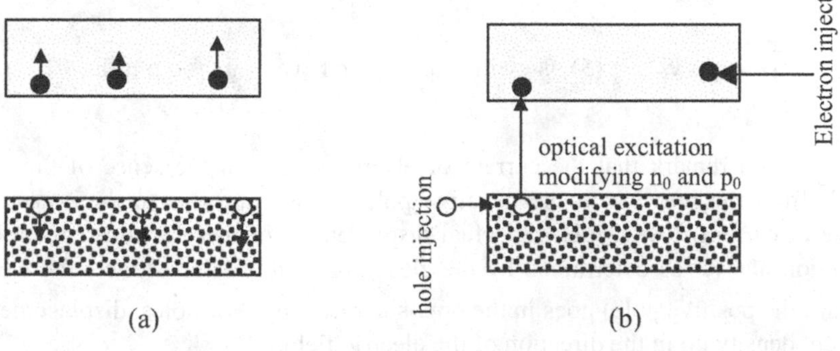

Figure 5.1. *(a) Intraband displacement of carriers under the effect of a weak perturbation and (b) effect of a strong perturbation with modification of the band population.*

Inversely, following a stronger perturbation (Figure 5.1b), the concentration of carriers can be modified inside the bands, for example, by injection of carriers into bands to produce electroluminescence in the domain of optoelectronics.

5.2. Properties of a semiconductor under an electric field

5.2.1. Ohm's law for a semiconductor

5.2.1.1. Classic form

Drude's classic theory (see Chapter 1, Volume 1) can be developed for a semiconductor and equally well for electrons as for holes. In the presence of an electric field ($\vec{E}$), their respective speeds are:

$$\vec{v}_n = -\mu_n \vec{E} \qquad (1)$$

$$\vec{v}_p = \mu_p \vec{E}, \qquad (2)$$

where μ_n and μ_p are the mobilities (defined as positive magnitudes) of the electrons and holes such that:

$$\mu_n = \frac{q\tau_n}{m_e^*} \text{ and } \mu_p = \frac{q\tau_p}{\left|m_h^*\right|}. \qquad (3)$$

Using this approach, τ_n and τ_p represent the average time between two successive collisions of electrons and holes. By denoting the concentrations of electrons and holes as n and p, respectively, and in accordance with the definition of the current density given by $\vec{j} = \rho\vec{v}$ (where $\rho = \rho_n = -q\,n$ for electrons and $\rho = \rho_p = q\,p$ for holes), we have as expressions for the corresponding current densities:

$$\vec{j}_n = -q\,n\,\vec{v}_n \quad (4), \text{ so with Eq. (1), we have } \vec{j}_n = q\,n\,\mu_n \vec{E} \qquad (4')$$

$$\vec{j}_p = q\,p\,\vec{v}_p \quad (5), \text{ so with Eq. (2), we have } \vec{j}_p = q\,p\,\mu_p \vec{E}. \qquad (5')$$

We can remark that the current of electrons goes in the sense of the electric field (from the positive to the negative pole). This conforms to the conventional sense of current flow, while the actual displacement of the electrons (given by the direction of $\vec{v}_n$ as determined by the electrostatic force that attracts the electrons toward the positive pole) goes in the opposite direction. For holes, displacement and current density go in the direction of the electric field.

Finally, the resultant current density (for electrons and holes) is:

$$\vec{j} = \vec{j}_n + \vec{j}_p = (\sigma_n + \sigma_p)\vec{E} \qquad (6)$$

where

$$\sigma_n = \frac{pq^2\tau_n}{m_e^*} = qn\mu_n \text{ and } \sigma_p = \frac{pq^2\tau_p}{\left|m_h^*\right|} = qp\mu_p . \qquad (7)$$

5.2.1.2. Conduction mechanism in a band energy scheme

In the absence of an electric field, the energy bands of a semiconductor are represented by the limits of the horizontal bands of value E_C (lowest point of the conduction band) and E_V (highest point on the valence band).

Take an electron that is situated at a point A which is the same as the origin O of an axis Ox along which the electron is transported (Figure 5.2). There is a cathode placed at x = 0 and an anode at x = OB = AB = x_B.

The origin of the potentials at the cathode are at point O (x = 0) and correspond to the E_C level of the semiconductor. A potential V > 0 is applied to the anode situated at x = x_B. The applied electric field ($\vec{E}$) goes from B toward O and the potential energy of a electron at x = x_B is decreased by qV following the application of this field (point B, pinpoints the position of the electron at x = x_B, thus dropped by qV to the position B').

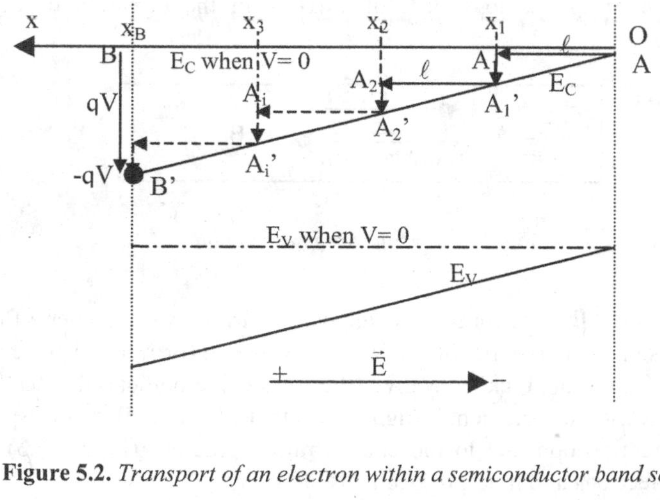

Figure 5.2. *Transport of an electron within a semiconductor band scheme.*

It is worth considering the transport of an electron from x = 0 to x = x_B at the scale of the lattice. Over a distance AA_1 equal to the mean free pathway (ℓ), the electron gains a kinetic energy denoted $\Delta\mathcal{E}_c$, which is equal to the work of the electric force and can be written in the form $\Delta\mathcal{E}_c = |q E \ell|$. At A_1, the electron undergoes a collision with the lattice as $AA_1 = \ell$, and its speed and also its kinetic energy return to zero. The electron has in effect lost $\Delta\mathcal{E}_c$ to the lattice, which is now warmed in accordance with Joule's law following the passage of current from A to A_1. Finally, following collision at A_1, the electron only has the potential energy for the position A_1', which is situated at the bottom of the conduction band.

At A_1', the electron subjected to $\vec{E}$ recommences the same trip which will take it through A_2' (with the intermediate position in A_2), then A_3', up to point B'. At this point the electron will cede an energy qV to the lattice which is either transformed into heat or radiation at each of the successive impacts at A_1, A_2, A_i...

5.2.2. Effect of a concentration gradient and the diffusion current

5.2.2.1. Fick's first law

Supposing that by using any one of several possible, for example, localized doping, we have created a concentration gradient. Figure 5.3 shows an example where there are many carriers in A and relatively few in B. This gradient will generate a movement of carriers from the high to the low concentration zone in an attempt to reestablish a uniform equilibrium. This phenomenon is that of diffusion which occurs in other physical systems such as with heat and with gases. The diffusion thus corresponds to a drift of carriers in the direction of decreasing concentration.

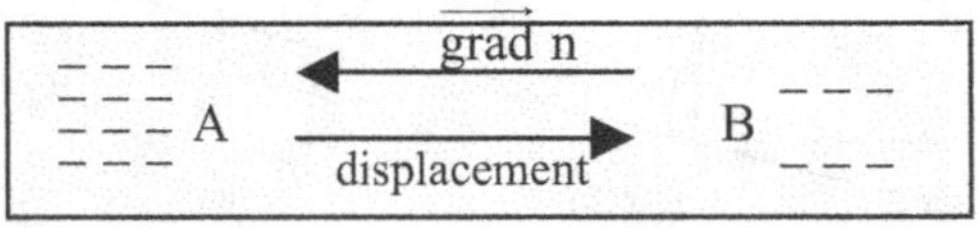

Figure 5.3. *Diffusion phenomenon.*

In order to establish the form of the diffusion current, we consider a flux (F) of carriers expressed as the number of carriers that traverse (or rather diffuse across) a unit surface per unit time. Fick's law gives the physical hypothesis that the flux is as large as the concentration gradient is high. Nevertheless, as the flux corresponds to a drift in the direction opposite to the concentration gradient (Figure 5.3), we can write that for electrons and holes, respectively, that:

$$\vec{F}_n = -D_n \overrightarrow{\text{grad}}\, n \qquad (8)$$

$$\vec{F}_p = -D_p \overrightarrow{\text{grad}}\, p\,. \qquad (9)$$

The proportionality coefficients D_n and D_p are positive and are termed diffusion constants, respectively, for electrons and holes.

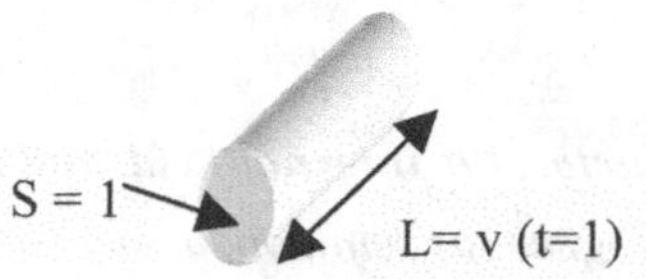

Figure 5.4. *Volume associated with a flux of particles.*

Given the definition of flux, the latter is represented by the number of carriers contained within a cylinder of cross–sectional area S (at unity) and length L corresponding to the transport of carriers moving at speed v for a space of time t = 1, (Figure 5.4). Hence, L = v x 1 = v , and the volume of the cylinder is given by :

$$V = L \times S = v.$$

The associated flux therefore is such that:

$\vec{F}_n = n\vec{v}_n$, for electrons of concentration n and speed $\vec{v}_n$;

$\vec{F}_p = p\vec{v}_p$, for holes with concentration p and speed $\vec{v}_p$.

By introducing these values for $\vec{F}_n$ and $\vec{F}_p$ into the Eqs. (4) and (5) for the current densities, we have the form of the diffusion currents of the electrons and the holes:

$$\vec{j}_n = -q\,\vec{F}_n = q\,D_n\,\overrightarrow{\text{grad}}\,n \qquad (10)$$

$$\vec{j}_p = q\,\vec{F}_p = -q\,D_p\,\overrightarrow{\text{grad}}\,p\,. \qquad (11)$$

5.2.2.2. Current in a semiconductor with a concentration gradient and subject to an electric field

Equations (4'), (5'), (10), and (11) give the resultant currents for electrons and holes:

$$\vec{j}_n = q\,n\,\mu_n\,\vec{E} + q\,D_n\,\overrightarrow{\text{grad}}\,n \qquad (12)$$

$$\vec{j}_p = q\,p\,\mu_p\vec{E} - q\,D_p\,\overrightarrow{\text{grad}}\,p\,. \qquad (13)$$

The density of the resultant current over the complete section of the semiconductor therefore is given by:

$$\vec{J} = \vec{j}_n + \vec{j}_p \,. \qquad (14)$$

5.2.3. Inhomogeneous semiconductor, the internal field, and Einstein's relation

5.2.3.1. Internal field in an inhomogeneous semiconductor

Consider an inhomogeneous n-type semiconductor that is isolated and at thermal equilibrium. There is no current of electrons and the resultant current given by Eq. (12) thus is zero. Also, we have:

$$\vec{j}_n = q\, n\, \mu_n\, \vec{E} + q\, D_n\, \overrightarrow{\text{grad}}\, n = 0\,,$$

from which it is possible to determine the expression for the internal field:

$$\vec{E}_{int} = -\frac{1}{n}\frac{D_n}{\mu_n}\overrightarrow{\text{grad} n}\,. \qquad (15)$$

5.2.3.2. Einstein's equation

- Concentration of carriers expressed as a function of n_i

The concentrations of electrons and of holes are given by Eqs. (23) and (25) of Chapter 4, which can be written:

$$n = N_c \exp\left(\frac{E_F - E_C}{kT}\right) \text{ and } p = N_v \exp\left(-\frac{E_F - E_V}{kT}\right).$$

On introducing the intrinsic level (E_I), which gives the position of the Fermi level of an intrinsic semiconductor, we can state that $(n)_{E_F=E_I} = (p)_{E_F=E_I}$, and hence

$$E_I = \frac{1}{2}\left(E_V + E_C\right) + \frac{kT}{2}\text{Ln}\frac{N_v}{N_c}\,.$$

It is thus possible to write that

$$E_F - E_C = \left(E_F - E_I\right) + \left(E_I - E_C\right) = \left(E_F - E_I\right) - \frac{1}{2}\left(E_C - E_V\right) + \frac{kT}{2}\text{Ln}\frac{N_v}{N_c}.$$

This equation plugged in to the expression for n gives:

$$n = N_c \exp\frac{(E_F - E_I) - \frac{1}{2}E_G + \frac{kT}{2}\mathrm{Ln}\frac{N_v}{N_c}}{kT} = \sqrt{N_c N_v}\exp\left(\frac{E_F - E_I}{kT}\right)\exp\left(-\frac{E_G}{2kT}\right).$$

Similarly, introducing n into the expression for n_i given in Eq. (27) of Chapter 4 finally gives:

$$n = n_i \exp\frac{E_F - E_I}{kT}. \qquad (16)$$

In the same manner, we find for the hole concentration that

$$p = n_i \exp\frac{E_I - E_F}{kT}. \qquad (17)$$

- Einstein's equation

By making $u_T = \frac{kT}{q}$ as the thermal potential and $U = \frac{E_F - E_I}{q}$ as the crystal potential, Eq. (16) can be written as $n = n_I \exp(\frac{U}{u_T})$.

In one dimension, we have $\frac{dn}{dx} = \frac{n_I}{u_T}\frac{dU}{dx}e^{U/u_T} = \frac{n}{u_T}\frac{dU}{dx}$.

In three dimensions, we thus have $\overrightarrow{\mathrm{grad}}\, n = \frac{n}{u_T}\overrightarrow{\mathrm{grad}}\, U$, and the electric field given by Eq. (15) can be rewritten:

$$\vec{E}_{int} = -\frac{1}{n}\frac{D_n}{\mu_n}\overrightarrow{\mathrm{grad} n} = -\frac{D_n}{\mu_n}.\frac{1}{u_T}\overrightarrow{\mathrm{grad} U}.$$

A comparison with $\vec{E}_{int} = -\overrightarrow{\mathrm{grad}}\, U$ yields Einstein's equation:

$$\frac{D_n}{\mu_n} = u_T = \frac{kT}{q} = \frac{D_p}{\mu_p}. \qquad (18)$$

5.2.4. Measuring the conductivity of a semiconductor and resistance squared

5.2.4.1. Principle

The contact between the classic metallic electrodes and a semiconductor can be non-Ohmic. This is where the work–function of the metal (electrode) is greater than that of the semiconductor. The electronic transport in the direction metal-to-semiconductor thus is partially blocked and there are mechanisms specific to the interface (such as thermoelectric effects, the Schottky effect, or tunnel injections) that determine the current (I) rather than the resistance of the material (which would give the classic Ohm's law as in V = RI from which R could be determined knowing V for a given value of I). Other effects such as the injection of minority carriers can perturb also the measurement of the current in a classic two-electrode configuration.

In order to get around the problem of the nature of the contact, we can use a system based on four electrodes, as illustrated in Figure 5.5.

- Two electrodes are used to inject a current directly measured, but for which the origin is not of importance (whether by a thermoelectric or a Schottky effect or whatever). The relation V = RI is not used, and as a consequence, the contact resistance is not taken into account.

- Two other electrodes which are independent from the first pair are used to measure the tension by a zeroing method. The current being zero in the circuit that measures V, the nature of the contact between these two electrodes and the semiconductor does not intervene and therefore does not determine the measurement of V.

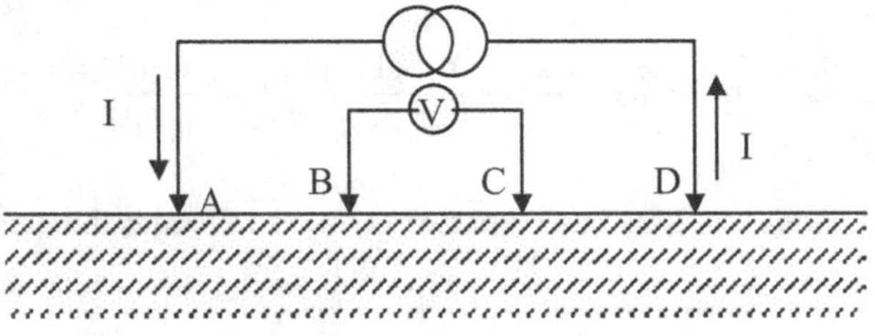

Figure 5.5. *Four point configuration for measuring conductivity.*

5.2.4.2. Determination of the resistivity of an infinitely thick film of semiconductor

Following an injection of current of intensity I at A, there is the appearance of current half-lines originating at O, as shown in Figure 5.6 and if the semiconductor is isotropic and homogeneous. As the current density vector

$\vec{j}$ (= $\sigma \vec{E}$ = -σ $\overrightarrow{\text{grad}}$ V) is collinear with the potential gradient (which is normal to the equipotentials), the equipotentials can be represented by spheres and the current lines as radii to these spheres.

For a potential difference (dV) between two points P and Q defined by the intersection of equipotentials V and V + dV (half-spheres of radius r and r + dr) along the surface of the semiconductor, we can state that with the resistance dR between two spheres being $dR = \rho \frac{dr}{2\pi r^2}$ (ρ is the semiconductor resistivity), in terms of moduli we have:

$$|dV| = I\,dR = \rho\, I \frac{dr}{2\pi r^2}.$$

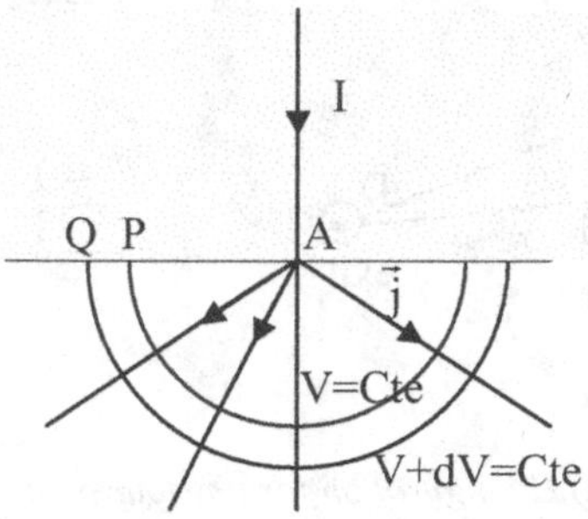

Figure 5.6. *Current injection into an infinity thick semiconductor film.*

Algebraically, there is a drop in the tension across dR of I dR, and therefore $dV = -\rho\, I \frac{dr}{2\pi r^2}$. From this can be determined that for two points B and C (Figure 5.5) defined with respect to A by B = d_B and AC = d_c (also with $\int \frac{dr}{r^2} = -\frac{1}{r}$) we have:

$$V_B - V_C = V_{BC} = \int_C^B dV = \int_{d_C}^{d_B} -\rho I \frac{dr}{2\pi r^2} = \frac{\rho I}{2\pi}\left(\frac{1}{d_B} - \frac{1}{d_C}\right).$$

Now consider the problem for point D where the current is extracted from the semiconductor. We can think that an electrode placed at D injects a current –I, so that if we make DB = d_B' and DC = d_C', the current created between B and C is the potential difference V'_{BC}, as in:

$$V'_{BC} = V'_B - V'_C = \int_C^B dV' = \int_{d'_C}^{d'_B} \rho I \frac{dr}{2\pi r^2} = \frac{\rho I}{2\pi}\left(\frac{1}{d'_C} - \frac{1}{d'_B}\right).$$

Finally, the two electrodes at A and D give rise to a potential difference (Figure 5.7a) whatever the position of the points:

$$V = V_{BC} + V'_{BC} = \frac{\rho I}{2\pi}\left(\frac{1}{d_B} + \frac{1}{d'_C} - \frac{1}{d'_B} - \frac{1}{d_C}\right). \qquad (19)$$

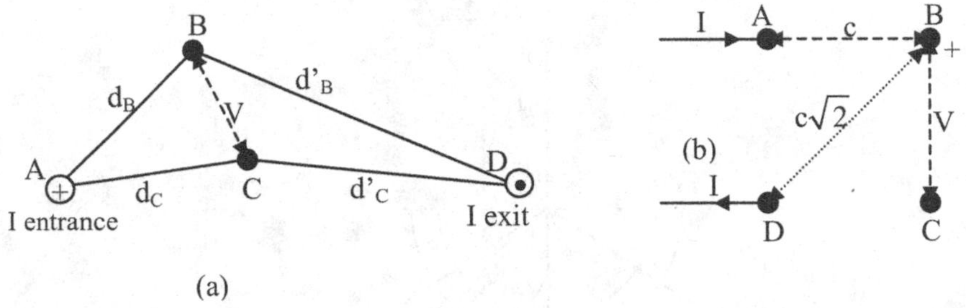

Figure 5.7. *(a) Configuration for any four points and (b) square configuration.*

- Equidistance points in a line (Figure 5.5)

Here $d_B = d$, $d_C = 2d$, $d'_B = 2d$, $d'_C = d$, which on plugging into Eq. (19) gives:

$$V = \frac{\rho}{2\pi d} I, \text{ hence } \rho = 2\pi d \frac{V}{I} \text{ where } \sigma = \frac{1}{\rho} = \frac{1}{2\pi d}\frac{I}{V} \qquad (20)$$

(with d in cm, σ is in Ω^{-1} cm^{-1}, and 1 Ω^{-1} $cm^{-1} = 10^2$ Ω^{-1} m^{-1}).

- Points in a square (Figure 5.7b)

With $d_B = d'_C = c$ and $d_C = d'_B = c\sqrt{2}$, we have from Eq. (19):

$V = \frac{\rho}{\pi c}\frac{\sqrt{2}-1}{\sqrt{2}} I$, from which:

$$\rho = 10.75\, c\frac{V}{I} \text{ and } \sigma = 0.093\frac{1}{c}\frac{I}{V}. \qquad (21)$$

5.2.4.3. Determination of the resistivity of an infinity thin film of semiconductor

The injection of a current I into a homogeneous plaque of infinite surface area and extreme thinness (e) results in current lines being parallel to the surface. Due to symmetry, and using the same arguments as those presented in the preceding section, the equipotentials can be represented by cylinders (which also are Gaussian surfaces) of height e and with an axis extending from the injected current. The current lines are the radii of the cylinders, as shown in Figure 5.8.

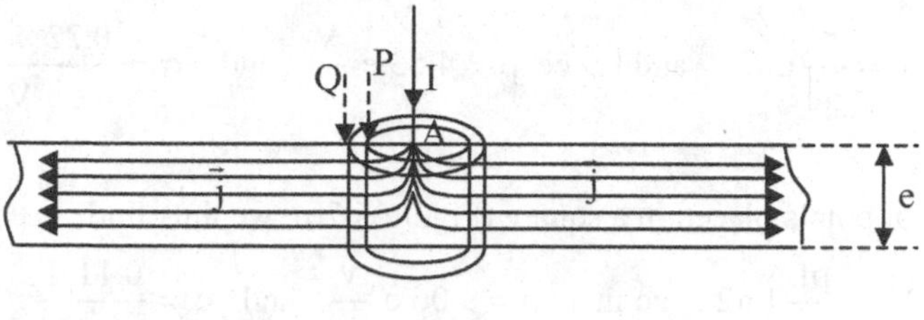

Figure 5.8. *Current and equipotential lines in an infinitely thin film.*

The resistance between two cylinders of radius r and r + dr is given by $dR = \rho \frac{dr}{2\pi re}$. The potential difference between these two cylinders of which the lateral surfaces are traversed by a current I (in other words between two points P and Q which are separated by a distance dr) thus is given by $|dV| = IdR = \rho I \frac{dr}{2\pi re}$, so that algebraically we have for the drop in potential

$$dV = -\rho I \frac{dr}{2\pi re}.$$

We thus obtain two points B and C located with respect to A by AB = d_B and AC = d_c, to give:

$$V_B - V_C = V_{BC} = \int_C^B dV = \int_{d_C}^{d_B} -\rho I \frac{dr}{2\pi re} = \frac{\rho I}{2\pi e} Ln \frac{d_C}{d_B}.$$

Similarly, the contribution to the voltage between B and C from the electrode D from which the current comes is:

$$V_B' - V_C' = V_{BC}' = \int_C^B dV = \int_{d'_C}^{d'_B} \rho I \frac{dr}{2\pi re} = \frac{\rho I}{2\pi e} Ln \frac{d_B'}{d_C'}$$

The two electrodes at A and D finally give rise to a potential difference V between B and C which is such that:

$$V = \frac{\rho I}{2\pi e} \mathrm{Ln} \frac{d_C d_B'}{d_B d_C'}. \quad (22)$$

- For aligned and equidistant points, we thus find:

$$V = \frac{\rho I}{\pi e} \mathrm{Ln}2, \text{ and hence } \rho = 4.53\, e\, \frac{V}{I} \text{ and } \sigma = \frac{0.22}{e} \frac{I}{V}. \quad (23)$$

- For points placed in a square (Figure 5.7b), we thus find:

$$V = \frac{\rho I}{2\pi e} \mathrm{Ln}2, \text{ so that } \rho = 9.06\, e\, \frac{V}{I} \text{ and } \sigma = \frac{0.11}{e} \frac{I}{V}. \quad (24)$$

5.2.5. Resistance per square or more simply put resistance squared and denoted $R^{\square}$

For parallelepiped (Figure 5.9a) with rectangular upper and lower surfaces of sides denoted a and ℓ and a lateral face of height e, the resistance between the two opposed lateral faces separated by ℓ is given by $R_{rect} = \rho \frac{\ell}{S} = \rho \frac{\ell}{e\, a}$.

If $\ell = a$, the upper and lower faces are squares of side length a (as in Figure 5.9b), and the resistance between two lateral faces becomes what is termed the resistance squared, of which the value is independent of the value of a because:

$$R_{square} = R^{\square} = \frac{\rho}{e}, \quad (25)$$

the unit of which is the ohm per square, i.e., $\Omega/\square$.

Between R_{rect} and $R^{\square}$ we have the relation $R_{rect} = R^{\square} \frac{\ell}{a}$.

The expression for the resistivity always being empirical and in the form [Eqs. (20,21,23,24)]

$$\rho = K\, e \frac{V}{I}, \quad (26)$$

where K is a fixed value for a given configuration (for a thin sample, K = 4.53, and for square points, K = 9.06).

The ratio of Eq. (26) to Eq. (25) gives:

$$R^{\square} = K\frac{V}{I}. \qquad (27)$$

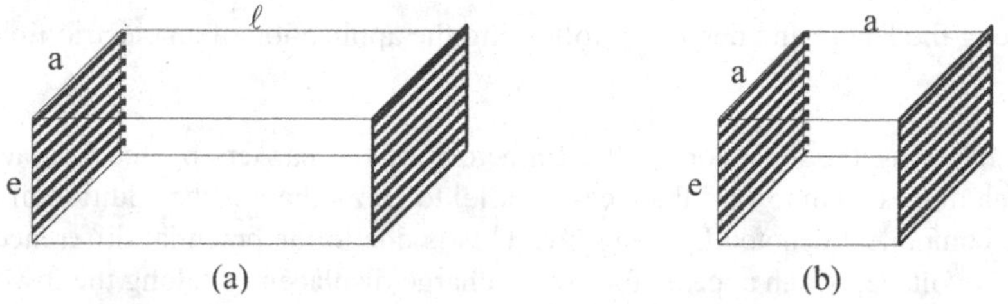

Figure 5.9. *Upper face as (a) a rectangle and (b) a square.*

Important comment

Equation (27) is often abusively interpreted by those working in electronics as they use it to deduce the square resistance (that is to say its value and therefore its measurement) independently of the size of the sample. This is false, because Eq. (25) shows that the ***value*** of $R^{\square}$ depends on the thickness (e) of the sample and the resistivity (ρ) only depends on the physical nature of the material. The only thing that is true is that the value of $R^{\square}$ does not depend on the value of the side (a) of the sample measured.

Also, it should be stated that the ***measurement*** of the square resistance does not necessitate knowledge of the thickness of the sample, as it is simply a measurement of V/I, within an order of K.

Therefore, a comparison of the square resistance is one at a macroscopic level for the resistance, which has geometric dimensions. A physical and microscopic comparison of materials necessitates knowledge of ρ for each material. In order to determine ρ we need to know the thickness of the material, in accordance with Eq. (26), for example.

To go even further and say that two samples have the same resistance squared is not the same as saying that they are made from the same material (and have the same physical constant ρ). This could only be true at a limit if the samples also have the same thickness (which is often unknown anyway due to ignorance of the doping thickness, hence the use of resistance squared!).

5.3. Magnetoelectric characterization of semiconductors

5.3.1. The Hall effect

5.3.1.1. Phenomenological study of the Hall field

The geometrical configuration chosen in Figure 5.10 shows a bar (parallelepiped) semiconductor with its edges parallel to the axes of the trihedral Oxyz. It is placed in a uniform magnetic field $\vec{B}_0 = B_0\vec{e}_z$ that is applied along Oz. A current is passed through along the length and denoted j_x following the application of an electric field ($\vec{E}_x$) along Ox.

Following the deviation of the trajectory of the carriers by the Laplace force (which moves them toward the faces parallel to Oxz), there is the addition of a transversal component denoted E_h along Oy. This is due to the potential difference, called Hall's voltage, which appears following charge displacement along the lower face of the parallelepiped (parallel to Oxz).

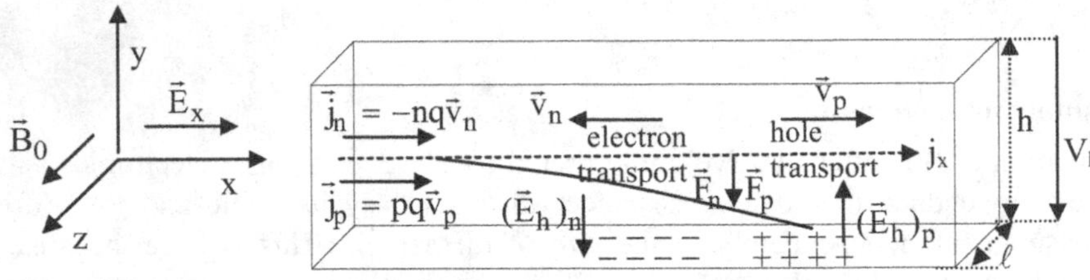

Figure 5.10. *Configuration of electric and magnetic fields.*

In effect, the Laplace forces

- $\vec{F}_n = -q\left(\vec{v}_n \times \vec{B}_0\right)$ for electrons, (28)
- $\vec{F}_p = q\left(\vec{v}_p \times \vec{B}_0\right)$ for holes, (29)

are both orientated along ($-\overrightarrow{Oy}$), and electrons and holes move toward the lower face. This gives rise to an electric field (termed the Hall field) $\left(\vec{E}_h\right)_n$ that is orientated toward $-\overrightarrow{Oy}$ for electron transport, along with a field denoted $\left(\vec{E}_h\right)_p$ following $\overrightarrow{Oy}$ for holes.

5.3.1.2. The Hall constant

The Hall field in turn generates a force due to the electron current $(\vec{F}_h)_n = -q(\vec{E}_h)_n$, and a force due to the holes given by $(\vec{F}_h)_p = q(\vec{E}_h)_p$. These forces are both directed along $\overrightarrow{Oy}$, and as a consequence are opposed to the Laplace forces $\vec{F}_n$ and $\vec{F}_p$.

Under a permanent regime, we have $m\left(\frac{d\vec{v}}{dt}\right) = 0 = \vec{F}_{Laplace} + \vec{F}_{Hall}$, so that $\vec{F}_{Laplace} = -\vec{F}_{Hall}$. There is no current flow between the upper and lower faces (unconnected by a closed circuit and therefore "open circuit"), and at equilibrium the Laplace and Hall forces compensate one another. We thus have:

- for electrons: $-q\left(\vec{v}_n \times \vec{B}_0\right) = q(\vec{E}_h)_n$, whereby using Eq. (4) gives:

$$(\vec{E}_h)_n = -\frac{1}{nq}\left(\vec{B}_0 \times \vec{j}_n\right) = R_n\left(\vec{B}_0 \times \vec{j}_n\right) \text{ where } R_n = -\frac{1}{nq} < 0. \quad (30)$$

- for holes: $q\left(\vec{v}_p \times \vec{B}_0\right) = -q(\vec{E}_h)_p$, where using Eq. (5) gives:

$$(\vec{E}_h)_p = \frac{1}{pq}\left(\vec{B}_0 \times \vec{j}_p\right) = R_p\left(\vec{B}_0 \times \vec{j}_p\right) \text{ where } R_p = \frac{1}{pq} > 0. \quad (31)$$

The Hall tension (V_h) between the lower and upper faces is, in moduli, in the form

$$|V_h| = E_h h = j\frac{B_0}{cq}h = \frac{I}{h\ell}\frac{B_0}{cq}h = \frac{1}{cq}\frac{I}{\ell}B_0,$$

where c is the concentration of carriers so that c = n or c = p, h is the height of the parallelepiped along Oy, and ℓ is the width of the parallelepiped along Oz.

Algebraically, for the electrons the potential difference is negative and:

$$V_h = -\frac{1}{nq}\frac{I}{\ell}B_0 = R_n\frac{I}{\ell}B_0 < 0. \quad (32)$$

For the holes the Hall potential difference is positive and:

$$V_h = \frac{1}{pq}\frac{I}{\ell}B_0 = R_p\frac{I}{\ell}B_0 > 0. \quad (33)$$

To conclude, the Hall voltage is:

- positive for a hole current and negative for an electric current;
- proportional to the magnetic field B_0 and the intensity I of the current; and
- inversely proportional to the concentration of carriers.

Numerically the value of V_h is of the order of 10 mV in semiconductors. In metals where the concentration of carriers is much higher, the same tension is considerably reduced and becomes around 10 μV.

5.3.1.3. Mobility and Hall mobility

For holes, according to Eq. (7) $\sigma_p = qp\mu_p$, so that $q\,p = \frac{\sigma_p}{\mu_p}$. From Eq. (31), we also have $q\,p = \frac{1}{R_p}$, from which we can pull out:

$$\mu_p = R_p\sigma_p. \qquad (34)$$

For the electrons, we have $nq = -\frac{1}{R_n} = \frac{\sigma_n}{\mu_n}$, so that

$$\mu_n = -R_n\sigma_n \qquad (35)$$

(as R_n is negative, we once again find μ_n being positive, as defined above).

In effect, the preceding physical theory is rather well simplified, as we have assumed that each of the carriers (electrons or holes) have the same single speed. This neglects any changes due to phonons or impurities in the lattice. In this case, a term needs to be introduced so that, for example, Eq. (34) can be rewritten as $R_p\sigma_p = r\,\mu_p = \mu_H$, where μ_H is called the Hall mobility (of the order of 20 to 90 % larger than the conduction mobility μ_p). This Hall mobility thus is given as a product of the conductivity and the Hall constant.

5.3.2. Magnetoresistance and magnetoconductance

5.3.2.1. Definitions

In general terms, if ρ(0) and σ(0), respectively, denote the resistivity and conductivity of a sample measured in the absence of a magnetic field, and if ρ(H) and σ(H), respectively, denote the resistivity and conductivity of the same sample measured in the presence of a magnetic field H, then:

• the magnetoresistance (MR) is defined by

$$MR = \frac{\rho(H) - \rho(0)}{\rho(0)} = \frac{\Delta\rho}{\rho_0}, \qquad (36)$$

• the magnetoconductance (MC) is defined by

$$MC = \frac{\sigma(H) - \sigma(0)}{\sigma(0)} = \frac{\Delta\sigma}{\sigma_0}, \qquad (37)$$

• the relation that exists between MR and MC is thus:

$$MR = \frac{\Delta\rho}{\rho_0} = \frac{\rho(H) - \rho(0)}{\rho(0)} \approx -\frac{\Delta\sigma}{\sigma_0} = -MC. \qquad (38)$$

Depending on whether the direction of the applied electric field is normal ($\vec{B}_0$ along Oz and $\vec{E}_x$ along Ox) or parallel ($\vec{E}_x$ and $\vec{B}_0$ along Ox) to the direction of the magnetic field, we respectively use the terms of transversal or longitudinal magnetoresistance (or of magnetoconductance). In accordance with the most generally studied configurations, the following calculations consider the transversal magnetoresistance shown in Figure 5.11 which adopts the same format as Figure 5.10.

The calculations are for a semiconductor that typically contains two types of charges with algebraic values q_1 and q_2, for which the densities are n_1 and n_2 and the effective masses are m_1^* and m_2^*. With the carriers being either electrons at the bottom of the conduction band or holes at the top of the valence band, these masses are positive. The relaxation times are denoted τ_1 and τ_2, and their mobilities are defined by $\mu_1 = |q_1|\tau_1/m^*_1$ and $\mu_2 = |q_2|\tau_2/m^*_2$ (positive magnitudes for this definition).

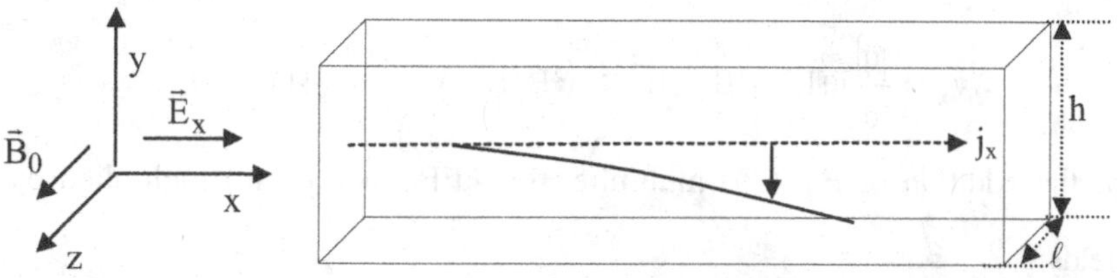

Figure 5.11. *Configuration of the electric and magnetic fields used in measuring transversal magnetoresistance.*

5.3.2.2. Carrier speeds (exampled calculation for one carrier type)

For a carrier with a mass m* and an algebraic charge q, taking into account the presence of a frictional force of the form $-m^*\vec{v}/\tau$, the dynamic fundamental relation is written:

$$m\frac{d\vec{v}}{dt} = q\vec{E} + q\left(\vec{v} \times \vec{B}_0\right) - m^*\frac{\vec{v}}{\tau}.$$

Under a permanent regime $(d/dt) = 0$, we obtain $\vec{E} = \dfrac{m^*\vec{v}}{q\tau} - \left(\vec{v} \times \vec{B}_0\right)$.

As $\mu = \dfrac{|q|\tau}{m^*}$, projections to the axes Ox and Oy yield:

$$E_x = \frac{m^* v_x}{q\tau} - v_y B_0 = \frac{|q|}{q}\frac{v_x}{\mu} - v_y B_0 \qquad (39)$$

$$E_y = \frac{m^* v_y}{q\tau} + v_x B_0 = \frac{|q|}{q}\frac{v_y}{\mu} + v_x B_0. \qquad (40)$$

The addition of Eq. (39) multiplied by $\mu\dfrac{|q|}{q}$ to Eq. (40) multiplied by $\mu^2 B_0$ yields

$$\mu\frac{|q|}{q}E_x + \mu^2 B_0 E_y = v_x\left(1 + \mu^2 B_0{}^2\right).$$

By assuming that the induction B_0 is sufficiently weak so that $\mu^2 B_0^2 \ll 1$, and by neglecting the terms for B_0 at a power greater than or equal to 3, we obtain:

$$v_x = \frac{|q|}{q}\mu(1 - \mu^2 B_0{}^2)E_x + \mu^2 B_0 E_y. \qquad (41)$$

Similarly, the addition of Eq. (39) multiplied by $-\mu^2 B_0$ to Eq. (40) multiplied by $\mu\dfrac{|q|}{q}$, yields:

$$v_y = \frac{|q|}{q}\mu(1 - \mu^2 B_0{}^2)E_y - \mu^2 B_0 E_x. \qquad (42)$$

5.3.2.3. Equations for two types of carriers

When there are two types of carriers (1 and 2), and in the absence of an induction B_0, there is no current along Oy. The conductivity $[\sigma(0)]$ is therefore of the form $\sigma(0) = |q_1| n_1\mu_1 + |q_2| n_2\mu_2$.

In the presence of B_0, the components of the current density are such that:

$$J_x = n_1q_1v_{1x} + n_2q_2v_{2x} = \sigma_{xx} E_x + \sigma_{xy} E_y$$
$$J_y = n_1q_1v_{1y} + n_2q_2v_{2y} = \sigma_{yx} E_x + \sigma_{yy} E_y,$$

where

$$\sigma_{xx} = \sigma_{yy} = |q_1| n_1\mu_1(1 - \mu_1^2B_0^2) + |q_2| n_2\mu_2(1 - \mu_2^2B_0^2)$$
$$\sigma_{xy} = -\sigma_{yx} = (q_1n_1\mu_1^2 + q_2n_2\mu_2^2)B_0 .$$

For a sample of limited size along Oy, the face of the sample being at open circuit (for example, with a voltammeter of great impedance for measuring the Hall effect), there is no current debit along Oy and j_y is zero. From this can be deduced that

$E_y = -\dfrac{\sigma_{yx}}{\sigma_{yy}} E_x = \dfrac{\sigma_{xy}}{\sigma_{xx}} E_x$. By moving this into j_x we obtain:

$$j_x = \frac{\sigma_{xx}{}^2 + \sigma_{xy}{}^2}{\sigma_{xx}} E_x = \sigma_\ell E_x , \qquad (43)$$

where σ_ℓ is the longitudinal conductivity (in the direction of the applied field E_x).

By substituting the expressions for σ_{xx} and σ_{xy} into that for σ_ℓ, we obtain:

$$\sigma_\ell = |q_1|n_1\mu_1 + |q_2|n_2\mu_2 - \frac{|q_1|n_1\mu_1|q_2|n_2\mu_2(\mu_1 \mp \mu_2)^2}{|q_1|n_1\mu_1 + |q_2|n_2\mu_2} B_0^2 . \qquad (44)$$

Note that in $(\mu_1 \mp \mu_2)$ the minus sign indicates that the charges are of the same sign, whereas the plus sign indicates the opposite.

5.3.2.3. Expressions for the magnetoconductance and the magnetoresistance

The transversal magnetoconductance (MC), which corresponds to a measure of the conductivity (σ_ℓ) along a direction Ox normal to the direction Oz of the applied magnetic induction, is defined by the ratio:

$$\mathrm{MC} = \frac{\Delta\sigma}{\sigma_0} = \frac{\sigma(H) - \sigma(0)}{\sigma(0)} = \frac{\sigma_1 - \sigma_0}{\sigma_0},$$

from which we have:

$$\frac{\Delta\sigma}{\sigma_0} = -\frac{|q_1| n_1\mu_1 |q_2| n_2\mu_2 (\mu_1 \mp \mu_2)^2}{(|q_1| n_1\mu_1 + |q_2| n_2\mu_2)^2} B_0^2 . \qquad (45)$$

When two types of charges are identical ($q_1 = q_2 = q$ and are either both holes or electrons) the expression can be simplified, as in:

$$\frac{\Delta\sigma}{\sigma_0} = -\frac{n_1 n_2 (\mu_1 - \mu_2)^2 \mu_1\mu_2}{(n_1\mu_1 + n_2\mu_2)^2} B_0{}^2$$

For charges with opposite signs ($q_1 = -q_2 = q$), we obtain

$$\frac{\Delta\sigma}{\sigma_0} = -\frac{n_1 n_2 (\mu_1 + \mu_2)^2 \mu_1\mu_2}{(n_1\mu_1 + n_2\mu_2)^2} B_0{}^2$$

Finally, for both cases, the law of evolution is given by:

$$\frac{\Delta\sigma}{\sigma_0} = -A\, B_0{}^2, \text{ where } A > 0 \qquad (46)$$

For its part, the magnetoresistance (MR) is such that:

$$\frac{\Delta\rho}{\rho_0} = \frac{\rho(H) - \rho(0)}{\rho(0)} \approx -\frac{\Delta\sigma}{\sigma_0} = A\, B_0{}^2 . \qquad (47)$$

This is an increasing function paired to B_0. Whatever the direction of B_0, this magnetoresistance represents the growth in the transverse resistance (that is to say $j_x \perp B_0$) when we apply a weak magnetic induction.

5.4. The Gunn effect and microwave emissions

5.4.1. Expressions for σ, j, and <v> for carriers in a semiconductor with a conduction band of two minima

It is worth recalling first of all from Section 5.1 that for electrons in a semiconductor conduction band the average drift speed of the electrons (which are noted here as $<v> = \vec{v}_d$) is given in the form

$$\vec{v}_d = -\frac{q\tau}{m^*}\vec{E} = -\mu_n \vec{E}, \qquad (1')$$

where τ is the relaxation time corresponding to the interval between two successive collisions.

For intense fields, the carriers reach speeds of the order of the thermal speed (v_{th}) which is such that $\frac{1}{2}m^* v_{th}^2 = \frac{3}{2}kT$, and the mobility μ tends toward zero along with τ. A representation is given in Figure 5.12.

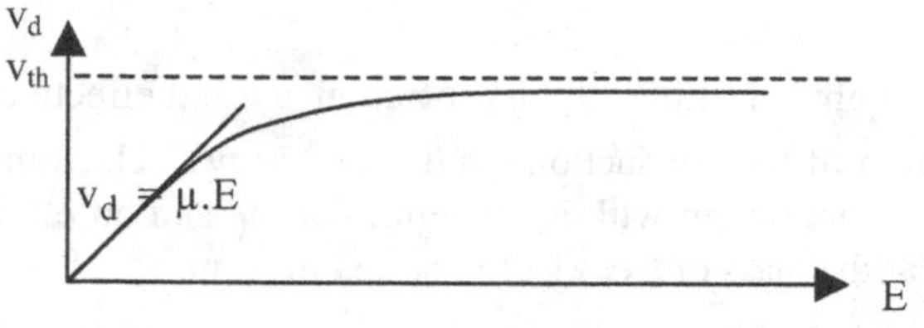

Figure 5.12. *Variation in the drift velocity with electric field.*

Thus, the electronic current density $\vec{j} = \rho_n \vec{v}_n$ is written for the linear part of the plot where $\vec{v}_n = \vec{v}_d = -\mu_n \vec{E}$. Hence $\vec{j} = -qn\vec{v}_d = -qn\mu_n\vec{E} = \sigma_n \vec{E}$, and the differential conductivity is given by:

$$\sigma = \frac{dj}{dE} = \sigma_n = qn\mu_n = \frac{nq^2\tau}{m^*} > 0. \qquad (48)$$

If we consider a semiconductor such as GaAs or InP, the conduction band has two minima in accordance with the representation given in Figure 5.13, which illustrates the case of a crystal of GaAs doped with As. The concentration of the ionized doping atoms is denoted N.

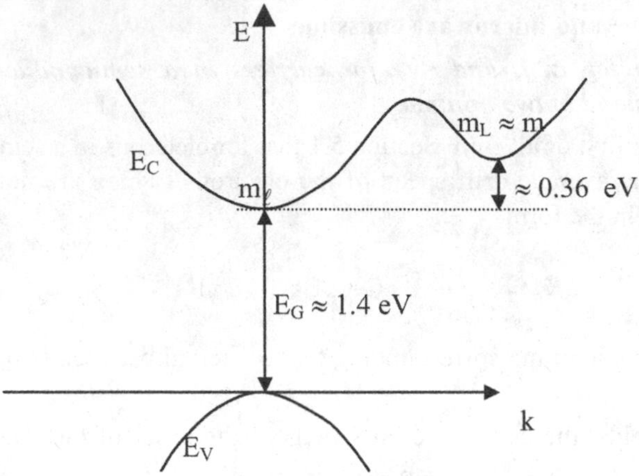

Figure 5.13. *Band scheme for GaAs, with two minima in the conduction band.*

The so-called "light" electrons of concentration n_ℓ and effective mass m_ℓ are situated at the bottom of the conduction band. The "heavy" electrons populate the bottom of the second minimum with a concentration n_L and an effective mass m_L more or less equal to the mass of free electrons, i.e., $m_L = m$.

In general terms, and for whatever electric field, it is these two types of electronic charges that participate in the conduction. Here, and as a first approximation, we can assume that the relaxation time (τ) for the two types of electron are identical. We thus have:

$$\mu_\ell = \frac{q\tau}{m_\ell} \qquad (49)$$

$$\mu_L = \frac{q\tau}{m_L}. \qquad (50)$$

By dividing Eqs. (50)/(49) we find:

$$\mu_L = \mu_\ell \frac{m_\ell}{m_L}. \qquad (51)$$

The conductivity is:

$$\sigma = q\left(n_\ell \mu_\ell + n_L \mu_L\right) = q\mu_\ell \left(n_\ell + n_L \frac{m_\ell}{m_L}\right). \qquad (52)$$

It is possible to hypothesize that the concentration (n_L) of heavy electrons increases with the electric field (to the detriment of the concentration of light electrons) in the following manner:

$$n_L = N\frac{E}{E_0}, \qquad (53)$$

where E_0 is the field for which $n_L = N$, and by as a consequence, $n_\ell = 0$ (as N is such that $N = n_\ell + n_L$).

Simultaneously, the law of variation for n_ℓ is

$$n_\ell = N - n_L = N\left(1 - \frac{E}{E_0}\right), \qquad (54)$$

- when $E < E_0$, substitution of Eqs. (53) and (54) into (52) gives:

$$\sigma = qN\mu_\ell\left[\left(1 - \frac{E}{E_0}\right) + \frac{E}{E_0}\frac{m_\ell}{m_L}\right], \qquad (55)$$

from which can be deduced that

$$j = \sigma E = qN\mu_\ell\left[1 - \frac{E}{E_0} + \frac{E}{E_0}\frac{m_\ell}{m_L}\right]E. \qquad (56)$$

The average drift velocity (v) for the sum of the electrons of concentration N is such that $j = qNv$, and for this average speed relative to all electrons, we have:

$$v = \frac{j}{Nq} = \mu_\ell E\left[1 - \frac{E}{E_0} + \frac{E}{E_0}\frac{m_\ell}{m_L}\right]; \qquad (57)$$

- when $E > E_0$, $n_\ell = 0$, $n_L = N$,

$$\sigma = qN\mu_\ell\frac{m_\ell}{m_L}, \quad j = qN\mu_\ell\frac{m_\ell}{m_L}E \qquad (58)$$

and:

$$v = \frac{j}{Nq} = \mu_\ell\frac{m_\ell}{m_L}E. \qquad (59)$$

Equations (56) and (58) make it possible to describe the function $j = f(E)$.

Figure 5.14 shows that it must go through a maximum at a field E_m which must be such that $\left(\frac{dj}{dE}\right)_{E=E_m} = 0$.

With Eq. (56), this condition gives rise to:

$$\left[1 - \frac{2E}{E_0} + \frac{2E}{E_0}\frac{m_\ell}{m_L}\right]_{E=E_m} = 0\text{, from which } E_m = \frac{E_0}{2}\frac{m_L}{m_L - m_\ell}.$$

When $m_\ell \ll m_L$ we find

$$E_m \approx \frac{E_0}{2}. \qquad (58)$$

Finally, when $E < E_m$ ($\approx E_0/2$), the differential conductivity (σ_{diff}) is positive. When E_m ($\approx E_0/2$) $< E < E_0$, then σ_{diff} is negative.

When $E > E_0$, we have a single type of carrier with a concentration $N = n_L$, so that $j = q\, n_L \mu_L E$ [linear law in contrast to Eq. (56)], and $\sigma_{diff} = q\, n_L \mu_L$.

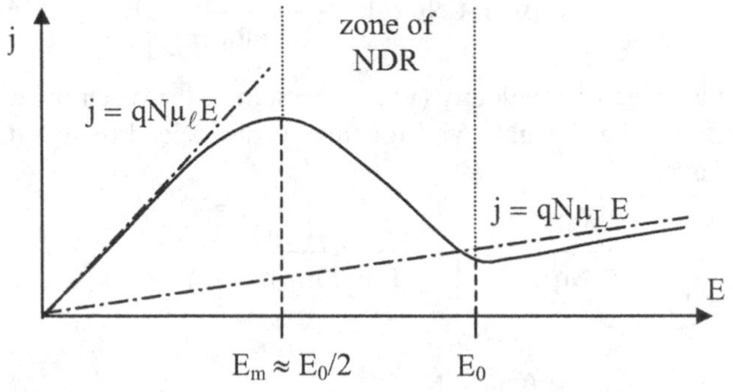

Figure 5.14. *Plot of j = f(E) and the zone of negative differential resistance (NDR) (or conductivity).*

As $v = \frac{j}{Nq}$, a plot of $v = g(E)$ as shown in Figure 5.15 has the same appearance as that of the plot of $j = f(E)$. It is only at very high fields, though, that the drift velocity approaches the thermal velocity.

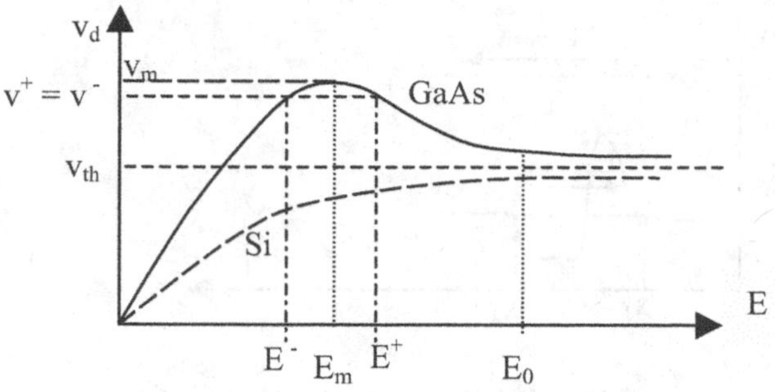

Figure 5.15. *Variation in the drift velocity with the electric field for GaAs.*

5.4.2. Emission of an electromagnetic wave in the microwave region

Between E_m and E_0, the differential resistance is negative and gives rise to unstable states in the semiconductor. There can arise therefore in the semiconductor various regions each characterized by different values of conductivity and field.

So if the semiconductor is placed in a field just above E_m, the current density is practically at a maximum. If there is an initial instability or a small heterogeneity in the semiconductor in a region that we suppose extends from x_1 to x_2, the electric field can be sufficiently greater than E_m so that the current, and hence the carrier velocity, is considerably lower than their maximum values j_m and v_m (see Figure 5.15). The carriers thus see their velocity reduced in the x_1 plane, and as a consequence they have a tendency to accumulate, whereas in x_2 they are late in arriving and as a consequence there is a positive charge formation (by non-compensation).

A dipolar layer is thus formed between x_1 and x_2, which contributes in the zone $[x_1 - x_2]$ to a localized supplementary field $E_{supplem.}$ (Figure 5.16). This field is in addition to the field E already in this zone, resulting in a more intense field, $E^+ = E + E_{supplem.}$, which further reinforces the slowing down of electrons, which now exhibit a velocity given by $v^+ < v_m$ (Figure 5.15). Inversely, given the polarity of the dipolar layer, outside the zone $[x_1 - x_2]$ the field is reduced toward a value $E^- < E_m$ much like the velocity which takes on a value $v^- < v_m$. Outside of the zone $[x_1 - x_2]$ where the field is $E^- < E_m$, any instabilities can no longer be formed (and remain localized in the initial zone). Because of the symmetry of the plot of $v = f(E)$ around E_m, the velocity of the electrons is in fact reduced to the same value $v^- = v^+ = v'$ both in the zone $[x_1 - x_2]$ and outside (with the evident notations, $v^- = \mu_{ext}E^-$ and $v^+ = \mu_{x1x2}E^+$).

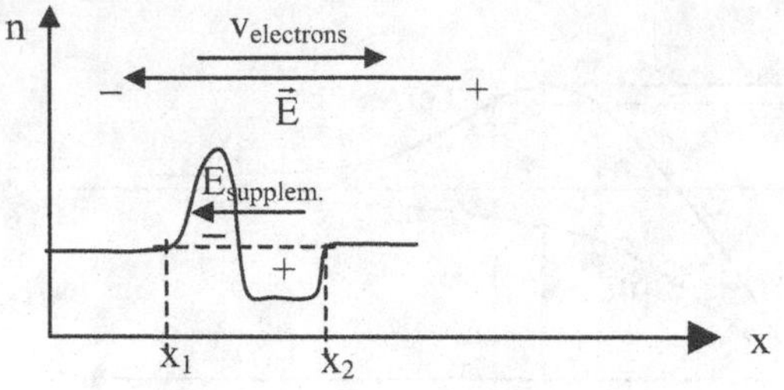

Figure 5.16. *Modification of a field by the presence of a dipolar layer.*

Taking the localization of defaults around the interfaces into account, the instabilities (at the layer $x_1 - x_2$) are in fact formed in the neighborhood of the cathode and thus are transported without modification at a velocity v' across the semiconductor. During this voyage, the density of the current in the sample is j' = q N v'.

When the domain reaches the anode where it is destroyed, the current in the sample once again takes on its initial and practically maximum value of j_m associated with a field just above E_m. The conditions to reform a new domain are thus reunited. The process in effect occurs periodically with a frequency associated with the velocity v' of the propagation of the domain. If the domain has a length L, the impulsions (layer $x_1 - x_2$) exhibit a frequency given by $\nu' = \dfrac{v'}{L}$.

Orders of scale

The frequency of the oscillations thus produced can be evaluated for GaAs. For example, $v' = v^- = \mu_{ext} E^-$ where μ_{ext} is the mobility in the volume of the semiconductor in which a field governs E^- which is slightly less that E_m. The μ_{ext} is then given by Eq. (57) where $E = E^-$. For GaAs, where $E_0 = 6$ kV cm^{-1}, we can take $E^- \approx E_m \approx E_0/2 = 3$ kV cm^{-1}, from which with $\dfrac{m_\ell}{m_L} \ll 1$, we have

$$v' = v^- \approx \mu_\ell E^- \left[1 - \frac{E^-}{E_0}\right].$$

From this can be deduced an order of scale of $v' \approx 10^5$ m s^{-1}.

For a sample of length $L \approx 10^{-5}$ m, i.e., L = 10 µm, we have $\nu' \approx 10$ GHz.

When $L = 10^{-4}$ m, i.e., L = 100 µm, we obtain $\nu' \approx 1$ GHz.

This is an emission in the microwave domain, and since its discovery in 1963, there have been numerous solid sources using the Gunn effect which have been developed. Their emission frequency can be adjusted simply by changing the length L (typically between 1 and 100 μm) of the GaAs sample. The homogeneity of the sample nevertheless must be carefully controlled so that domains are not formed in various parts around the crystal.

5.5. Problems

5.5.1. Problem 1. Hall constant

1. Establish the expression for the Hall constant for a material with two types of carriers.
2. Study the particular cases where the two types of carriers are identical ($q_1 = q_2 = q$), the signs different ($q_1 = - q_2 = q$), where there is in effect only one type of carrier ($n_2 = 0$).
3. Show how from measurements of conductivity, the magnetoresistance and the Hall constant, it is possible to determine expressions for the mobilities μ_1 and μ_2 of the carriers as well as their concentration n if the two carriers (of opposing charges) have the same concentration.

Answers

1. The Hall constant (R_H) can be defined by $\frac{E_y}{j_x} = B_0 R_H$,

where $j_x = \sigma_\ell E_x$ and $E_y = \frac{\sigma_{xy}}{\sigma_{xx}} E_x$.

For a material with two carriers, we have:

$$\frac{E_y}{j_x} = \frac{E_y}{\sigma_1 E_x} = \frac{\sigma_{xy}}{\sigma_{xx}} \frac{1}{\sigma_\ell} = \frac{\sigma_{xyx}}{\sigma_{xx}} \frac{\sigma_{xx}}{\sigma_{xx}{}^2 + \sigma_{xy}{}^2} = \frac{\sigma_{xy}}{\sigma_{xx}{}^2 + \sigma_{xy}{}^2}.$$

By neglecting the small terms (which is the same as keeping the 1st degree term for B_0 on the numerator and the 0 degree term on the denominator) we end up with:

$$\frac{E_y}{j_x} = \frac{(q_1 n_1 \mu_1{}^2 + q_2 n_2 \mu_2{}^2) B_0}{(|q_1| n_1 \mu_1 + |q_2| n_2 \mu_2)^2} = B_0 R_H \text{, soit } R_H = \frac{(q_1 n_1 \mu_1{}^2 + q_2 n_2 \mu_2{}^2)}{(|q_1| n_1 \mu_1 + |q_2| n_2 \mu_2)^2}.$$

2. Specific cases

a. The two charges are identical and $q_1 = q_2 = q$, so

$R_H = \dfrac{1}{q}\dfrac{(n_1\mu_1^2 + n_2\mu_2^2)}{(n_1\mu_1 + n_2\mu_2)^2}$. If $q > 0$ (holes), $R_H > 0$ while $q < 0$ (electrons) gives $R_H < 0$. If in addition, $n_1 = n_2 = n$,

$$R_H = \frac{1}{qn}\frac{(\mu_1^2 + \mu_2^2)}{(\mu_1 + \mu_2)^2}.$$

b. The two charges have different signs $q_1 = -q_2 = q$, so

$R_H = \dfrac{1}{q}\dfrac{(n_1\mu_1^2 - n_2\mu_2^2)}{(n_1\mu_1 + n_2\mu_2)^2}$, and if in addition, $n_1 = n_2 = n$,

$$R_H = \frac{1}{qn}\frac{(\mu_1^2 - \mu_2^2)}{(\mu_1 + \mu_2)^2}.$$

c. If there is only one type of carrier, for example, $n_2 = 0$, then by making $n_1 = n$, $q_1 = q$ and $\mu_1 = \mu$, we have $R_H = \dfrac{1}{qn}$.

R_H clearly has the same sign as q.

3. Applied determination of the characteristics n_1, μ_1, n_2, μ_2 of a semiconductor

It is assumed that we have at our disposal the following characteristics:

- continuous conductivity (in the absence of B_0) corresponding to $\sigma(0) = \sigma_0$;
- the magnetoresistance (which according to our hypotheses has a small effect) as MR or σ_ℓ ; and
- the Hall constant R_H.

From this it is *a priori* possible to determine the four magnitudes listed above. If the material is an intrinsic semiconductor, we can assume that $n_1 = n_2 = n$, and the signs are opposed, as in $q_1 = -q_2 = q$.

This leaves but three unknowns, namely n, μ_1, and μ_2, which can be derived from three independent equations:

$$\sigma(0) = |q|\, n(\mu_1 + \mu_2),$$

$$\sigma_\ell = |q| n(\mu_1 + \mu_2)(1 - \mu_1\mu_2 B_0{}^2),$$

$$R_H = \frac{1}{qn}\frac{(\mu_1{}^2 - \mu_2{}^2)}{(\mu_1 + \mu_2)^2}.$$

It is possible to state that for the Hall effect, it is the carriers with the greatest mobility that give their sign to R_H (if $\mu_1 > \mu_2$, $R_H \propto 1/q = 1/q_1$, and if $\mu_1 < \mu_2$, then $R_H \propto -1/q = 1/q_2$).

5.5.2. Problem 2. Seebeck effect

1. Preliminary question on the electric current in a semiconductor

This concerns an *inhomogeneous* n-type semiconductor at thermal equilibrium and not subject to an external electric field.

a. Give the general expression for the electronic current density (which contains two components, one being tied to the concentration gradient).

b. From this, deduce the expression for the internal field.

2. General expression for the Seebeck coefficient

This concerns an initially *homogeneous* n-type semiconductor that is one-dimensional along Ox onto which is applied a temperature gradient at two surface points A and B. The hottest point is at A, and typically $[T_A - T_B] \approx 5$ K is small.

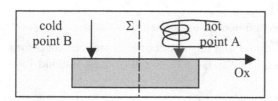

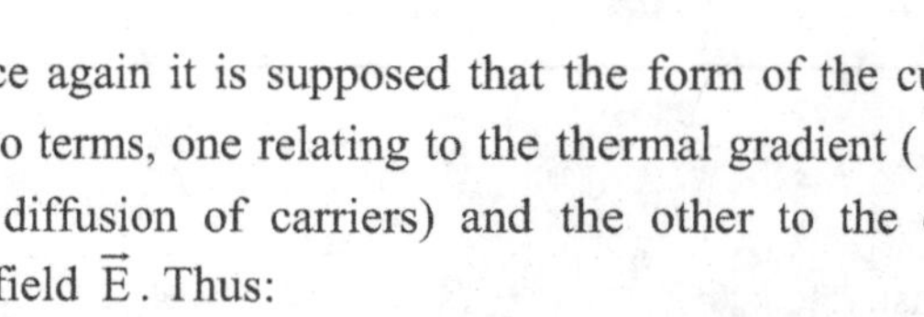

Once again it is supposed that the form of the current density can be written using two terms, one relating to the thermal gradient ($\overrightarrow{\text{grad}T}$ which gives rise to the internal diffusion of carriers) and the other to the conduction generated by an internal field $\vec{E}$. Thus:

$$\vec{j_n} = \sigma \vec{E} + M\, \overrightarrow{\text{grad}T}.$$

At equilibrium, the resultant current $\vec{j}_n$ is zero and the internal field can be written in the form $\vec{E} = S\,\overrightarrow{\text{grad}T}$ in an equation that defines S, the Seebeck coefficient.

a. Express S as a function of M and of σ.

b. Express S as a function of $\Delta V = V_A - V_B$ and of $\Delta T = T_A - T_B$.

It is often the latter equation that serves in defining Seebeck's coefficient, and it will be used in the rest of this problem.

3. Expression for S in a nondegenerate n-type semiconductor

a. Indicate the physical method and calculation required to find out the electronic concentration (n) for a semiconductor with the form $n = n(T) = AT^{3/2}\exp\left(-\frac{E_c - E_F}{kT}\right)$.

b. A diffusion current j_D is created by a concentration current which is itself formed by a temperature gradient (which we can simplify to $\frac{dT}{dx}$ for a one–dimensional problem).

• The fact that $[T_A - T_B] \approx 5$ K is typically small indicates what with respect to the E_F of the semiconductor?

• Express j_D as a function of the diffusion constant D_n and of $\frac{dT}{dx}$.

c. Following the diffusion, an internal electric field appears and generates an antagonistic electric current j_C.

•. Give the equilibrium condition for the system.

• Recall Einstein's relation. From this deduce the expression for S for a n-type semiconductor as a function of $(E_C - E_F)$, – e (electronic charge), k, and T.

4. Show how knowing both the conductivity and S make it possible to determine the carrier mobility.

Answers

1.

a. Electronic current density:

$$\vec{j}_n = \sigma_n \vec{E} + qD_n\,\overrightarrow{\text{grad}n} = qn\mu_n \vec{E} + qD_n\,\overrightarrow{\text{grad}n}\,.$$

b. At equilibrium, no current goes through the semiconductor, so $j_n = 0$ and $qn\mu_n\vec{E} = -qD_n\overrightarrow{\text{grad}n}$, from which $\vec{E} = -\frac{1}{n}\frac{D_n}{\mu_n}\overrightarrow{\text{grad}n}$. (1)

2.

a. We thus have $\overrightarrow{j_n} = \sigma\vec{E} + M\,\overrightarrow{\text{grad}T}$, and at equilibrium we also have $j_n = 0$.

From this can be deduced that $\vec{E} = -\frac{M}{\sigma}\overrightarrow{\text{grad}T}$. (2)

As S is defined by $\vec{E} = S\,\overrightarrow{\text{grad}T}$ we therefore have $S = -\frac{M}{\sigma}$.

b. From $\vec{E} = -\overrightarrow{\text{grad}V}$ it is possible to deduce algebraically with respect to Ox that

$$E = -\frac{\Delta V}{\Delta x} = -\frac{V_A - V_B}{x_A - x_B}. \quad (3)$$

Similarly, from Eq. (2) we can state that in one dimension

$$E = S\frac{\Delta T}{\Delta x} = S\frac{T_A - T_B}{x_A - x_B}. \quad (4)$$

From Eqs. (3) and (4) we deduce that $S = -\frac{V_A - V_B}{T_A - T_B} = -\frac{\Delta V}{\Delta T}$. (5)

3.

a. We have $n = \int_{E_C}^{\infty} Z(E)F(E)dE$, where $Z(E) \propto \sqrt{E - E_C}$, and for a non-degenerate semiconductor [$(E - E_F) > 2$ or 3 kT]] $F(E) \simeq \exp\left(-\frac{E - E_F}{kT}\right)$. We also have [where C_n is a constant for the proportionality factor in Z(E)]:

$$n = C_n \int_{E_C}^{\infty} \sqrt{E - E_C}\,\exp\left(-\frac{E - E_F}{kT}\right)dE.$$

By making $u = \frac{E - E_C}{kT}$,

and by writing $\exp(-\frac{E-E_F}{kT}) = \exp(-\frac{E-E_C}{kT})\exp(-\frac{E_C-E_F}{kT})$

we find that $n = C_n (kT)^{3/2} \exp\left(-\frac{EC-EF}{kT}\right)\int_0^{\infty}\sqrt{u}\exp(-u)du$.

As $\int_0^{\infty}\sqrt{u}\exp(-u)du = \frac{\pi}{2}$, we finally have:

$$n = AT^{3/2}\exp\left(-\frac{E_C-E_F}{kT}\right). \quad (6)$$

b.

- As $T \approx$ constant across the whole semiconductor, $E_F \approx$ constant with respect to T across the semiconductor.
- We have $j_D = qD_n \mathrm{grad}n = qD_n\frac{dn}{dx} = qD_n\frac{dn}{dT}\frac{dT}{dx}$.

Taking Eq. (6) into account, we have:

$\frac{dn}{dT} = AT^{3/2}\exp\left(-\frac{E_C-E_F}{kT}\right)\left(\frac{3}{2T}+\frac{E_C-E_F}{kT^2}\right) = n\left[\frac{E_C-E_F}{kT^2}+\frac{3}{2T}\right]$, from which

$$j_D = qD_n n\left[\frac{E_C-E_F}{kT^2}+\frac{3}{2T}\right]\frac{dT}{dx}. \quad (7)$$

c.

- From $\vec{j}_n = \vec{j}_c + \vec{j}_D = 0$ at equilibrium, the result is that $\vec{j}_D = -\vec{j}_c$, so that with $\vec{j}_c = qn\mu_n\vec{E}$, that is to say, $-j_c = qn\mu_n\frac{dV}{dx}$, we have

$$D_n\left[\frac{E_C-E_F}{kT^2}+\frac{3}{2T}\right]\frac{dT}{dx} = \mu_n\frac{dV}{dx}. \quad (8)$$

- Einstein's relation can be written as $\frac{D_n}{\mu_n} = \frac{kT}{q} = \frac{kT}{e}$, where $e > 0$, and the electronic charge is denoted as $-e$.

Finally we have

$$S \overset{(5)}{=} -\frac{dV}{dT} \overset{(7)}{=} -\frac{D_n}{\mu_n T}\left[\frac{E_C - E_F}{kT} + \frac{3}{2}\right],$$

so that with Einstein's relation, $\frac{D_n}{\mu_n T} = \frac{k}{e}$, and

$$S = S_N = -\frac{k}{e}\left[\frac{E_C - E_F}{kT} + \frac{3}{2}\right]. \qquad (9)$$

As $E_C > E_F = E_{Fn}$ (n-type semiconductor), we have $(E_C - E_{FN}) > 0$, and $S = S_N > 0$.

Comment 1. In more general terms, the thermoelectric power (S_N) has the form

$$S = S_N = -\frac{k}{e}\left[\frac{E_C - E_F}{kT} + B\right], \qquad (9')$$

and here B = 3/2. However, in more general theories, we have $B = \frac{5}{2} - s$, where s takes on different values depending on the theoretical model used for conduction.

Comment 2. For a p-type semiconductor, we obtain $S = S_P = \frac{k}{e}\left[\frac{E_F - E_V}{kT} + \frac{3}{2}\right]$, and $S_P > 0$. Hence the Seebeck coefficient can be used to characterize the p or n-type character of the semiconductor.

4. If we know the conductivity, so that $\sigma_n = q\, n\, \mu_n$, then for a n-type semiconductor, we have

$$\mu_n = \frac{\sigma}{qn}. \qquad (10)$$

With n given by Eq. (6), which also can be written as

$$n = AT^{3/2}\exp\left(-\frac{E_C - E_F}{kT}\right) = N_C \exp\left(-\frac{E_C - E_F}{kT}\right),$$

then by taking Eq. (9’) into account, we have

$$n = N_C \exp\left(\frac{qS}{k} + B\right),$$

and Eq. (10) yields:

$$\mu_n = \frac{\sigma_n}{qN_C}\exp\left(-\frac{qS}{k} - B\right). \qquad (11)$$

Chapter 6

Introduction to Nonlinear Effects

This chapter details the effects of the intensity of electromagnetic waves on materials. When the intensity reaches a sufficiently high level, the material's response is no longer linear. Of particular interest is the expression for polarization under such conditions, along with the applications that result from the electrooptical effects.

6.1. Context

The previous volume, entitled "*Basic Electromagnetism and Materials*", describes nonlinear properties of materials. They were considered to be singularly due to the materials themselves, for example, in the case of ferroelectric and ferromagnetic materials, which give rise to electric polarization and a spontaneous magnetic intensity, respectively. The electric and magnetic dipole moments of these materials thus are not proportional to the magnetic or electric excitation field.

The nonlinear properties provoked by an electric excitation must be such that the electric field gives rise to nonlinear displacement forces in the material. In general, we can suppose that a material submitted to one or more external electric forces undergoes deformations which are typically ion displacements in minerals—where ionic bonds dominate—or electronic displacements in organic materials—where electronic and molecular orbitals dominate through covalent atomic bonds.

With appropriate external fields, the recall force that is exercised on displaced charges (ionic in minerals and electrons in organics) no longer follows the classic expression $f_{recall} = - kx$. The recall force is a simple elastic force that pulls displaced electrons back to their equilibrium positions. The polarizations are no longer linear.

The first part of this chapter takes as an example the application of an intense external field (E^{ω}) that is associated with an optical wave and affects the movements of electrons in a material. The second-order form of the electronic polarization is deduced, and the nonlinear response to the electric field, produced by the electric field in terms of polarization, is considered.

The second part looks at the optical properties of materials subject to a field signal denoted (E_s) of static or low frequency. Within this is also propagating an optical wave with a field E^{ω}. This gives rise to a study of the index of the material and the intensity or dephasing of the optical wave following its propagation in the material. The use of this system in electrooptical modulators then will be detailed.

6.2. Mechanical generation of the second harmonic (in one dimension)

6.2.1. Effect of an intense optical field (E^{ω})

For an electronic polarization, nonlinear effects result when the recall force, which exercises a field E (denoted here $E = E^{\omega}$), no longer takes on a simple linear form:

$$f_r = - k\,x = - m\,\omega^2\,x.$$

In the case where f_r is reduced to this simple form, then it is the coulombic field, caused by the action of the nuclei on the electrons, which is responsible for the recall force. Typically, the internal electric fields are of the order of 10^8 to 10^9 V cm^{-1}, while the optical field is only around 10^4 V cm^{-1} for an optical field of several MW cm^{-1}.

When a field, such as a laser, is no longer negligible with respect to the internal field, then the recalling force should be changed to:

$$f_r = - k\,x + d\,x^2 = - m\,\omega^2\,x + m\,\delta^2\,x^2,$$

where f_r, as a consequence, is derived from a potential, or more precisely, an anharmonic potential energy (W or rather more conventionally denoted V):

$$\vec{f}_r = -\overrightarrow{\text{grad}}\,W = -\overrightarrow{\text{grad}}\,V \Rightarrow \quad V = \frac{1}{2}kx^2 - \frac{1}{3}dx^3.$$

6.2.2. Putting the problem into equations

The anharmonic potential given in the form $V(x)=\frac{1}{2}kx^2-\frac{1}{3}dx^3$, along with the corresponding recall force $f_r=-kx+dx^2$, are schematized in Figure 6.1.

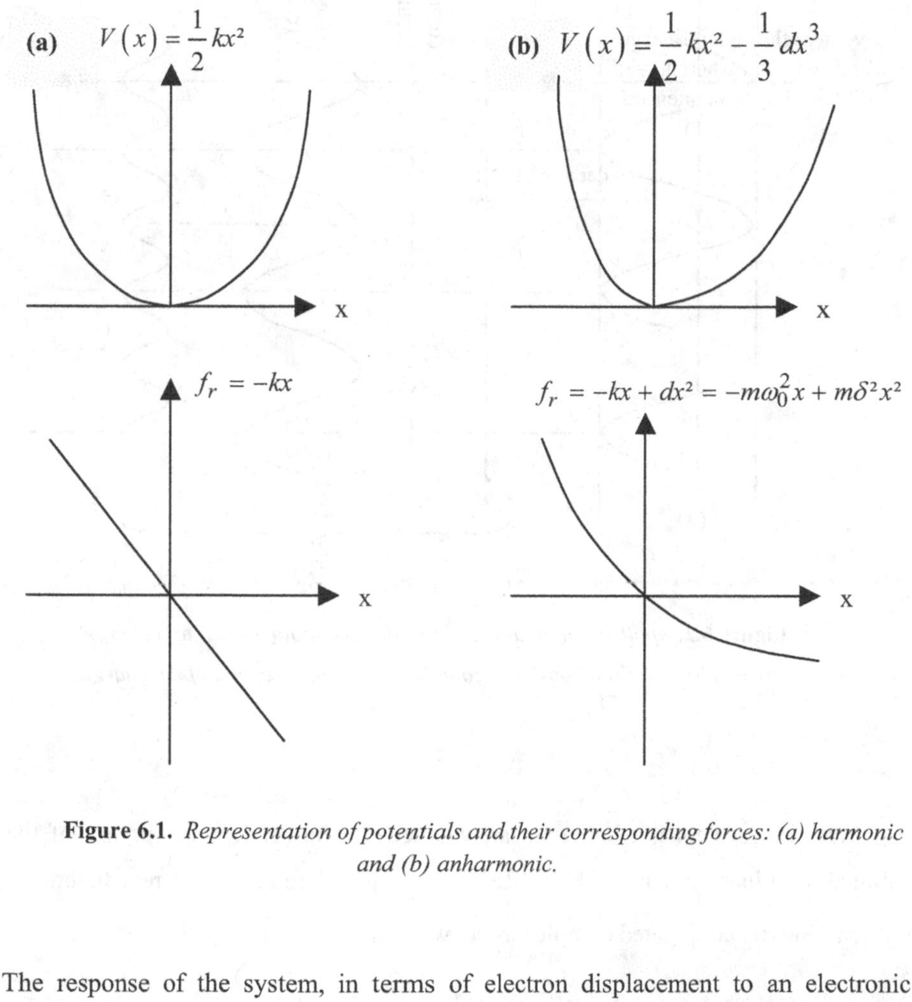

Figure 6.1. *Representation of potentials and their corresponding forces: (a) harmonic and (b) anharmonic.*

The response of the system, in terms of electron displacement to an electronic polarization, to the optical excitation which is an electric field of the form $E^{\omega}=E_0\cos\omega t$ is governed by a fundamental dynamic law in which the displacement x(t) is varied in response to the different harmonic components (Figure 6.2).

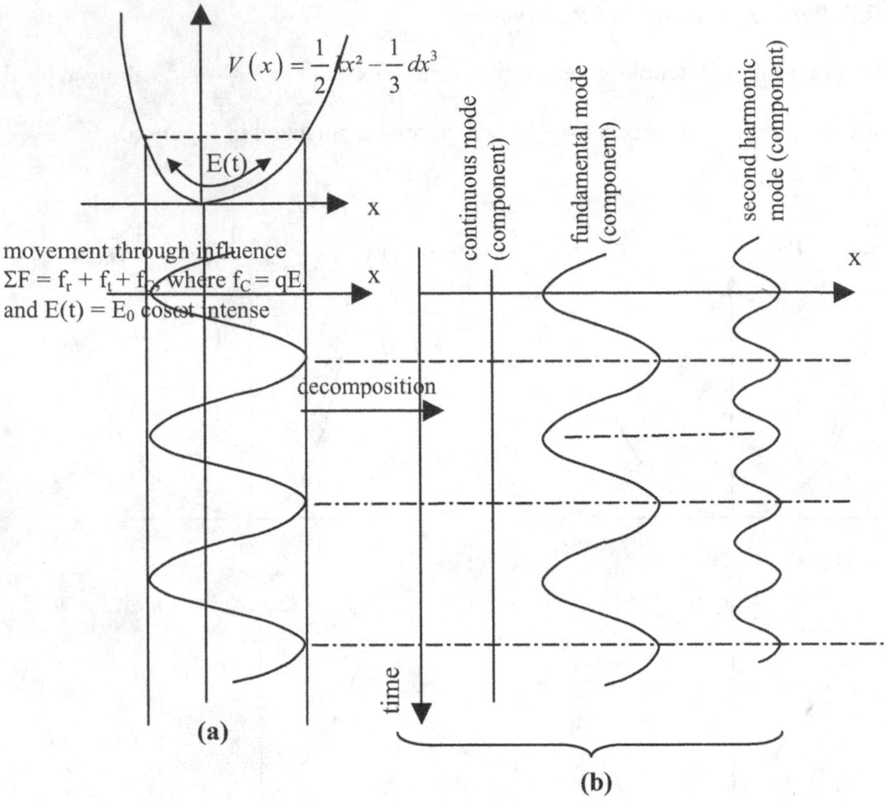

Figure 6.2. *Application of an anharmonic potential (zone a) to a charge, giving rise to a decomposition (zone b) of the displacement of the charge.*

With the system not being linear, we need to be prudent in our use of the habitual imaginary notation $\left(E^{\omega} = \mathrm{Re}\left(E_0 e^{j\omega t}\right)\right)$ and remain with real functions. By denoting the conjugated complex as cc, we can write that:

$$E^{\omega} = E_0 \cos \omega t = \frac{E_0}{2}\left(e^{j\omega t} + cc\right).$$

When using a classic study of the movement of electrons in a material, where there is a linear regime with the recall force being given as $f_r = -k\,x$, we bring in all of the forces applied to the system, i.e.,

- coulombic force $\vec{f}_C = q\,\vec{E}^{\omega}$ (where here q is the charge under consideration, and q = – e for electrons);
- frictional force f_t, which is such that $f_t = -m\Gamma v_x = -m\Gamma\dot{x}$; and
- a recalling force $f_r = -kx$, which is further detailed in Chapter 8, Volume 1.

The same is applicable to a nonlinear regime, as long as f_r is defined by its new equation, as in $f_r = -kx + dx^2 = -m\omega_0^2 x + m\delta^2 x^2$.

The fundamental dynamic equation with respect to x is written in the form:

$m\,\ddot{x} = \sum F = f_r + f_t + f_C$, so that by simplifying with m we have:

$$\ddot{x} + \Gamma\dot{x} + \omega_0^2 x - \delta^2 x^2 = \frac{qE_0}{2m}\left(e^{j\omega t} + cc\right). \qquad (1)$$

Under a permanent regime, x is now related to the various anharmonic components (Figure 6.2), and hence:

$$x(t) = \frac{1}{2}\left(x_0 + x_1 e^{i\omega t} + x_2 e^{i2\omega t} + \ldots + cc\right). \qquad (2)$$

In order to simplify the calculation, we can take $x_0 = 0$.

6.2.3. Solution to the problem of displacement terms

The resolution of the problem is through a long series of calculations, which involve plugging the equation for x(t) into Eq. (1).

We thus obtain Eq. (3):

$$-\frac{\omega^2}{2}\left(x_1 e^{i\omega t} + cc\right) - 2\omega^2\left(x_2 e^{i\omega t} + cc\right) + \frac{i\omega\,\Gamma}{2}\left(x_1 e^{i\omega t} + cc\right)$$
$$+ i\omega\,\Gamma\left(x_2 e^{i\omega t} + cc\right) + \frac{\omega_0^2}{2}\left(x_1 e^{i2\omega t} + cc\right)$$
$$+ \frac{\omega_0^2}{2}\left(x_2 e^{i\omega t} + cc\right) + \frac{\omega_0^2}{2}(x_2 e^{i2\omega t} + cc)$$
$$-\frac{\delta^2}{4}\left(x_1^2 e^{2i\omega t} + 2x_1 x_2^* e^{-i\omega t} + x_1 x_1^* + x_2 x_2^* + 2x_1 x_2 e^{3i\omega t} + x_2^2 e^{4i\omega t} + cc\right)$$
$$= \frac{qE_0}{2m}\left(e^{i\omega t} + cc\right). \qquad (3)$$

6.2.3.1. First-order terms (linear approximation)

The identification of coefficients of the terms to $e^{i\omega t}$ in Eq. (3) gives rise to:

$$x_1 = \frac{qE_0}{m}\frac{1}{\left(\omega_0^2 - \omega^2\right) + i\omega\Gamma} \approx \frac{qE_0}{2\omega m}\frac{1}{\left(\omega_0 - \omega\right) + i\Gamma/2}. \qquad (4)$$

if resonance at $\omega = \omega_0$

3.2.3.2. Second-order terms

Similarly, the identification of coefficients for terms in $e^{i2\omega t}$ makes it possible to determine x_2, which verifies:

$$x_2\left(-4\omega^2 + 2i\omega\,\Gamma + \omega_0^2\right) = \frac{\delta^2}{2} x_1^2 .$$

We thus can see that it is the term in x_1^2 that generates the movement in 2ω (term d x^2 in the equation $f_r = -m\omega_0^2 x + m\delta^2 x^2 = -kx + d\,x^2$).

By substituting Eq. (4) for x_1 in our last equation, we obtain:

$$x_2 = \frac{q^2\delta^2 E_0^2}{2m^2}\frac{1}{\left[\left(\omega_0^2 - \omega^2\right) + i\omega\,\Gamma\right]^2\left[\left(\omega_0^2 - 4\omega^2\right) + 2i\omega\,\Gamma\right]}. \qquad (5)$$

6.2.4. Solution to the problem in terms of polarization

6.2.4.1. Polarizabilities and polarizations of different orders

The generalized formula for the dipole moment (p) of a molecule with a permanent dipole (μ_0) submitted to a local field denoted E_l^{ω} is:

$$p = \mu_0 + \varepsilon_0 ([\alpha] E_l^{\omega} + [\beta] E_l^{\omega} E_l^{\omega} + [\gamma] E_l^{\omega} E_l^{\omega} E_l^{\omega} + \ldots). \qquad (6)$$

Just as α represents the first-order polarizability, β (units 10^{-4} m V^{-1}) gives the molecular polarizability to a second order, which is also called the first-order hyperpolarizability. Similarly, γ denotes the third-order polarizability. In more general terms, [α], [β], and [γ] are tensors.

The corresponding macroscopic polarization, such that $\vec{P} = n\vec{p}$ where n is the number of molecules per unit volume, when subject to a wave $\vec{E}_{\omega}$ is in one dimension of the form:

$$\vec{P} = \vec{P}_0 + \chi_1^{(\omega)} \vec{E}^{\omega} + \chi_2^{(2\omega)} \vec{E}^{\omega} \vec{E}^{\omega}, \qquad (7)$$

where χ is the absolute dielectric susceptibility such that $\chi = \varepsilon_0 \chi_r$ and $\chi_r = (\varepsilon_r - 1)$ is the relative dielectric susceptibility.

In even more general terms, we should write:

$$\vec{P} = \vec{P}_0 + [\chi_1] \vec{E}^{\omega} + [\chi_2] \vec{E}^{\omega} \vec{E}^{\omega} + [\chi_3] \vec{E}^{\omega} \vec{E}^{\omega} \vec{E}^{\omega} + \ldots,$$

where the $[\chi_i]$ terms are tensors. See also Comment 3 of Section 6.2.5.3 for more detailed remarks.

When the molecules are ordered in such a way in a system that there is a center of inversion, which results in centrosymmetric dispositions, then strong microscopic nonlinearities (high β) become inoperable at a macroscopic scale ($\chi_2^{(2\omega)}$ small).

6.2.4.2. First-order susceptibility

The linear polarization due to a movement $x_1(t)$ of a charge q [where x_1 is given by Eq. (4)] is such that:

$$P^{(1)}(t) = Nqx_1(t) = Nq\frac{x_1}{2}\left(e^{i\omega t} + cc\right).$$

By identification with $P^{(1)}(t) = \frac{P_0}{2}\left(\chi_1^{(\omega)} E e^{i\omega t} + cc\right)$, we obtain

$$\chi_1^{(\omega)} = \frac{Nq^2}{m\varepsilon_0}\frac{1}{\left(\omega_0^2 - \omega^2\right) + i\omega\Gamma}. \qquad (8)$$

As discussed in Volume 1, we once again find the expression for the dielectric susceptibility for a linear material.

6.2.4.3. Second-order susceptibility

This appears in the second-order term of polarization, which is written by analogy to the linear polarization, while taking Eq. (7) into account, so:

$$P^{(2)}(t) = \frac{\varepsilon_0}{2}\left[\chi_2^{(2\omega)} E_0^2 e^{i2\omega t} + cc\right] = Nq\frac{x_2}{2}\left(e^{i2\omega t} + cc\right).$$

From Eq. (5), we thus can deduce $\chi_2^{(2\omega)}$:

$$\chi_2^{(2\omega)} = \frac{Nq^3\delta^2}{2m^2\varepsilon_0}\frac{1}{\left[\left(\omega_0^2 - \omega^2\right) + i\omega\Gamma\right]^2\left[\left(\omega_0^2 - 4\omega^2\right) + 2i\omega\Gamma\right]}. \qquad (9)$$

Just as $\chi_1^{(\omega)}$, in fact $\chi_2^{(2\omega)}$ also exhibits real and imaginary parts, but its denominator is doubly resonant at $\omega = \omega_0$ and $\omega = \frac{\omega_0}{2}$.

6.2.5. Comments

6.2.5.1. Comment 1

As the charge under study is the electron, $q = -e$ and $q^3 = -e^3$. Equation (9) in fact can be written as:

$$\chi_2^{(2\omega)} = -\frac{Ne^3\delta^2}{2m^2\varepsilon_0}\frac{1}{\left[\left(\omega_0^2-\omega^2\right)+i\omega\Gamma\right]^2\left[\left(\omega_0^2-4\omega^2\right)+2i\omega\Gamma\right]}. \qquad (9')$$

6.2.5.2. Comment 2

In addition, we can establish the following relation between the linear $\chi_1^{(\omega)}$ and nonlinear $\chi_2^{(2\omega)}$ susceptibilities:

$$\frac{\chi_2^{(2\omega)}}{\left[\chi_1^{(\omega)}\right]^2\chi_1^{(2\omega)}\varepsilon_0^2} = \frac{m\delta^2}{2N^2q^3} = \xi^{(2\omega)}.$$

The parameter $\xi^{(2\omega)}$, called Miller's parameter, is in fact practically identical for all materials: $\xi^{(2\omega)} \cong 3$ to 8×10^9 SI for InAs, GaSb, GaAs, CdTe, ZnTe, and ZnSe.

6.2.5.3. Comment 3

The problem so far has been treated in just one dimension. If in place of just one component the incident wave has three, then the second-order polarization has components denoted P_x, P_y, and P_z. In the most general of cases, these can be obtained from all possible quadratic components of E_x, E_y, and E_z, as in:

$$\begin{bmatrix} P_x \\ P_y \\ P_z \end{bmatrix} = \varepsilon_0 \underline{\underline{\chi}}_2^{(2\omega)} \begin{bmatrix} E_x^2 \\ E_y^2 \\ E_z^2 \\ E_zE_y \\ E_zE_x \\ E_xE_y \end{bmatrix}, \quad \text{where } \underline{\underline{\chi}}_2^{(2\omega)} \text{ is a tensor.}$$

6.3. Electrooptical effects and the basic equations

6.3.1. Excitation from two pulsations and an introduction to the Pockels effect

The above classic study showed that in the mechanics of the second harmonic, an excitation at a photon E^{ω} can generate displacements at ω frequencies [movement $x_1(t)$] and 2ω [displacement $x_2(t)$]. The frequency of the photons emitted by the charges moving along x_1 (at frequency ω) or along x_2 (at a frequency of 2 ω) will occur with a dipolar radiation of frequency ω or 2ω.

We thus can generalize this emission mechanism to that of the excitation of a system with two photons of angular frequencies ω' and ω''. The emissions detailed in Figure 6.3 thus are expected.

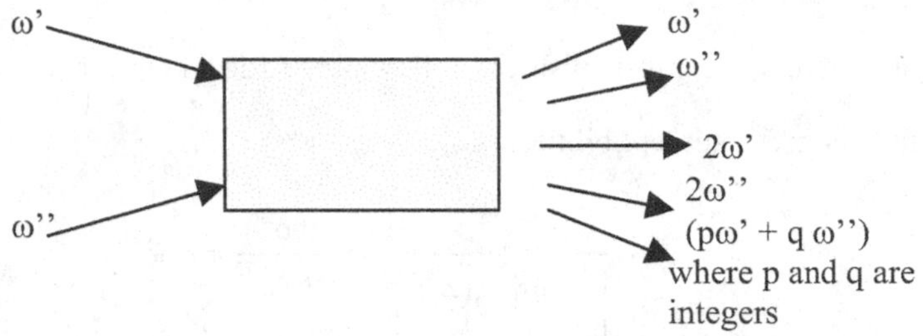

Figure 6.3. *Possible emissions from a system excited by pulsations ω' and ω''.*

We can use the classic expression [Eq. (7) from Section 6.1.2] for the second-order (relative) susceptibility established for a system excited by one pulsation ω. For an excitation pulsation ω_i, by making $D(\omega_i) = \omega_0^2 - \omega_i^2 + i\omega_i\Gamma$, this equation can be rewritten as:

$$\chi_2^{(2\omega_i)} = \frac{Nq^3\delta^2}{2m^2\varepsilon_0} \frac{1}{D^2(\omega_i)D(2\omega_i)}.$$

Similarly, for an excitation based on two waves (pulsations ω' and ω''), the harmonic of the pulsation ω_T = ω' + ω'' is associated with a second-order susceptibility in the form:

$$\chi_2^{(\omega'+\omega'')} = \frac{Nq^3\delta^2}{2m^2\varepsilon_0} \frac{1}{D(\omega')D(\omega'')D(\omega'+\omega'')}.$$

Thus there is an interesting and specific case when one of the waves is an optical wave (of angular frequency ω' = ω), and the other is a low-frequency wave which could have a zero frequency of a static field (ω'' = 0). The preceding formula thus becomes:

$$\chi_2^{(\omega+0)} = \frac{Nq^3\delta^2}{2m^2\varepsilon_0}\frac{1}{D(0)D(\omega)D(\omega)}.$$

This second-order susceptibility no longer depends only on the optical pulsation ω. Nevertheless, a simplification to a single term for a polarization varying with a single pulsation ω can be misrepresentative, just as the direct calculation for the polarization showed. This calculation is actually rather complicated, and in order to study these electrooptical effects, it is preferable to use a phenomenological study (see Problem 2 of this chapter). This simplifies the calculation when using a small variation in the index for a static or low-frequency field (for further details see Section 6.3.2).

Comment. If the expression for the polarization is directly written in its classical from, then $P(t) = P_0 + \chi_1 E(t) + \chi_2 E^2(t) + \ldots$, and here

$$E(t) = E_{(0)} + E_{(\omega)}e^{j\omega t} + cc \text{ (and } \chi \text{ is the absolute susceptibility)},$$

while the dependence of the susceptibility on two pulsations (static and optical) can bring in rather difficult calculations.

In effect, we should write (with first and second-order susceptibilities, χ_1 and χ_2):

$$P(t) = \left\{\chi_1\left[E_{(0)} + E_{(\omega)}e^{j\omega t} + cc\right] + \chi_2\left[E_{(0)} + E_{(\omega)}e^{j\omega t} + cc\right]^2 + \ldots\right\}.$$

Straight away we can see that this text is ambiguous, as in fact, for the first term, for example, χ_1 carries both the static term $E_{(0)}$ and the optical term $E_{(\omega)}e^{j\omega t}$. In order to avoid these difficulties, it is preferable to present the more simple and empirical examples, as in Problem 1 and Section 6.3.2.

6.3.2. *Basic equations for nonlinear optics*

6.3.2.1. The classic equations

When subject to an electric force [static of low frequency electric field (E_s)], a material undergoes deformations, which typically correspond to ionic displacements in minerals and electronic displacements in organics (deformation of electronic/molecular orbitals). The resulting variations in polarizations induce a modification in the measured index at a given value of ω (see Figure 6.4).

When the index varies proportionally with the applied field, the electrooptical effect is linear in what is called the Pockels effect. When the index varies quadratically with the applied field (proportional to the square of the applied field), the electrooptical effect is called the Kerr effect.

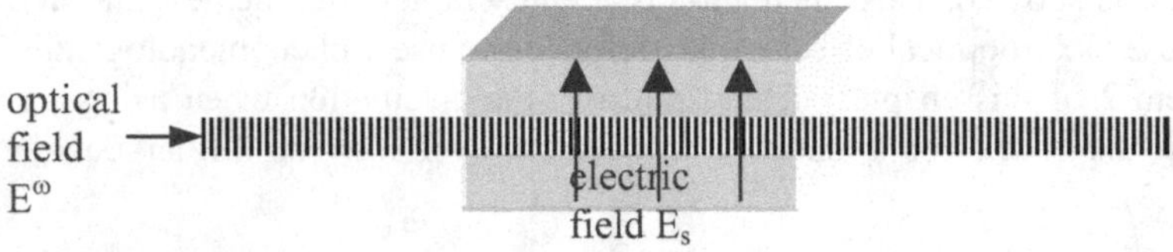

Figure 6.4. *Coupling of the electric field E_s and the optical field E^{ω} through an electro-optical material.*

In the hypothesis that the variation in the index is small with respect to the applied field E_s (≡ E to simplify the following notation), a Taylor development about the values of the index for E = 0 can be carried out.

By notation $n(E=0) = n(0) = n$, $a_1 = \left.\frac{dn}{dE}\right)_{E=0}$, $a_2 = \left.\frac{d^2n}{dE^2}\right)_{E=0}$,

we thus find

$$n(E) = n + a_1 E + \frac{1}{2} a_2 E^2 + \ldots \quad (1)$$

For reasons of notation that will be made apparent in Section 6.3.2.2, we introduce the electrooptical coefficients by making:

$$\begin{cases} r = -\dfrac{2a_1}{n^3} \\ s = -\dfrac{a_2}{n^3} \end{cases}$$

so that:

$$n(E) = n - \frac{1}{2} rn^3 E - \frac{1}{2} sn^3 E^2 + \ldots \quad (2)$$

Typically the second and third-order terms are small (by several orders of size) with respect to n.

6.3.2.2. Form of the electric impermeability

The electric impermeability is useful when describing anisotropic media (ellipsoidal indices) and is defined by:

$$\eta = 1/\varepsilon_r = 1/n^2.$$

As in one part:

$$\eta = \frac{1}{n^2}, \text{ so that } \frac{d\eta}{dn} = \frac{-2}{n^3},$$

and in another:

$$\Delta n = n(E) - n = -\frac{1}{2} rn^3 E - \frac{1}{2} sn^3 E^2,$$

we thus can write:

$$\Delta\eta = \left(\frac{d\eta}{dn}\right)\Delta n = \left(-\frac{2}{n^3}\right)\left(-\frac{1}{2} rn^3 E - \frac{1}{2} sn^3 E^2 + \ldots\right) = rE + sE^2.$$

With $\Delta\eta = \eta(E) - \eta$, we finally arrive at:

$$\eta(E) = \eta + rE + sE^2 + \ldots \quad (3)$$

We should immediately remark that the coefficients r and s were initially defined so as to arrive at this relatively simple equation which introduces into Eq. (3) for $\Delta\eta$ the coefficient r or s in front of E or E^2.

6.3.2.3. Pockels effects and order of scale

In a lot of materials, the term for the Kerr effect (second order) is small with respect to the Pockels effect (first order), so that:

$$n(E) \approx n - \frac{1}{2} rn^3 E. \quad (4)$$

Typically, $r \approx 10^{-12}$ to 10^{-10} m V^{-1}, which is around 1 to 100 pm V^{-1}.

When $E = 10^6$ V m^{-1}, the term $\frac{1}{2} rn^3E$ is of the order of 10^{-6} to 10^{-4}, which is a very small variation in the index induced by E (often used minerals include KH_2PO_4, $LiNbO_3$, $LiTaO_3$, and CdTe).

6.4. Principle of electrooptical modulators

As a generalization, there are two types of modulators. They are:

- phase modulators which make it possible to transmit information optically by modifying the phase of the optical wave [using a signal V(t) or $\vec{E}(t)$]; and
- amplitude modulators in which the signal modulates the intensity of the optical wave (parallel monochromatic luminous beams with a laser beam).

6.4.1. Phase modulator

The phase modulator does not require a particularly complicated structure. It is constructed from a single optical waveguide which makes up the arm of the modulator and into which the incident optical wave is injected. The modulation is produced by a transversal tension (V) associated with an electric field ($E_s \equiv E$) that induces a dephasing ($\Delta\varphi$) dependent on the frequency of the electric field. The continuous information in this frequency thus is inscribed upon the optical wave in the form of a phase modulation, as schematized in Figure 6.5. It should be noted that the figure does not show the positions of the electrodes used to apply E.

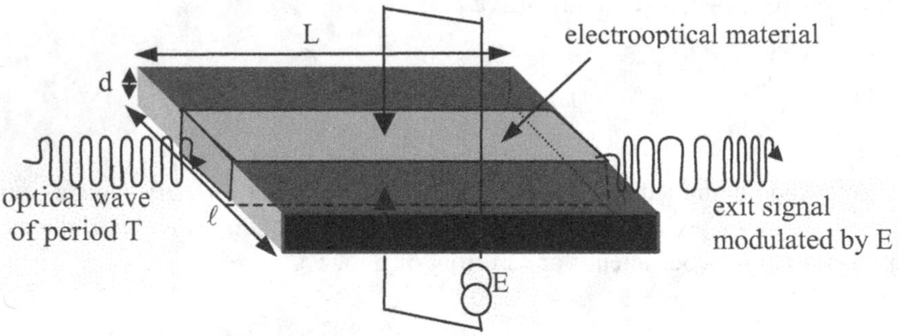

Figure 6.5. *Schematization of a phase modulator.*

In fact there are two possible dispositions: one is that of a sandwich structure with the electrodes sandwiching the electrooptical material and the other uses coplanar electrodes, that is to say they are deposited on the same surface of the electrooptical material and are separated by a narrow nonmetallic band which has parallel sides of dimension ℓ. In the latter, the form of the field lines is such that the interaction with the electrooptical material is reduced. The covering factor (T) between the optical and electrical fields diminishes and results in an increase in the half-wave tension (V_π), which is defined below.

When an optical beam traverses a Pockels cell of length L and is submitted to a field E, the wave undergoes a dephasing:

$$\varphi = n(E)k_0L = \frac{2\pi}{\lambda_0}n(E)L, \qquad (5)$$

where λ_o is the wavelength in a vacuum.

The introduction of Eq. (4) into Eq. (5) yields:

$$\varphi = \frac{2\pi}{\lambda_0}nL - \frac{2\pi}{\lambda_0}L\frac{1}{2}r\,n^3E = \varphi_0 - \pi\frac{r\,n^3EL}{\lambda_0}, \qquad (6)$$

where $\varphi_0 = \frac{2\pi nL}{\lambda_0}$ represents the dephasing when E = 0, meaning without an applied field E (or tension V).

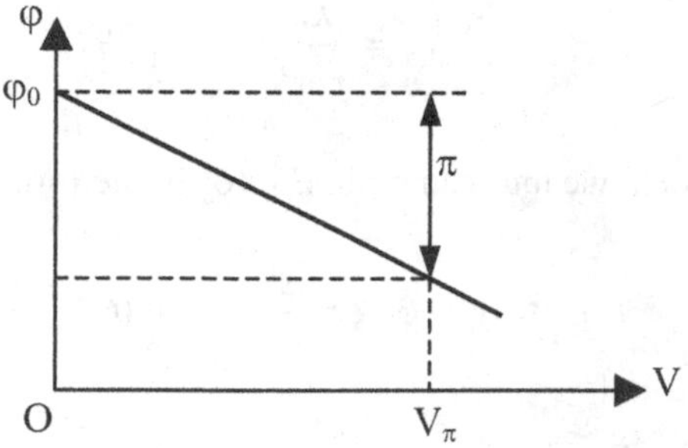

Figure 6.6. *Variation in the dephasing (φ) as a function of the polarization tension (V), along with a definition of V_π (for a transversal cell).*

The variation in φ (dephasing in the presence of a polarization field E) as a function of the polarization tension (V) is represented in Figure 6.6. From this can be defined the so-called half-wave tension which is such that its application leads to a dephasing by π with respect to a wave not subject to E:

$$\varphi_0 - (\varphi)_{V=\text{tension half-wave}} = \pi\,. \qquad (7)$$

In a transversal cell, this half-wave tension denoted V_π, (which verifies $\varphi_0 - (\varphi)_{V=V\pi} = \pi$), is such that:

$$E = (V/d)_{V=V\pi} = V_\pi/d\,.$$

This value introduced into Eq. (6) gives rise to Eq. (7), which can be written in the form:

$$\varphi_0 - (\varphi)_{V=V\pi} = \pi = \frac{\pi r n^3}{\lambda_0}\frac{V_\pi L}{d}\,.$$

From which can be deduced:

$$V_\pi = \frac{\lambda_0}{r\,n^3}\frac{d}{L}\,. \qquad (8)$$

For a longitudinal Pockels cell, the field is in the form E = V/L, and the half-wave tension, which is commonly denoted $V^l_{P,\pi}$, becomes:

$$V^l_{P,\pi} = \frac{\lambda_0}{r\,n^3}\,. \qquad (9)$$

For a transversal cell, we thus can write Eq. (6) in the form:

$$\varphi = \varphi_0 - \pi\frac{V}{V_\pi}\,. \qquad (6')$$

For a longitudinal cell, we can write:

$$\varphi = \varphi_0 - \pi\frac{V}{V^l_{P,\pi}}\,. \qquad (6'')$$

The dephasing thus evolves linearly with the applied tension (Figure 6.6), so that effective control over the modulation of an optical wave traversing the cell can be made. The value of $V^{l}_{P,\pi}$ is of the order of 1 to several kV for longitudinal modulators. The transversal modulators make it possible to reduce V_π by varying the ratio $\frac{d}{L}$ (d ≈ several microns, L ≈ 1 cm), so that $V_\pi \approx 1$ to 100 volt.

6.4.2. Amplitude modulator

The dephasing does not affect the intensity of an optical wave. Nevertheless, the insertion of a phase modulator into the arm of an interferometer, for example, of the Mach–Zehnder type, can be used to give an intensity modulator, as shown in Figure 6.7.

If I_i is the amplitude of the incident optical wave, split equally into two waves, then each has an intensity $\frac{1}{2} I_i$ in each of the arms, which are not subject to E.

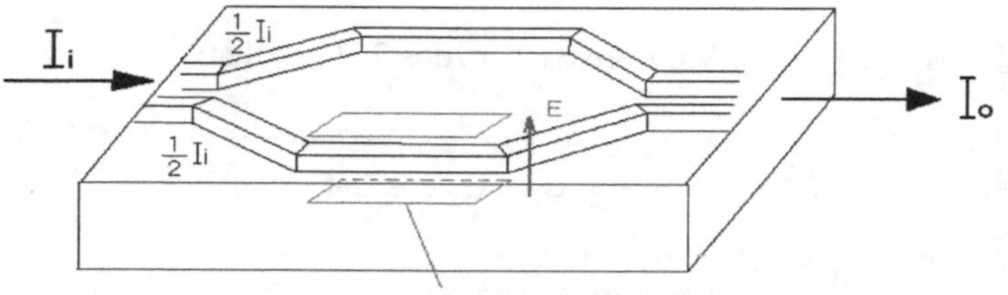

Figure 6.7. *Schematization of a Mach–Zehnder amplitude modulator.*

The application of E to the arms modifies the transmitted intensity which then takes on the form $\frac{1}{2} I_i \cos\Delta\varphi$, where $\Delta\varphi$ is the dephasing induced by E in the "modulated" arm with respect to the arm not subject to E. The exit intensity thus is

$$I_0 = \frac{1}{2} I_i + \frac{1}{2} I_i \cos\Delta\varphi = I_i \cos^2\frac{\Delta\varphi}{2} \qquad (10)$$

where $\Delta\varphi = |\varphi - \varphi_0| = \pi\frac{V}{V_\pi}$ [from Eq. (6) of Section 3.4.1].

As $\cos^2 x = \dfrac{1+\cos 2x}{2}$ and $\cos 2x \approx 1 - \dfrac{4x^2}{2}$, we have:

$$\cos^2 x \approx \frac{1}{2}\left(1 + 1 - \frac{4x^2}{2}\right) = 1 - x^2 . \qquad (11)$$

By introducing $x = \dfrac{\Delta\varphi}{2}$ into Eq. (11) and substituting into Eq. (10), we have:

$$I_0 = I_i \left[1 - \frac{\Delta\varphi^2}{4}\right] = I_i \left[1 - \frac{\pi^2}{4V_\pi^2} V^2\right] = I_i - K I_i V^2 , \qquad (12)$$

so that in addition

$$\frac{I_0}{I_i} = 1 - K V^2 . \qquad (12')$$

If $V = V_0 \cos \omega t$, we have $V^2 \propto \cos^2 \omega t \propto \cos 2\omega t$, so that:

$$\frac{I_0}{I_i} \propto C_1 - C_2 \cos 2\omega t .$$

The intensity I_0 is modulated by the frequency 2ω.

6.4.3. The merit factor

As indicated in Section 6.4.1, the half-wave tension takes on the form $V_\pi = \dfrac{\lambda_0}{r\, n^3} \dfrac{d}{L}$ (transversal polarization modulator) and should be as low as possible (around 1 volt in principle). The value of V_π is often taken as a reference as to the performance of a modulator.

Taking the expression for V_π into account, in order to make it smaller, it is of interest to decrease the coefficient $\dfrac{d}{L}$, which means increasing the length—within all reason—of the arm length. A value of $L \approx 3$ cm can be considered normal (with d of the order of several microns).

In addition, the factor $r\,n^3$ should be as large as possible so as to decrease the value of V_π. This factor also can be considered one of technical specification; $M = r\,n^3$ is expressed in pm V^{-1}.

We thus have (at $\lambda \approx 1000$ nm) FM $\approx 10^4$ pm V^{-1} in $BaTiO_3$, and FM $\approx 7 \times 10^2$ pm V^{-1} in $LiNbO_3$. In optimized organic materials, we can reach FM $\approx 10^3$ pm V^{-1}.

6.5. Problems

6.5.1. Problem 1. Second-order susceptibility and molecular centrosymmetry

Show that the second-order susceptibility (χ_2) is nonzero uniquely for noncentrosymmetric molecules.

Answer

If a molecule is symmetric, then changing $\vec{E}$ to $-\vec{E}$ is the same as moving the charges in exactly the same way symmetrically, so that:

$$P^{(2)}\left(\vec{E}\right) = -P^{(2)}\left(-\vec{E}\right). \qquad (1)$$

In addition and by definition for $P^{(2)}$ we have

$$\left.\begin{array}{l} P^{(2)}(\vec{E}) = \chi_{a2}\,\vec{E}.\vec{E} \\ P^{(2)}(-\vec{E}) = \chi_{a2}\,(-\vec{E}).(-\vec{E}) \end{array}\right\} \Rightarrow P^{(2)}(\vec{E}) = P^{(2)}(-\vec{E}) \qquad (2)$$

This finally gives:

$$\left.\begin{array}{ll} \text{from Eq. (1)} & \chi_{a2}\,\vec{E}.\vec{E} = -\chi_{a2}\,(-\vec{E}).(-\vec{E}) \\ \text{from Eq. (2)} & \chi_{a2}\,\vec{E}.\vec{E} = \chi_{a2}\,(-\vec{E}).(-\vec{E}) \end{array}\right\} \Rightarrow \chi_{a2} = -\chi_{a2}\text{, so that } \chi_{a2} = 0$$

The conclusion therefore is that the second-order dielectric susceptibility is zero for symmetrical molecules. As a consequence, for this susceptibility to be nonzero, the molecules need to be noncentrosymmetric.

6.5.2. Problem 2. Phenomenological study of the Pockels effect

This problem considers a material in which the displacement of electrons is particularly facile (for example, systems with delocalized electrons such as in π conjugated organic materials). In addition, the system is noncentrosymmetric and has a center of gravity for positive charges ($G_A \equiv O$) which does not coincide with the center of gravity of negative charges (G_E) which is such that $OG_E = x$, and $x > 0$.

Such a situation can be obtained by a method called "poling" which involves applying in a controlled manner an electric field ($\vec{E}_P$) directed along Ox at a temperature in the neighborhood of the glass transition temperature (T_g) of the material to an ambient temperature (T_a). This is because there is an orientation of orbitals that takes place at the T_g which then becomes stuck at T_a. The consequence of this is that the localization of the center of gravity of the electronic charges at G_E is such that $OG_E = x$, where $x > 0$ if $\vec{E}_P$ is antiparallel to $\overrightarrow{Ox}$, as shown in Figure a.

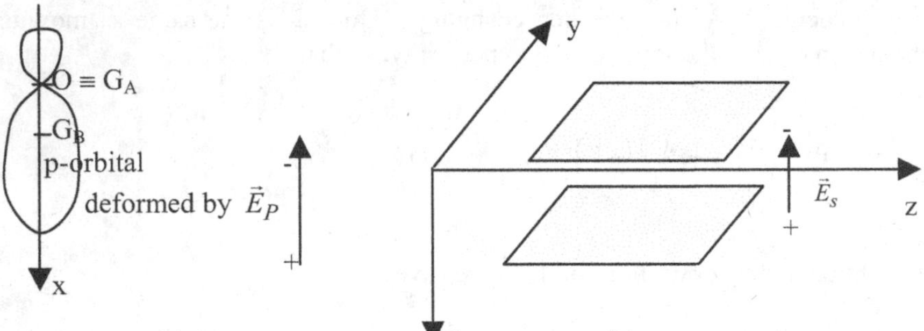

(a) (b)

An optical wave (E^{ω}) is polarized along Ox and propagates along Oz. The above-described active material is placed between two electrodes placed as detailed in Figure b. These permit the application of a tension to the material and thus a signal characterized as a static field ($\vec{E}_s$) that is directed along –Ox.

Overall, we take into account the effect of the optical field E^{ω} by the intermediate and only coulombic force given by $f_C = q\,E^{\omega}$ which it generates. The effect of the signal generated by the signal characterized by the field E_s is only taken into account by the recalling force f_r which is overall expressed in the form $f_r = -kx + Dx$. On making $kx = m\omega_0^2 x$, the component Dx due to E_s is written $Dx = m\,\delta^2 x$. Physically speaking, this term introduces a variation in harmonicity with respect to the classic case where f_r is taken in its simple form $f_r = -k\,x$. Then given that δ^2 is physically much greater than E_s is high, it is possible to write δ^2 in the form $\delta^2 = a\,E_s$.

1. Under a permanent regime, we neglect the frictional forces $f_t = -m\,\Gamma\,\dot{x}$ by assuming that $\Gamma << |\omega - \omega_0|$ [far from the resonance, $(\omega - \omega_0)$ is large and $\Gamma \rightarrow 0$].

a. Give the dynamic fundamental relationship with respect to Ox.

b. With $E_\omega = E_0 \cos \omega t$, the solution for a forced regime is in the form $x = x_0 \cos \omega t$ $(= x_0 e^{j\omega t})$. From this deduce the expression for x_0 (and hence for x).

2.

a. Here $\omega_p^2 = \dfrac{Nq^2}{\varepsilon_0 m}$, where N is the electronic density. Determine the polarization $\vec{P}$ produced by the displacement by x of the electronic charges along Ox, which is written in the form $\vec{P} = \varepsilon_0 \chi \vec{E}$.

b. Assuming that δ is small with respect to $|\omega - \omega_0|$, then we are once again considering a domain outside that of the zone of resonance (where $|\omega - \omega_0|$ is large). Express for this case the susceptibility χ calculated above as a function of a classic value as in $\chi_\omega = \dfrac{\omega_p^2}{\omega_0^2 - \omega^2}$.

3.

a. With $\varepsilon_r = n_x^2 = 1+\chi$, determine the form of the index n_x following the direction Ox in the presence of E_s ($n_x = f(\omega, \omega_0, \delta, \chi_\omega)$).

b. By introducing $n_\omega = \sqrt{1 + \chi_\omega}$, the index with respect to Ox in the absence of E_S, express n_x as a function of n_ω.

4.

a. Give the value of $\Delta n = n_x - n_\omega$. Eliminate χ_ω so that Δn can be expressed as a function of n_ω (index in the absence of E_s).

b. By using the fact that we have made $\delta^2 = a\,E_s$, and assuming that $n_\omega^2 >> 1$ (for a medium consisting of π-conjugated polar chromophores of high dielectric permittivity), show that we once again find the expression characteristic of the Pockels effect, $\Delta n = \frac{1}{2} r\, n^3 E_S$.

Answers

1.

a. Taking the hypotheses into account, and for an electron with a charge q (= – e), the dynamic fundamental relation with respect to Ox can be written as:

$$m\ddot{x} = -m(\omega_0^2 - \delta^2)x + qE^{\omega}. \qquad (1)$$

b. With $E^{\omega} = E_0 \cos \omega t$, the solution for the forced regime $x = x_0 \cos \omega t$ ($= x_o e^{i\omega t}$, hence $\ddot{x} = -\omega^2 x$) leads to:

$$([\omega_0^2 - \delta^2] - \omega^2)x = -\frac{eE^{\omega}}{m}.$$

2.

a. If N is the electronic density, the polarization $\vec{P}$ produced by the displacement x of the electronic charges along $\overrightarrow{Ox}$ is $\vec{P} = -Ne\, x\, \vec{u}_x$, so that with $\omega_p^2 = \frac{Nq^2}{\varepsilon_0 m}$, we have

$$\vec{P} = \varepsilon_0 \frac{\omega_p^2}{\left[\omega_0^2 - \delta^2\right] - \omega^2} \vec{E}^{\omega} = \varepsilon_0 \chi \vec{E}^{\omega}. \qquad (2)$$

Typically, we make

$$\chi_{\omega} = \frac{\omega_p^2}{\omega_0^2 - \omega^2}. \qquad (3)$$

This is the term for the linear polarization of the same system but not subjected to a supplementary static signal (E_S). In effect, the term δ^2 no longer appears in f_r, so that the corresponding polarization $P^{(\omega)}$ is in the form:

$$\vec{P}^{(\omega)} = \varepsilon_0 \frac{\omega_p^2}{\omega_0^2 - \omega^2} \vec{E}^{\omega} = \varepsilon_0 \chi_{\omega} \vec{E}^{\omega}.$$

b. It is assumed that δ is small with respect to $|\omega - \omega_0|$, which means working outside of the resonance zones. We therefore use Eq. (2) and introduce Eq. (3):

$$\chi = \frac{\omega_p^2}{\omega_0^2 - \omega^2 - \delta^2} = \frac{\omega_p^2}{\left(\omega_0^2 - \omega^2\right)\left(1 - \dfrac{\delta^2}{\omega_0^2 - \omega^2}\right)} \approx \chi_\omega\left(1 + \frac{\delta^2}{\omega_0^2 - \omega^2}\right).$$

3.

a. With $\varepsilon_r = n_x^2 = 1+\chi$, we have for the index with respect to the direction Ox and in the presence of E_S:

$$n_x = \sqrt{1+\chi} \approx \left(1 + \chi_\omega + \frac{\chi_\omega \delta^2}{\omega_0^2 - \omega^2}\right)^{1/2} = \left\{\left(1+\chi_\omega\right)\left[1 + \frac{\chi_\omega \delta^2}{\left(\omega_0^2 - \omega^2\right)\left(1+\chi_\omega\right)}\right]\right\}^{1/2}$$

$$= \left(1+\chi_\omega\right)^{1/2}\left[1 + \frac{\chi_\omega \delta^2}{\left(\omega_0^2 - \omega^2\right)\left(1+\chi_\omega\right)}\right]^{1/2}.$$

b. By introducing $n_\omega = \sqrt{1+\chi_\omega}$, the index with respect to Ox in the absence of E_S is given by:

$$n_x \approx n_\omega\left[1 + \frac{1}{2}\frac{\chi_\omega \delta^2}{\left(\omega_0^2 - \omega^2\right)\left(1+\chi_\omega\right)}\right] = n_\omega + \frac{\chi_\omega \delta^2}{2n_\omega\left(\omega_0^2 - \omega^2\right)}. \qquad (5)$$

4.

a. Finally, we have:

$$\Delta n = n_x - n_\omega = \frac{\chi_\omega \delta^2}{2n_\omega\left(\omega_0^2 - \omega^2\right)}. \qquad (6)$$

With $\chi_\omega = \dfrac{\omega_p^2}{\omega_0^2 - \omega^2}$, so that $\dfrac{1}{\omega_0^2 - \omega^2} = \dfrac{\chi_\omega}{\omega_p^2}$, we obtain:

$$\Delta n = n_x - n_\omega = \frac{\chi_\omega^2 \delta^2}{2n_\omega \omega_p^2}.$$

As $n_\omega = \sqrt{1+\chi_\omega}$, and hence $\chi_\omega = n_\omega^2 - 1$, as a definitive equation we have:

$$\Delta n = \frac{\left(n_\omega^2 - 1\right)^2}{2n_\omega} \left(\frac{\delta}{\omega_p}\right)^2 . \qquad (7)$$

b. As indicated above, it is possible to set $\delta^2 = aE_S$, even though in the first approximation 1 is neglected with respect to n_ω^2 (with chromophores in the medium, the medium is intrinsically well polar, so that its permittivity $\varepsilon_r = n_\omega^2$ can be quite high—at least to an order of three or four places). This means that we have

$$\Delta n \approx \frac{n_\omega^4}{2n_\omega \omega_p^2} aE_s ,$$

so that Δn is in the form:

$$\Delta n = \frac{1}{2} r\, n^3 E_s , \qquad (8)$$

where $r \equiv r_p \approx \dfrac{a}{\omega_p^2}$ and is the electrooptical coefficient and $n_\omega \equiv n$ by notation. This is an index for the direction along the polarization of the optical wave, and is measured in the absence of the polarization field E_S.

Finally, Δn is proportional to E_S and Eq. (8) is identical to Eq. (4) in Section 3.3.3. and well characterizes the Pockels effect.

Comment. Alternative form for representation of the Pockels effect

As in general terms $d\left(\dfrac{1}{x^2}\right) = -2x^{-3}dx$, we can go on to write that

$$\Delta\left(\frac{1}{n^2}\right) = \left|d\left(\frac{1}{n^2}\right)\right| = 2n^{-3}\Delta n = 2\frac{\Delta n}{n^3} ,$$

so that by using Eq. (8), we have

$$\Delta\left(\frac{1}{n^2}\right) = rE_s . \qquad (9)$$

This is the classic and defining equation for the electrooptical effect, developed from the deformation of an ellipsoid of indices by a low-frequency electric field. This gives rise to a linear variation of coefficients $\left[\frac{1}{n^2}\right]_i$ in the ellipsoid of indices as a function of the electric field.

Chapter 7

Electromagnetic Cavities

This chapter details the effect of the form of the material environment on the electromagnetic field. It is more specifically concerned with three directions, limited dimension guides (cavities) which can be around a micron in size, hence the name microcavities. Particular attention is paid to the evolution of an electromagnetic field in a confined volume.

7.1. Definition

An electromagnetic cavity is a volume, which is empty or filled with a dielectric, and is limited in three directions by walls. When these walls are metallic, we have a metallic cavity, and when they are made from a dielectric, we have a dielectric resonator.

Typically, metallic cavities are obtained by closing a metallic waveguide with two metallic plates perpendicular to the axis Oy, which is longitudinal along the guide. If the guide section is rectangular, the cavity is a parallelepiped; however, if the section is circular, then the cavity is cylindrical.

7.2. Resonance conditions for a cavity and proper resonance modes

Figure 7.1 shows a rectangular guide closed at $y = 0$ and $y = d$ by metallic plates perpendicular to the axis Oy. The metallic cavity thus is defined as a parallelepiped of volume $V = a\,b\,d$.

As detailed in Section 12.4.4.1 of Volume 1 "*Basic Electromagnetism and Materials*", a wave TE_{mn} can propagate within a guide (or rather a cavity without the plates at y = 0 and y = d). For this there is a dispersion equation, given in that same Section as Eq. (15'), and detailed again here:

$$\frac{\omega^2}{c^2} = k_g^2 + \frac{m^2\pi^2}{a^2} + \frac{n^2\pi^2}{b^2}. \qquad (1)$$

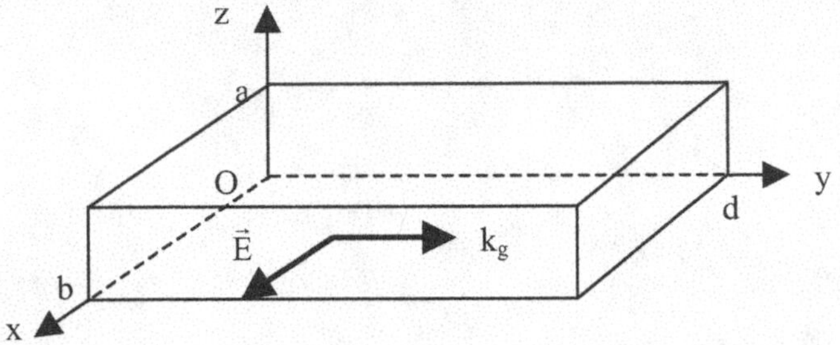

Figure 7.1. *Parallelepiped cavity.*

In the metallic cavity and following reflections of the wave on the metallic planes at y = 0 and y = d, a system of stationary waves is established. The resultant field thus varies with sin $k_g y$. This result was elaborated in Section 5.3.2 of this volume and obtained for a wave propagating along Oz. The limiting conditions imposed by the perpendicular planes at y = 0 and y = d in the y direction result in the formation of a node in the electric field at these planes. This can be represented by the equation $\left[\sin k_g y\right]_{y=d} = 0$, so that it is also possible to state:

$$k_g d = p\,\pi. \qquad (2)$$

The resonance modes, which are possible in the cavity, are termed the proper modes and denoted TE_{mnp}.

By substituting Eq. (2) into Eq. (1), we obtain the proper angular frequencies for the cavity, which are such that:

$$\omega^2_{m,n,p}\varepsilon_0\mu_0 = \frac{m^2\pi^2}{a^2} + \frac{n^2\pi^2}{b^2} + \frac{p^2\pi^2}{d^2}. \qquad (3)$$

So, the number of stationary waves is infinitely high, but the corresponding frequencies form a discontinuous group.

Comment. Equation (3) is a simple generalization in three dimensions of the result obtained for one dimension in Section 6.2.5 of Volume 1. When the cavity is excited by a wave, which has a frequency that verifies Eq. (3), the oscillation amplitudes become increasingly large, up to a maximum, just as in an oscillating circuit with a resonance frequency. The calculation thus shows how the overtension coefficient (Q) for the cavity at resonance is such that:

$$Q \approx \sqrt{\frac{a^2 + b^2}{\delta^2}},$$

where $\delta = \sqrt{2/\omega\mu\sigma}$ and is the depth to which the wave penetrates the metal.

7.3. Fabry–Perot-type optical cavities

7.3.1. Generalities and the Fabry–Perot resonator

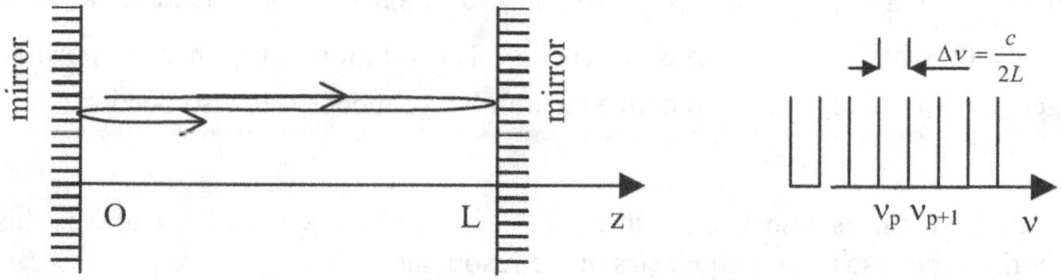

Figure 7. 2. *(a) A Fabry–Perot resonator (with flat mirrors) and (b) a system of resonance frequencies.*

Beyond those cavities with a classic geometry, there are other forms, such as those shown in Figure 7.1. Optical cavities have been widely studied, as they are one of the principal elements in lasers. In effect, a laser is essentially made up of three elements:

- an active medium capable of amplifying an incident signal by stimulated emission;

- a pumping system which permits a population inversion in the active medium; and

- an optical resonator into which is introduced the active medium. In effect it increases the amplification and sets the frequency and spatial distribution of the laser beam.

The most widely used resonator is a one-dimensional resonator termed a Fabry–Perot standard. It is made up of two plane parallel mirrors, which are highly reflective and separated by a distance L, as shown in Figure 7.2a.

7.3.2. Form of stationary wave system: resonance modes

The stationary wave system varies with sin kz. The condition for cancellation at z = 0 and z = L means that we also have k L = p π, where k is limited to values defined by:

$$k_p = \frac{p\pi}{L}. \qquad (4)$$

The negative values of p leave the (amplitude) form of the stationary waves invariant, as $\sin k_{-p}z = -\sin k_p z$. When $p = 0$, then the corresponding wave is $\sin k_0 z = 0$, which does not transport energy. The solutions for p thus are positive integers, i.e., p = 1, 2, 3, 4... and define the number of modes (see also Section 6.2.5, Volume 1).

The frequencies associated with Eq. (4) are such that ν = ck/2π, and it is this that defines the resonance frequencies of the resonator:

$$\nu_p = \frac{ck_p}{2\pi} = p\frac{c}{2L}, \text{ where } p = 1,2... \qquad (5)$$

Two successive frequencies thus are separated by a constant separation, as indicated in Figure 7.2b, and defined by:

$$\Delta\nu = \frac{c}{2L}. \qquad (6)$$

The resonance wavelengths are of the form:

$$\lambda_p = \frac{c}{\nu_p} = \frac{2L}{p},$$

and at resonance, the length of the resonator is such that

$$L = p\frac{\lambda_p}{2}, \qquad (7)$$

and is equal to an integer multiple of half-wavelengths.

7.3.3. An alternative point of view and the Fabry–Perot interferometer

The resonance modes also can be determined by using an interferometer (with a laser), which passes the light through a multiple system of mirrors, as shown in Figure 7.3a. Each mode results from a wave, which has traveled back and forth across the length (L) twice (2L). The corresponding dephasing thus is in part given by $\varphi = k2L$, the other part being associated with the reflections at the two mirrors which each give rise to a dephasing of π. We can deduce from this that:

$$\varphi = k2L = p2\pi, \text{ where p is an integer.} \qquad (8)$$

Once again we find $kL = p\pi$, an equation resembling Eq. (4).

Equation (8) can be thought of as a condition for a constructive counterreaction, where the waves are in phase (system exit and entering waves). The resultant of the constructive waves is given by $V = V_0 + V_1 + V_2 + ...$

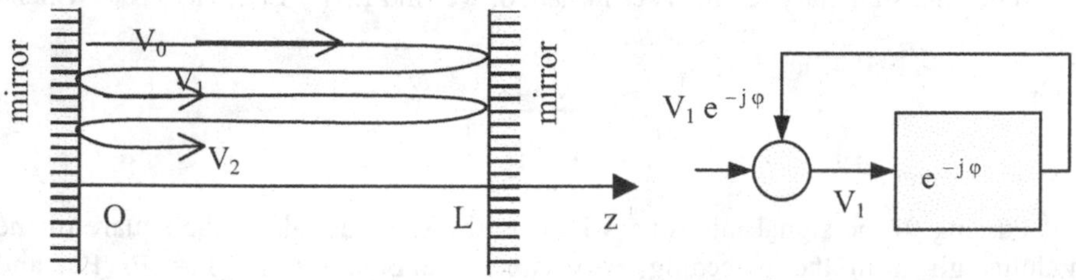

Figure 7.3. *(a) Back and forth pathway of waves in between mirrors in a resonator and (b) a diagram of the counterreaction optical system.*

7.4. The Airy laser formula

It is possible to determine the structure of a wave that leaves a laser quite rapidly. First we consider the simple example of an interferometer with an optical wavelength L, which is empty and externally lit by a monochromatic field given by $E_{ex} = E_{oex} e^{i\omega t}$.

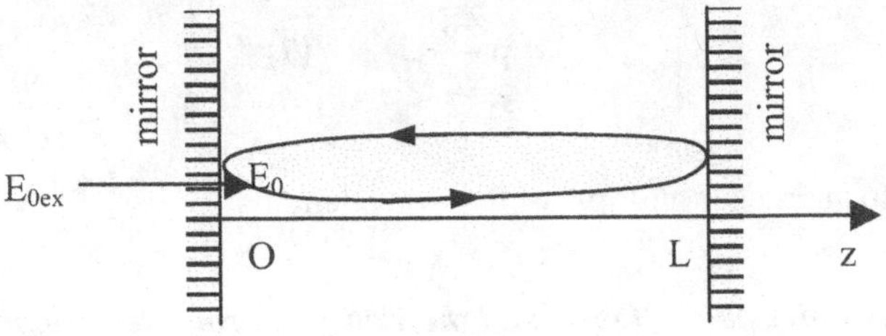

Figure 7.4. *Field modulus of following a single cycle.*

In order to find the field of the general form $E = E_0 e^{i\omega t}$ in a cavity, the preceding back and forth method (Figure 7.4) is used. If $E_0(t)$ represents the modulus of the field at an instant t for a point M in the cavity, then after one return passage the field is given by:

$$E_0(t+\Delta t) = E_0(t)\, r_1 r_2\, e^{ik2L} + \sqrt{T}\, E_{0ex} = E_0(t)\, r_1 r_2\, e^{i\omega 2L/c} + \sqrt{T}\, E_{0ex},$$

where r_1 and r_2 are the reflection coefficients for the mirrors, $\sqrt{T}$ is the transmission for the field given by E_{ex}, and $\varphi = k2L = \omega 2L/c$ is the dephasing in E following its return journey.

On making $\Delta t = 2L/c$, we can suppose that the source term for E_0 is $E_{0ex}\sqrt{T}$.

When the stationary regime is established, we find $E(t) = E(t + \Delta t)$, from which

$$E = \frac{\sqrt{T} E_{ex}}{1 - r_1 r_2 e^{i\varphi}}.$$

Passing to the signal intensity, which is the same as taking the square of the modulus given in the preceding expression, we obtain (with $I = E_0 E^*_0$ and $I_{ex} = E_{0ex} E^*_{0ex}$) the Airy equation that is characteristic of interferometers:

$$I = \frac{T I_{ex}}{[1 - r_1 r_2]^2 + 4 r_1 r_2 \sin^2(\varphi/2)}. \qquad (9)$$

According to Eq. (8), $\varphi = k2L = \frac{4\pi\nu L}{c}$, and from Eq. (6) $\Delta\nu = \frac{c}{2L}$, so we have:

$$\frac{\varphi}{2} = \pi\frac{\nu}{\Delta\nu}.$$

This makes it possible to write Eq. (9) in the form:

$$I = \frac{TI_{ex}}{\left[1 - r_1 r_2\right]^2 + 4 r_1 r_2 \sin^2\left(\pi\frac{\nu}{\Delta\nu}\right)}. \quad (9')$$

By making

$$I_{max} = \frac{TI_{ex}}{\left[1 - r_1 r_2\right]^2} \text{ and } F = \frac{\pi\left(r_1 r_2\right)^{1/2}}{1 - r_1 r_2} \text{ (smoothness of the resonator),}$$

Eq. (9') can be written in the "spectral response" form of the resonator:

$$I = \frac{I_{max}}{1 + \left(\frac{2F}{\pi}\right)^2 \sin^2\left(\pi\frac{\nu}{\Delta\nu}\right)}. \quad (10)$$

The maximum intensity ($I = I_{max}$) thus is obtained for the resonance frequency: $\nu = p\,\Delta\nu$ (where p = 1,2....), as indicated in Figure 7.5.

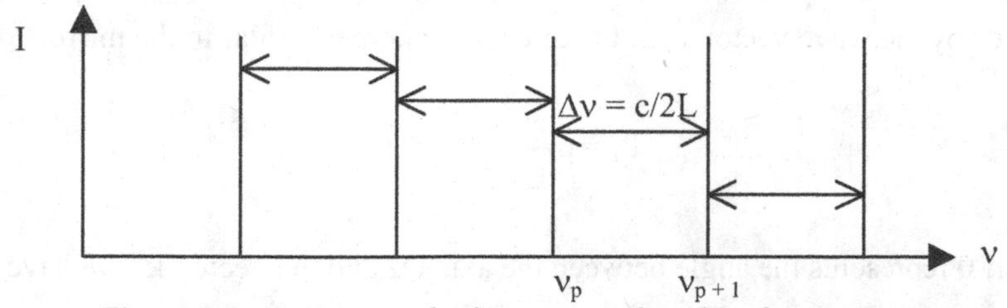

Figure 7.5. *Emission at the frequency ν_p for a loss-free resonator.*

For resonators that have losses, then we can use $r_1r_2 = e^{-L}$ where L represents the losses. When F >> 1, it is found that the resonance peaks exhibit a half-height width given by $\delta\nu = \Delta\nu/F$. The generated intensity is not totally zero outside of the resonance peaks, but there is a rapid decrease close to the maximum. It is comparable to the change in intensity with respect to the distance between two mirrors.

The emission spectrum of a laser therefore takes on the form given in Figure 7.6. It is constituted of a series of fine and equidistant peaks situated at the Fabry–Perot transmission maximal. They have as an envelope the emission curve of the active medium.

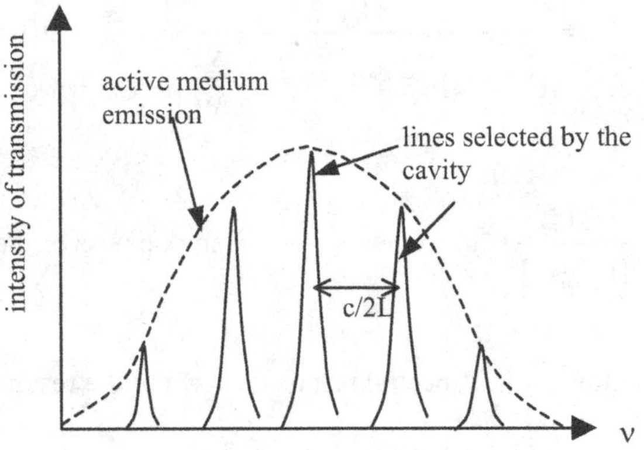

Figure 7.6. *Laser emission spectrum.*

7.5. Modification of spontaneous emission in a planar cavity and the angular diagram

When a source is placed in a planar cavity formed from two parallel mirrors (Fabry–Perot cavity), the emission modes are controlled by the limiting factors of the mirrors. These are the Fabry–Perot modes. They accord to the quantification defined by the wave vector $\vec{k}$ in the direction z perpendicular to the mirrors deduced from Eq. (4):

$$k_z = p\frac{\pi}{L}. \qquad (4')$$

If θ represents the angle between the axis Oz and the vector $\vec{k}$, we have $k_z = k\cos\theta$, where $k = \dfrac{2\pi}{\lambda}$.

The upshot is that condition (4') can be written as

$$k \cos \theta = p\frac{\pi}{L}, \qquad (11)$$

which shows that the field resonance occurs along directions that are selected at the angle θ_p which is such that

$$\theta_p = \arccos\left(\frac{p\pi}{kL}\right). \qquad (12)$$

Therefore, emission cones with an angle at their peak of θ_p can be found. The cones are only ideal when the mirrors themselves have perfect surfaces and reflection coefficients of 1. When this is not the case, the cones become more like cornets revolved around θ_p.

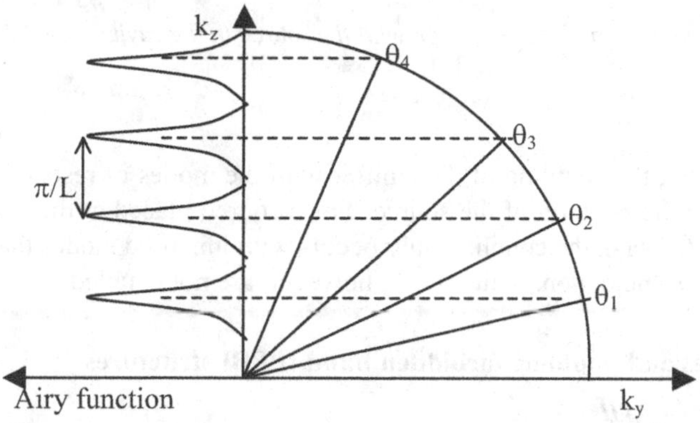

Figure 7.7. *Representation of the quantification of k_z. Each Airy function maximum corresponds to the direction of an emission from the cavity. The maximal of the emissions are separated by π/L.*

Within the k space (Figure 7.7), the extreme point of $\vec{k}$ describes a circle of radius ω/c, and the emission of the cavity evolves as a function of k_z, just as the Airy function studied in the preceding paragraph.

Figure 7.8 represents a Fabry–Perot cavity with the source S along with the system of stationary waves for four proper modes ($p_c = 4$) for a cavity of length L and such that [Eq. (7)]:

$$L \approx p_c\frac{\lambda}{2}. \qquad (13)$$

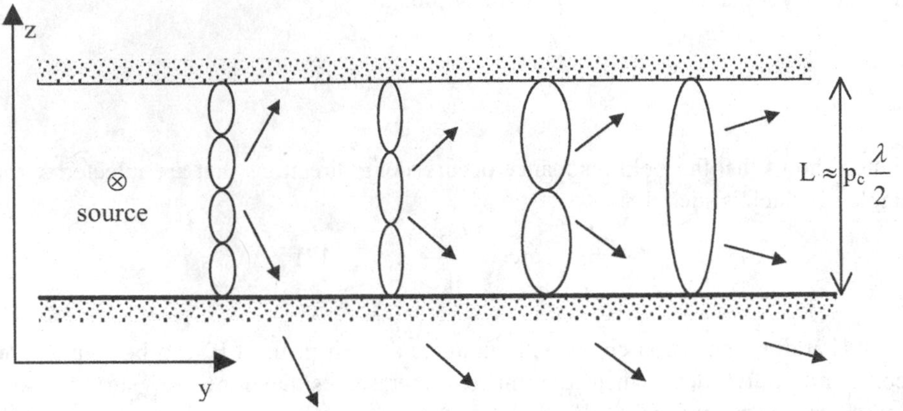

Figure 7.8. *Representation of four modes and the emitted rays for a planar cavity of width* $L \approx p_c \frac{\lambda}{2}$ *(and* $p_c = 4$*), with the source in the cavity.*

Comment. In effect, the coupling of the emitter with the modes of resonance of the cavity depends on the position of the source. For a source placed at the center of a symmetric cavity ($r_1 = r_2$), the coupling only occurs with impaired modes that exhibit a maximum in the same region. Pair modes, however, are not coupled.

7.6. Microcavities and photonic forbidden band (PFB) structures

7.6.1. Exampled scale effects

From the preceding results, it is possible to consider the effects of reducing the size of the cavity. Taking Eq. (13) into account, if L decreases, then p_c becomes small so that the emitter is now coupled to a reduced number of resonant modes. At the limiting case, where $p_c = 1$, the emission is coupled to a single mode, and the initially isotropic source becomes a directional source. Indeed, the finer the resonance peak, the more selective in direction the source becomes.

We also can state that for a cavity with a given length (L), the direction of the emission is dependent on the frequency. In effect, for a given mode, the product of k cosθ should remain constant, as indicated by Eq. (11). The result is that if the system is blue-shifted (λ decreases and k increases), then cosθ must decrease, which means that θ increases. This means that the emission spreads out from the normal.

7.6.2. PFB structures

7.6.2.1. Origin of PFBs

The propagation of a wave in a vacuum with the form (see also Section 4.1, Chapter 4, this volume) $s = A(x,y,z)e^{i\omega t} = A(x,y,z)e^{i2\pi\nu t}$ has a propagation equation that can be written as:

$$\Delta A + \frac{\omega^2}{c^2} A = 0 . \qquad (14)$$

This takes on the form

$$\Delta\psi + \frac{2m}{\hbar^2}(E - V)\psi = 0 , \qquad (15)$$

for a Broglie wave associated with a wavelength given by the equation:

$$\lambda = \frac{h}{mv} = \frac{h}{\sqrt{2m(E - V)}} . \qquad (16)$$

This relation is for a crystal where the potential V is periodic and there are solutions for the energy that arise as permitted (valence and conduction) and forbidden (the “gap” for semiconductors) bands.

The wave equation written in Eq. (14) for a vacuum needs to be rewritten for a medium with a given permittivity (ε_r) in accordance with Eqs. (7’) or (7”) from Section 7.1.3, Volume 1.

$$\Delta A + \frac{\omega^2}{c^2} \varepsilon_r A = 0 . \qquad (17)$$

Intuitively, it seems logical that if ε_r exhibits periodic values as does V in Eq. (15), then the photons must exhibit forbidden energies. They can be rigorously characterized, for example, by replacing A by a vector such as that of the electric field. As $E = \hbar\omega$, this forbidden property of the energy is represented here by the presence of forbidden domains of angular frequencies, or rather, frequencies.

Materials that exhibit a periodicity defined by ε_r and from which are produced photons exhibiting forbidden frequencies are called photonic forbidden band (PFB) materials. The term “photonic crystal” also is used for systems made up of dielectrics or periodic metals.

7.6.2.2. *Basic structures*

In general terms and in one dimension, the basic structure is made up of a periodic column of thin dielectric layers, generally called Bragg mirrors, and shown in Figure 7.9a.

In two dimensions, an example of photonic crystals is that of a square latticework of pores or branches as shown in Figure 7.9b.

In three dimensions, an example is that of the structure which looks rather like a pile of criss-cross sticks shown in Figure 7.9c. In practical terms, this sort of structure can be obtained by drilling a cube in two, orthogonal directions.

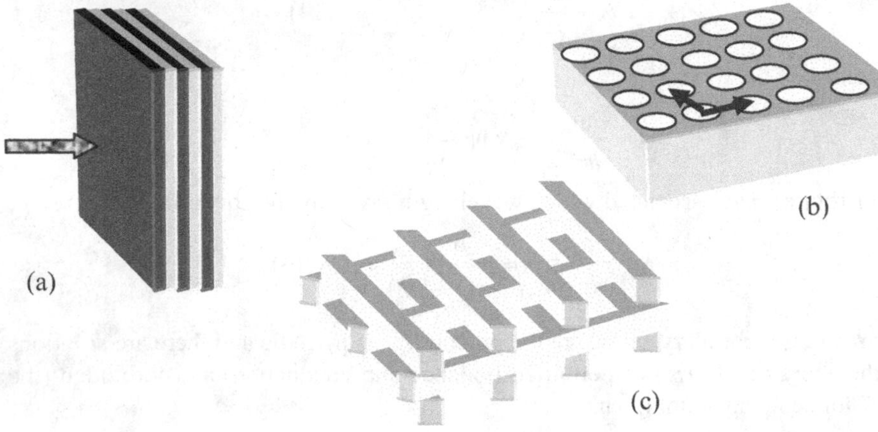

Figure 7.9. *Some basic structures of photonic crystals: (a) one-dimensional Bragg mirror, (b) two-dimensional square pore lattice, and (c) three-dimensional "pile of sticks" also termed "yablonovite".*

In the optical domain, the one-dimensional photonic crystals can be considered Bragg mirrors. They reflect the waves due to the interfaces between different layers in the material. Each layer has an optical thickness equivalent to a quarter of the wavelength of a given central wave (λ_0) which is termed the Bragg wavelength (Figure 7.10a).

The efficiency of these mirrors always depends on the incidence angle (θ) with respect to the dielectric layers. The existence of an omnidirectional forbidden band is impossible in these one-dimensional structures, as the reflectivity decreases on going further from the normal incidence (a classic study can be carried out on the processes involved in multiple interferences).

In addition, a wave also may be guided by a forbidden photonic band in a periodic medium, such as a photonic crystal fiber. Similar to refractive guides, the region, which effects the guiding, is called the gain and provides the forbidden photonic band (Figure 7.10b).

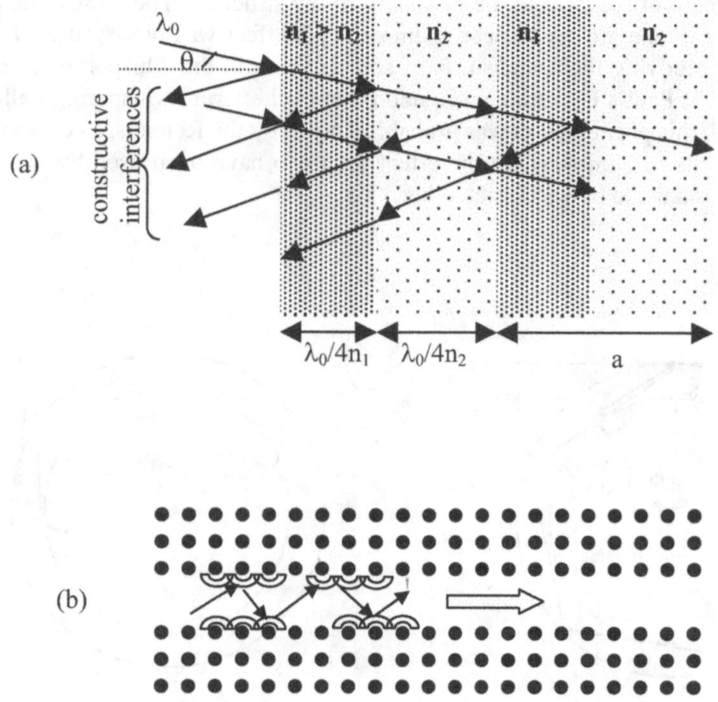

Figure 7.10. *(a) Structure of a Bragg mirror ($\theta \approx 0$) and (b) wave-guide through forbidden photonic bands, where the interferences produced by the periodic motifs generate reflections.*

7.7. Microcavities using whispering gallery modes

7.7.1. Generalities

To end this chapter, we look at the symmetrical cavities developed at the end of the 20th century. It also should be noted that there are asymmetric cavities that were developed to favor signal extraction.

With microdisk or microsphere cavities made using only slightly absorbing dielectrics, the gallery modes correspond to a guided propagation by total internal reflection at the peripheral of the cavity. Thus, the energy is spread only at the edges of the resonator.

The glancing propagation of the ray around the walls of the cavity results in a strong confinement of the field to the periphery of the structure. The more glancing the incident ray is, the greater the confinement. This effect was observed by Lord Rayleigh while studying the propagation of sound waves in the whispering gallery of the dome of St Paul's Cathedral in London, hence the term "whispering gallery modes". Indeed, this phenomenon was probably known by the Romans, as exampled by the arena walls in Nimes where the effect seems to have been used by various "actors" to communicate between each other.

7.7.2. Principle

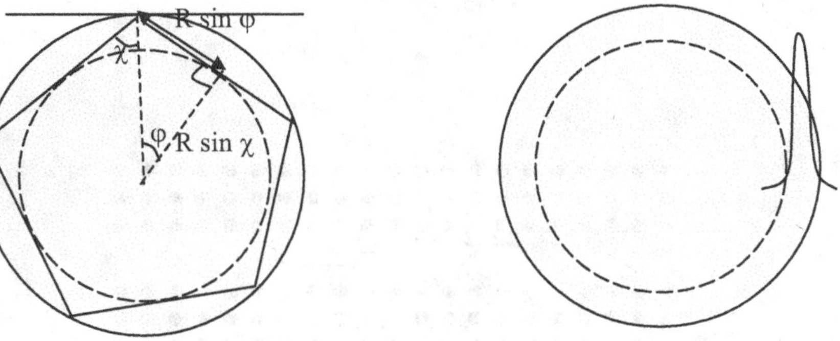

Figure 7.11. *(a) Propagation of whispering gallery modes at the periphery and with respect to the azimuth direction and (b) schematization of the localization of energy at the edges of the cavities.*

This study is limited to that of cylindrical cavities of a known radius (R) and termed 'microdisks'. Typically, the dimension of the cavities is such that R is equal to 10 to 200 μm, a large dimension with respect to the optical wavelengths. As a first approximation, the optical function of the system can be made using the principles of geometrical optics. These cavities are made from a material with an index (n) higher than that of the surrounding medium, for example, air which is the simplest and has an index close to unity.

There are two groups of rays, depending on the angle of incidence (χ) shown in Figure 7.11. They are:

- when $\chi < \chi_l$ where χ_l is the limiting angle (here defined as n sin $\chi_l = 1$) beyond which total reflection is obtained, then following several successive reflections the rays are completely refracted and leave the cavity. This corresponds to the radiation mode; and
- when $\chi > \chi_l$, then the condition of total internal reflection applies, and by symmetry the angle is conserved throughout the successive glancing (for which $\chi \approx \pi/2$) reflections around the cavity. This is in effect the system's guided mode.

The successive rays thus describe a polygon (as each of the reflections is in the same plane as the initial incidence, according to optical laws) of which the internal circle that is tangent to the sides of the polygon describe a radius $\rho = R \sin \chi$.

For rays of the second group, which describe several turns inside the cavity, it is those which have a phase varying on each turn by an integer of 2π which will give rise to constructive interferences characteristic of whispering gallery modes.

7.7.3. Basic equations for whispering gallery modes

7.7.3.1. Resonance condition leading to a determination of the gallery modes within an approximation of the optical geometry

If the polygon described by the rays has a number (C) of length 2R sin φ (Figure 7.11a), the angle at the center through which the sides are seen is given by $2\varphi = 2\pi/C$, hence $\varphi = \pi/C$.

The condition that a ray returns to its original phase after a turn inside the cavity is equal to a whole number of wavelengths, so that

$$n\, C\, 2R \sin \varphi = m\, \lambda\,, \qquad (18)$$

where m is a whole number at least equal to C so that each side of the polygon has a length (2R sin φ) at least equal to λ/n, the limit of the use of optical geometry.

In terms of geometry, if C increases, then m also increases, and the radius $\rho = R \sin \chi$ does likewise (as χ tends toward $\pi/2$ and ρ toward R). The result is that the confinement also increases.

With an approximation based on circular trajectories, for which φ tends toward zero where $\sin \varphi \approx \varphi$ while C becomes infinitely large, the resonance condition can be written simply as:

$$n\, 2\, \pi\, R = m\, \lambda\,. \qquad (19)$$

This is a resonance condition equivalent to that of a Fabry–Perot cavity where a return journey (2L) taken by the ray is now replaced by an effective length for the gallery mode, as in $2\pi R$.

The approximation for circular trajectories is justified by the fact that C is generally very large, so that for example with $R = 30\ \mu m$ and $\lambda = 610$ nm, Eq. (18) gives $C = 433$.

Comment. Just as in a Fabry–Perot cavity, the whispering gallery modes depend on the direction of the polarization of the light because the Fresnel reflection coefficients also depend on this. Thus, the resonance condition (of in-phase return) is sensitive to any dephasing introduced at each reflection. A generality that arises from this is that there are two resonance conditions: one is termed TE designating a field (E) normal to the plane of incidence (these waves are also termed "s" type for "senkrecht" which translates as "perpendicular" in German); while the other is for "TM" type waves where E is parallel to the plane of incidence (these waves also are termed polarized or "p" type waves).

7.7.3.2. Free spectral interval

If D denotes the diameter of the microdisk, then Eq. (19) yields $n\pi D = m\lambda_m$, where λ_m is the wavelength associated with the m mode.

Therefore, $\frac{1}{\lambda} = \frac{m}{\pi n D}$, which on differentiating gives:

$$d\left(\frac{1}{\lambda}\right) = -\frac{d\lambda}{\lambda^2} \text{ and } d\left(\frac{m}{\pi n D}\right) = \frac{dm}{\pi n D}.$$

By going into absolute values, $\frac{\Delta\lambda}{\lambda^2} = \frac{\Delta m}{\pi n D}$, and hence:

$$\Delta\lambda = \frac{\lambda^2}{\pi n D}\Delta m.$$

Thus for two neighboring modes λ_m and λ_{m+1} there is a separation equal to a spectral distance (also called the free spectral interval, or FSI) which is of the form:

$$\Delta\lambda = \frac{\lambda^2}{\pi n D}.$$

In numerical terms, when $D = 68\ \mu m$, $n = 1.49$, and $\lambda = 617$ nm, we find $\Delta\lambda \approx 1.2$ nm.

7.7.4. Photon lifetimes and extraction of the radiation

7.7.4.1. Symmetric cavities

Even in a perfect, symmetric cavity, with a homogeneous medium and a perfectly transparent isotropy, the lifetime of photons trapped in a gallery mode by internal total reflection is in fact finite. This is due to the presence of an evanescent wave which exponentially decreases in intensity on going from the interface of the cavity. Additionally, because of a curve in the wall of the cavity, the evanescent wave once again becomes propagating beyond a certain distance. An optical tunneling effect thus is generated with a propagation tangential to the cavity wall. The losses resulting from this are called diffraction losses.

A second type of loss, also called diffusion losses, results either from localized variations in the refraction index, or surface asperities at the dielectric surface.

Therefore radiation trapped in microdisk systems essentially is extracted through localized defaults around the cavity, including variations in the verticality of the lateral walls, and—most of all—roughness of the lateral walls.

7.7.4.2. Asymmetric cavities

In order to extract the light from a cavity, other than by diffusion through increasing wall roughness, it is possible to use asymmetric cavities so that the light is extracted at the point of the dissymmetry.

Classically, cavities with spiral forms or even deformed circles are used in which the extraction of the radiation at the level of the dissymmetry is all the more intense when the excitation of the cavity is adapted to the dissymmetry ("ring pumping"), as detailed in the following exercise.

7.8. Problem

Simplified study of an asymmetric cavity

This questions turns around the problem of an asymmetric cavity for which the circular section of a symmetrical microdisk is replaced with a section that has a spiral form.

1. The section of the cavity is described by a system of cylindrical coordinates (r, φ, z) where $r(\varphi) = r_0(1+\varepsilon \dfrac{\varphi}{2\pi})$ is for a given side z.

Determine the value D_A of the large axis, as well as the dimension e of the flat part. Use $\varepsilon = 0.1$ and $r_0 = 90$ μm. Calculate D_A and e.

2. The cavity is excited from above with a uniformly pumped beam.

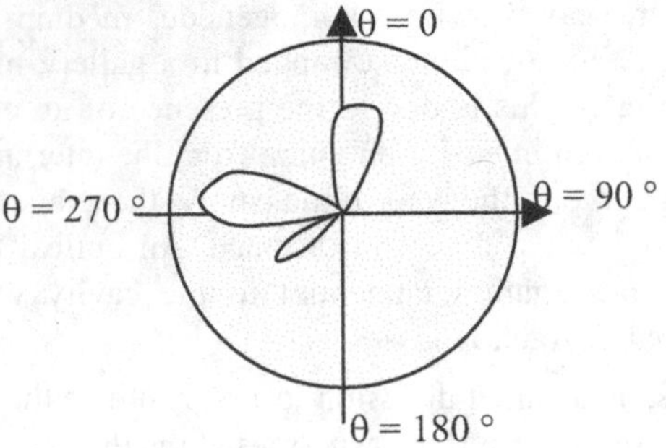

The spatial radiation diagram (radiation intensity) of the cavity has a form that is shown above. There are two principal lobes of emission that have been shown by experiment to be centered about the directions θ ≈ 5 ° and 275 °. A smaller lobe also appears at θ ≈ 240 °. Give a qualitative explanation for this distribution.

3. Which excitation geometry might reinforce the emission lobe at θ ≈ 5 °; reduce that at θ ≈ 275 °?

Answer

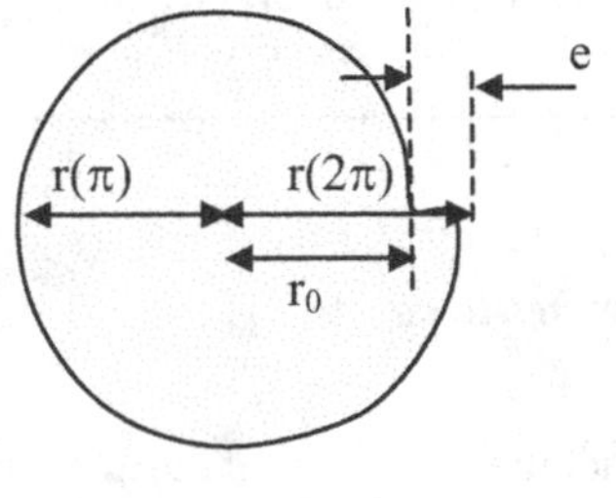

1. We have (see figure):

$$D_A = r(\pi) + r(2\pi) = r_0(1+\varepsilon\,\frac{\pi}{2\pi}) + r_0(1+\varepsilon\,\frac{2\pi}{2\pi}) = r_0(2+\frac{3}{2}\varepsilon)$$

$$e = r(2\pi) - r(0) = r_0\,(1+\varepsilon) - r_0 = \varepsilon\, r_0\,.$$

In numerical terms, we obtain $D_A \approx 195$ μm and e ≈ 9 μm.

2. The appearance of an intensity peak in a certain direction indicates that the cavity acts more strongly in that direction. It is possible to imagine that rays propagating in an anticlockwise direction (termed "propagative" and thus follow the trigonometric sense) will meet the flat surface of the spiral, and on doing so will have an incidence close to the normal and thus below the limiting angle χ_ℓ. The result according to Fresnel's law, and shown in Figure a, is that a part of this radiation will be refracted through a small angle in the neighborhood of that observed at $\theta \approx 5$ °. It equally can be supposed that a small contribution in the direction $\theta \approx 5$ ° could arise from losses due to optical tunneling effects due to rays generating an evanescent wave tangential to the wall of the cavity just before the flat part (as a drawn by the ray).

In addition, a part of the radiation incident to the flat surface is reflected. This gives rise to counterpropagative rays. When these come onto the wall of the cavity at an angle below the limiting angle, they can contribute to the small lobe at $\theta \approx 240°$ indicated in Figure b.

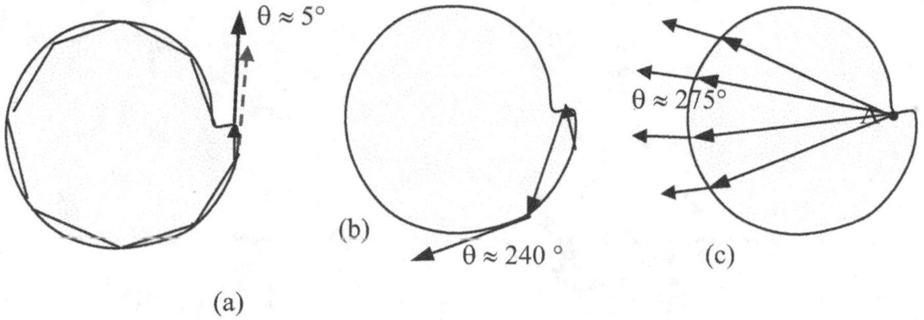

Now consider rays that are diffracted by the geometrical default, which makes up the angular corner of the cavity and is situated at the intersection of the flat surface and the beginning of the spiral (point A in Figure c).

The rays are diffracted in all directions. Those rays that hit the walls at angles greater than the limiting angle (χ_ℓ) are trapped in the cavity and then can go on to contribute to other emission lobes (depending on their angle and ulterior incidences). In contrast, rays that are refracted such that $\chi < \chi_\ell$ leave the cavity by refraction. As they traverse the central zone, which is illuminated by the excitation beam, they can be amplified by this pumping beam to give a sizable lobe in the direction $\theta \approx 275°$.

3.

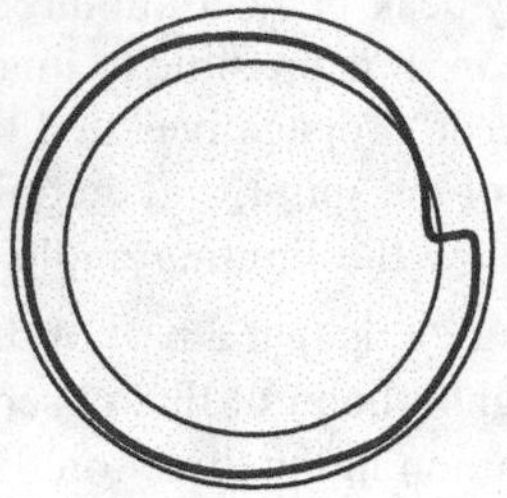

If an excitation is performed with a ring shaped beam, such as that represented in the above figure, then only the gallery modes confined to the periphery of the cavity are excited. The upshot is that the lobe at 5° is reinforced. The center of the cavity is no longer excited and the radiation at $\theta \approx 275°$ is well reduced.

Chapter 8

Particles in Electromagnetic Fields: Ionic and Electronic Optics

8.1. Mechanics of particles in an electromagnetic field

8.1.1. Introduction to mechanics: an aide-memoire

8.1.1.1. Forces acting on charged particles

In general terms, gravitational forces ($\vec{P} = m\vec{g}$) are assumed to be negligible with respect to the forces generated by the fields $\vec{E}$ and $\vec{B}$. Respectively, these forces produce a coulombic force as in $q\vec{E}$ and a Laplace force given by $q\vec{v} \times \vec{B}$. Under the action of these forces, a particle of charge q is subject to an acceleration $\vec{\gamma} = \dfrac{\overrightarrow{dv}}{dt}$ which is such that:

$$\vec{F} = m\frac{\overrightarrow{dv}}{dt} = q\vec{E} + q\vec{v} \times \vec{B} = q\left(\vec{E} + \vec{v} \times \vec{B}\right). \qquad (1)$$

As detailed in the preceding chapters, the magnetic force is often negligible with respect to the coulombic force.

8.1.1.2. Energy and the quantities of movement

For the elementary displacement denoted $\overrightarrow{dl}$ and of the form $\overrightarrow{dl} = \vec{v}\,dt$, the elementary work dW of the electromagnetic force $\vec{F}$ given by Eq. (1), is found in the form:

$$dW = \vec{F}.\vec{dl} = m\frac{\overrightarrow{dv}}{dt}\vec{v}\ dt = q\vec{E}.\vec{v}dt\ , \qquad (2)$$

so that :

$$dW = m\vec{v}\ \overrightarrow{dv} = q\vec{E}.\vec{v}dt\ .$$

From this it can be said that the work done by the magnetic force is zero, and it is only the electric force that intervenes.

Taking the limits on the speed to be:

- initial speed at O is given by $v = v_0$; and
- a speed v at point M, then:

the integration of Eq. (2) along the trajectory OM, as in

$$W = \int dW = \int_O^M m\ v\ dv\ = q\int_O^M \vec{E}.\vec{dl} = q\int_O^M -dV\ ,$$

yields (denoting V_0 and V_M the electrical potentials at O and M, respectively):

$$W = \frac{1}{2}m(v^2 - v_0^2) = q(V_0 - V_M)\,. \qquad (3)$$

If at O we have $v_0 = 0$ and with $(V_0 - V_M) = V$, the preceding equation can be simply written as (with $V > 0$ if $q > 0$, or $V < 0$ if $q = -e < 0$)

$$W = \frac{1}{2}mv^2 = qV\,. \qquad (4)$$

Considering the problem in numerical terms, we find that for an electron, where $q = -e$), we have:

$$v = \sqrt{\frac{2e}{m}|V|}\,,$$

to give $v = 5.93 \times 10^5\sqrt{|V|}$.

This means that when $|V| = 10$ kV $= 10^4$ V, we have $v = 5.93 \times 10^7$ m s^{-1}, which is a nonnegligible with respect to the speed of light. This explains the necessity of using relativistic masses for electron particles.

The speed of ions is approximately 40 to 600 times less than that of electrons due to the variation in $1/\sqrt{m}$. The correction to account for relativistic effects is only necessary for tensions above 10^6 V. In addition, this simplifies the work here, which will be performed only using nonrelativistic mechanics.

8.1.2. Movement of a charged particle in an electric or magnetic field

8.1.2.1. Deviation of a particle by an electric field

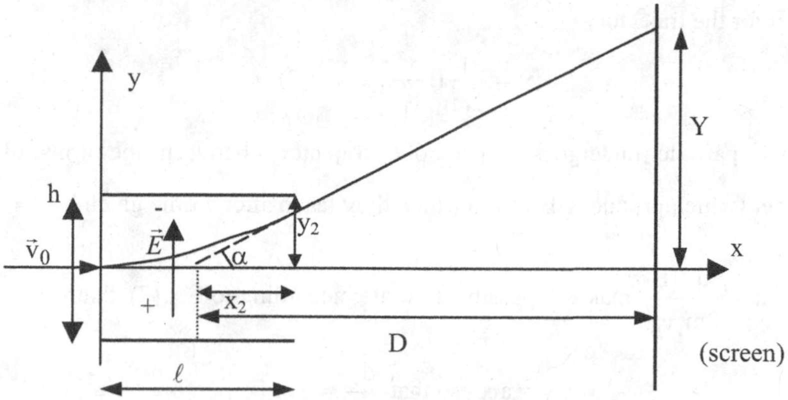

Figure 8.1. *Deviation of a particle (q) by a E.*

There is an incident, monokinetic flux of particles (each of charge q+) with uniform speed ($\vec{v}_0$) with respect to the x axis. From O onward, the particles are subject to an electric field ($\vec{E}$), which is directed along Oy and acts over a length ℓ in the x axis, as shown in Figure 8.1. With respect to this, we have $F_y = qE$, and the equations for movement (which rest in the plane of the above figure) are:

- along Ox, $m\dfrac{d^2x}{dt^2} = 0$;
- along Oy, $m\dfrac{d^2y}{dt^2} = qE$.

With time and space having origins given by the particle's position at O (i.e., $t_0 = 0$, $x_0 = 0$, $y_0 = 0$), we obtain by integration:

- along Ox we have $\dfrac{dx}{dt} = v_0$, so that $x = v_0 t$; (5)
- and with respect to Oy, we have:

$$\frac{dy}{dt} = \frac{q}{m}Et, \text{ so that } y = \frac{q}{2m}Et^2. \qquad (6)$$

From Eq. (5) we can pull out that $t = \frac{x}{v_0}$, which when placed into Eq. (6) gives the equation for the trajectory:

$$y = \frac{q}{2m} E \frac{x^2}{v_0^2} . \qquad (7)$$

Each particle undergoes a parabolic trajectory between the arms of the condenser (which produces $\vec{E}$) from which they leave after a time given by $t = \frac{\ell}{v_0}$.

Setting $a = \frac{q}{2m} \frac{E}{v_0^2}$ makes it possible to write, according to Eq. (7), that:

$$y = a\, x^2, \text{ so that } \frac{dy}{dx} = 2ax .$$

In addition, the angle α at which the particles emerge is defined by (see Figure 8.1),

$$\tan\alpha = \left(\frac{dy}{dx}\right)_{x=\ell} = 2a\ell . \qquad (8)$$

As we also have

$$\tan\alpha = \frac{y_2}{x_2} = \frac{a\ell^2}{x_2}, \qquad (9)$$

then given that equations (8) and (9) are equal, then:

$$x_2 = \frac{\ell}{2} . \qquad (10)$$

This indicates therefore that the exit tangent goes through the center of the condenser.

With the notations given in Figure 8.1, we also have $\tan\alpha = \frac{Y}{D}$, which with Eq. (8) can give:

$$Y = 2a\ell\, D = \frac{q}{m} \frac{E}{v_0^2} \ell\, D .$$

If the uniform speed (v_0) is produced by an accelerating tension V_0, such that $\frac{1}{2} m v_0^2 = qV_0$, then we also have:

$$Y = \frac{E\ell D}{2V_0}.$$

By taking E in the form $E = \frac{V}{h}$, we finally have

$$Y = \frac{\ell D}{2h}\frac{V}{V_0}. \qquad (11)$$

The deflection (Y) is proportional to the potential (V) of the condenser, and the proportionality coefficient (f_s) is given by $f_s = \frac{\ell D}{2hV_0}$.

8.1.2.2. Deviation of a particle by a magnetic field

8.1.2.2.1. The trajectory

Once again we can use a similar layout to that in the previous figure where there is a beam of monokinetic particles that has a uniform speed ($\vec{v}_0$) along Ox. At the point O they penetrate a uniform magnetic induction ($\vec{B}$) normal to the plane of Figure 8.2. The force applied to the particles, given by $q\vec{v} \times \vec{B}$, is normal to $\vec{B}$; that is to say it is in the plane Oxy.

The resulting acceleration, collinear with the force therefore also is in the plane of the figure.

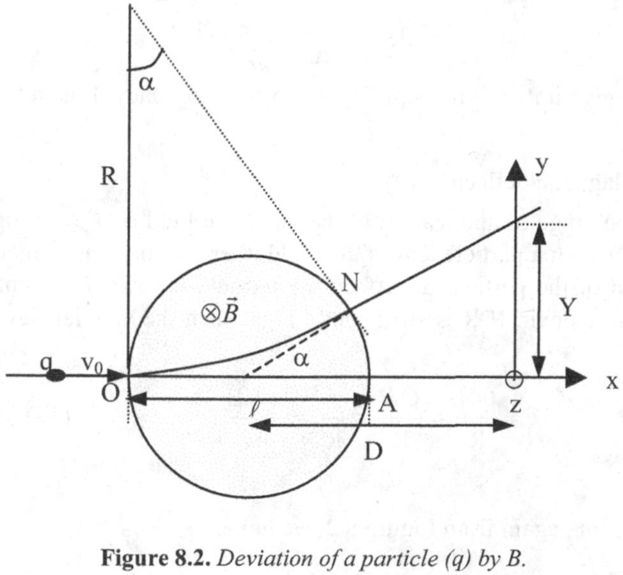

Figure 8.2. *Deviation of a particle (q) by B.*

In addition, this force is normal to the velocity ($\vec{v}$) of the particles and therefore does not have a component tangential to the trajectory. The tangential acceleration therefore is also zero. There is only a normal acceleration.

Using a Frenet triad, we find that:

$$\begin{cases} \gamma_t = \dfrac{dv}{dt} = 0 & \Rightarrow v = \text{constant as a modulus} = v_0 \\ \gamma_n = \dfrac{v^2}{R} \neq 0. & \end{cases}$$

With respect to the normal to the trajectory, we thus can write that $F_N = qv_0B = m\gamma_n$, from which

$$R = \frac{mv_0}{qB}, \qquad (12)$$

where R is the radius of the trajectory curve and is independent of the point along the trajectory, which therefore is a circle (i.e., R = constant). For a given frequency f_c (which gives the number of cycles per unit time) and in one second, a particle will cover a length $l = v_0$ while the length of one turn is $2\pi R$. From this we have

$$f_c = \frac{v_0}{2\pi R}.$$

Taking Eq. (12) into account, we can deduce that:

$$f_c = \frac{qB}{2\pi m}, \qquad (13)$$

where f_c, to give it its full name, is the cyclotron frequency. It is independent of v_0.

8.1.2.2.2. Magnetic deflection

Denoting the range of application of the magnetic field as ℓ (see Figure 8.2) and N the point at which a particle leaves this field, there is only one tangential component to the speed of the particle. In effect, the particle follows a tangent to the circular trajectory on its exit. If R is sufficiently high, then the angular deviation α is such that:

$$\alpha = \frac{ON}{R} \approx \frac{OA}{R} = \frac{\ell}{R} \overset{(12)}{=} \frac{q\ell B}{mv_0}. \qquad (14)$$

In addition, again from Figure 8.2, we have:

$$\alpha \approx \tan\alpha = \frac{Y}{D}. \qquad (15)$$

From Eq. (14) and (15), it is possible therefore to deduce that

$$Y = \frac{q}{m}\frac{\ell DB}{v_0}. \qquad (16)$$

If the uniform speed (v_0) here is again the result of an accelerating tension V_0, then it is such that $\frac{1}{2}mv_0^2 = qV_0$. On taking v_0 from this last equation, then from Eq. (16) we have:

$$Y = \sqrt{\frac{q}{m}}\frac{\ell DB}{\sqrt{2V_0}}, \qquad (17)$$

(from the equation for Y, we can see that we now have access to characterizations in terms of $q\,m^{-1}$).

8.1.2.3. Crossed $E \times B$ *fields: the magnetron effect and ionic pumping*

8.1.2.3.1. Basic setup for the analytical study

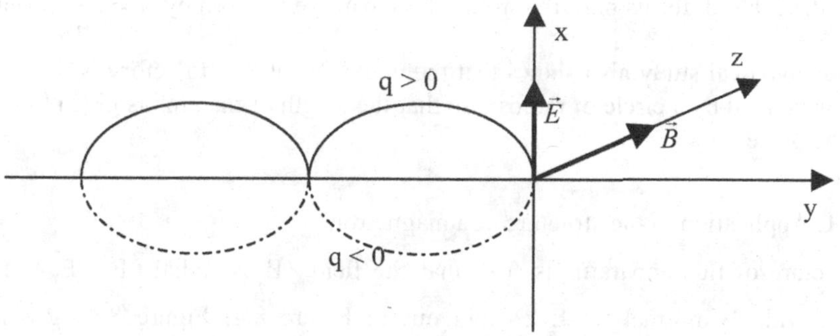

Figure 8.3. *Cycloid trajectory of a charge (q) subject to a field* $E \times B$.

The study detailed in problem 2 of this chapter shows that a charge q subject to a pair of crossed fields given by $\vec{E} \times \vec{B}$ (where $\vec{E}$, for example, is directed along Ox and $\vec{B}$ along Oz) has a resultant trajectory which is a cycloid in the plane Oxy. By making

$$\begin{cases} \omega = \dfrac{qB}{m} \\ a = \dfrac{qE}{m\omega^2} \end{cases}$$

the coordinates of a current point M (x,y,z) are (parametric equations):

$$\begin{cases} x = a(1 - \cos \omega t) \\ y = -a(\omega t - \sin \omega t) \\ z = 0. \end{cases}$$

It is worth noting that for a charge $q > 0$, we have $x > 0$, while for a negative charge we have $x < 0$, which gives the behavior shown in Figure 8.3.

In qualitative terms, the sense of the cycloid can be determined from the Laplace force $\vec{F} = q(\vec{E} + \vec{v} \times \vec{B})$ applied at O. The axis of the arch depends on the product of $\vec{E} \times \vec{B}$, so that the displacement along Ox is parallel to $\vec{E}$ if $q > 0$ and antiparallel to $\vec{E}$ if $q < 0$.

On the cycloid, for its part, the speed of the particle is given by $v = 2\dfrac{E}{B}\sin \omega t$.

The analytical study also shows that the equation for the trajectory is that of a cycloid generated by a circle of radius a so that the length of the arch is given by $L = 2\pi a$ in the plane Oxy.

8.1.2.3.1. Application to the structure of a magnetron

The structure of this apparatus is such that the field $\vec{E}$ is radial ($E = E_r$), and therefore with $\vec{B}$ normal to $\vec{E}$ as laid out in Figure 8.4. Figure 8.4a gives a segmental view.

If the field is applied simply along Ox, the trajectory of the particle (with charge $q > 0$, i.e., an ion) can be represented as in Figure 8.4b. In effect, the electric field displays a rotational symmetry as shown in Figure 8.4c, and the axis of the trajectory at each point (A) must be normal to E_r (and to B). The axis of the trajectory at each point is no longer parallel to Oy as indicated in Figure 8.4b, but deformed to a circle (shown as a dotted line in Figure 8.4d) and takes on the trajectory shown in Figure 8.4d.

The result is that the amplitude of the oscillations (related to the height 2a of the arch and in the form $2a = \frac{2mE}{qB^2}$) is sufficiently small so that the ion does not reach the cathode. The ions remain trapped within the structure. This happens when the B field is sufficiently large so that the parameter a is sufficiently small. This configuration is in fact often used with electrons—where the length of the trajectory is considerably increased. As an example, this also can be used to increase their likelihood of colliding with gas atoms in order to yield ions.

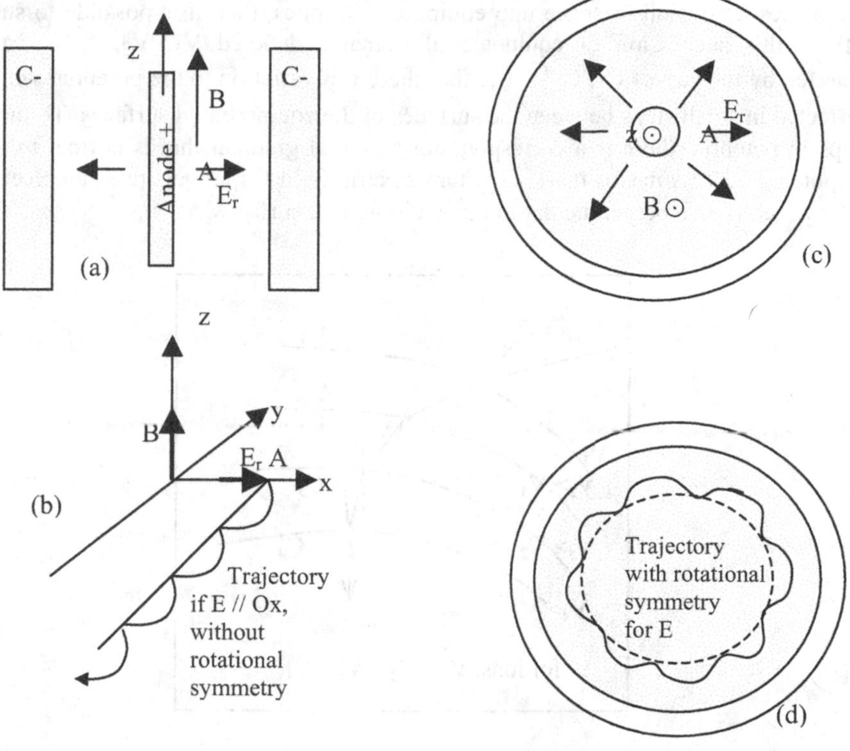

Figure 8.4. *Trajectories in an E × B field, when E is radial (see text for further details).*

Inversely, for sufficiently weak fields, 2a will be great enough to allow the collection of ions at the cathode. If the latter is capable of absorbing the ions, by chemisorptions, for example, then we obtain what is in effect an ion pump. This process is widely used for making ultrahigh vacuums (10^{-7} to 10^{-10} mm Hg) with pumps made out of titanium.

8.2. Ionic or electronic optics: the electrostatic lens

8.2.1. The analogue to the refractive index: trajectorial refraction of a charged particle placed in a succession of equipotential zones

An electron (or ion) is moving through space and meets a sudden change in an electric potential localized on an otherwise equipotential surface; as shown in Figure 8.5 the curves denoted C_i show a range of changes in the trajectories.

If the equipotential surfaces are sufficiently close to one another to assume that each makes a division of space into equipotential zones, then it is possible to state that for this succession of equipotential domains denoted V_1, V_2, ...V_i, each separated by the curves C_1, C_2, ...C_i, that there is a variation in the potential which is effected in small steps between the surfaces of the equipotential surfaces. To these jumps in potential there is a corresponding potential gradient that is normal to the equipotentials. This means that there is an electric field E that generates an electric force, which is also perpendicular to the equipotential surfaces.

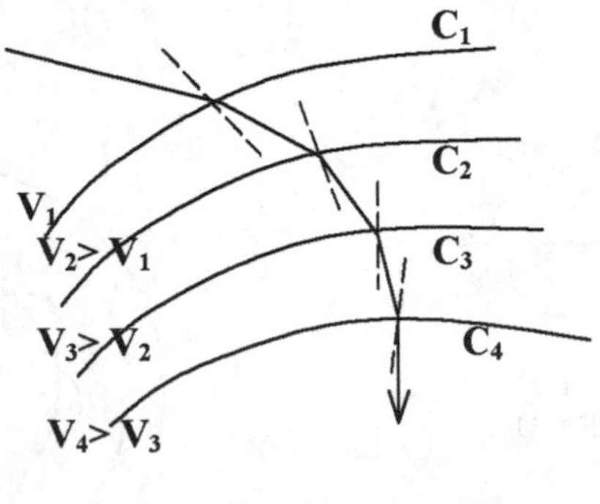

for ions, $V_1 > V_2 > V_3 > V_4$

Figure 8.5. *Refraction of the trajectory of an electron (moving in an acceleration field) over equipotential surfaces.*

With respect to the normal of the equipotentials, the particle is submitted to an electric force given by

$$F_N = m\left(\frac{dv}{dt}\right)_N = qE.$$

However, along the tangent the field and therefore the electric force also are zero, so that:

$$F_t = m\left(\frac{dv}{dt}\right)_t = 0.$$

Finally, the speed varies with respect to the normal (N) of the equipotentials (as $dv_N \neq 0$), but remains constant along the tangent (as $dv_t = 0$).

For example, at the interface of C_1 (in Figure 8.6), the conservation of the tangential component of the speed makes it possible to state that $v_{1t} = v_{2t}$, so that:

$$v_1 \sin i_1 = v_2 \sin i_2.$$

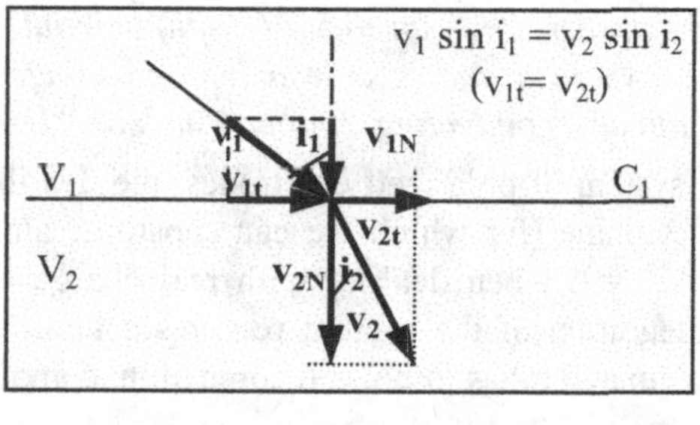

Figure 8.6. *Refraction of trajectories along the equipotential curve C_1.*

In addition, the speed v_i of the particle of charge q and mass m placed in a potential denoted V_i is such that $\frac{1}{2}mv_i^2 = qV_i$, so that by making $K = \sqrt{2\frac{q}{m}}$, we have:

$$v_i = K\sqrt{V_i}\ .$$

On substituting $v_1 = K\sqrt{V_1}$ and $v_2 = K\sqrt{V_2}$ into $v_1 \sin i_1 = v_2 \sin i_2$, we obtain:

$$\sqrt{V_1} \sin i_1 = \sqrt{V_2} \sin i_2 . \qquad (1)$$

This is a Snell–Descartes-type law, where here $\sqrt{V}$ plays the role of the refractive index.

Finally, the trajectory of a particle placed in a varying electric field (associated with the successive equipotential zones V_1, V_2, ... V_i) also can be constructed as if it were a ray of light passing through materials with successive indices given by $n_1 = \sqrt{V_1}$, $n_2 = \sqrt{V_2}$,... $n_i = \sqrt{V_i}$, as in Figure 8.5.

Comment. The law of similar trajectories

It is notable that the refraction of the trajectory lines is independent of the value of the q and m of the particles. In effect, the geometric path followed does not depend on the type of charge (vis-à-vis its mass or charge). Only the kinetics (speed along the trajectory and thus time required for the passage) depend on the type of particle.

8.2.2. *Practical determination of equipotential surfaces (and thus field lines)*

8.2.2.1. Equivalence between the spread of equipotential lines produced by electrodes in a vacuum and in a weakly conducting medium (such as an electrolytic bath or a conducting "paper" such as "Teledeltos")

Under the effect of a system of polarized electrodes, the distribution of the electric potential in an empty volume (for which we can substitute air) is governed by the Laplace law, where $\Delta V = 0$ when dealing with real charge volume densities (ρ) equal to zero. The integration of the system requires a knowledge of the limiting conditions of the problem and it has to be said some quite complicated calculations.

An alternative to trying to theoretically resolve the problem is by empirical determination of the equipotential lines. In order to do this, a weakly conducting medium (of conductivity σ) is used which can support the passage of current lines between the electrodes. It thus can be seen that the field lines and the equipotentials exhibit the same distribution in a vacuum.

The law of charge conservation under a permanent regime (where $\frac{\partial \rho}{\partial t} = 0$) for a weakly conducting medium gives: div $\vec{j} = 0$, and with $\vec{j} = \sigma \vec{E}$, we have σ div $\vec{E} = 0$, so that div $\vec{E} = 0$.

As there is no voltage or field generated by induction, then $\frac{\partial \vec{A}}{\partial t} = 0$ and $\vec{E} = -\overrightarrow{\text{grad}}V$, so that in a weakly conducting medium we find that $\Delta V = 0$ remains true (as $0 = \text{div}\ \vec{E} = -\text{div}\,\overrightarrow{\text{grad}}V = -\Delta V$) when we have the same limiting conditions given the geometric similarity of the two systems (electrodes in a vacuum or in a weak conductor). The distribution of the potential, measured with the help of a probe moved around the weak conductor, makes it possible to determine the equipotential lines (and therefore also the field) between the electrodes as if they were placed in a vacuum.

8.2.2.2. Symmetrical systems

For an example of cylindrical electrodes (although this remains valid for all systems presenting a symmetry), the electric field has no components outside of those in the plane of symmetry that passes through the coaxial axis and the normal to the electrodes. The field does not give rise to any components perpendicular to the plane of symmetry, just as the electric current ($i = \sigma E$) is zero in the plane of symmetry. The plane of symmetry is that of the "Teledeltos" paper onto which are traced the electrodes (Figure 8.5). The equipotential surfaces in the space then can determined from these equipotential lines through a rotation about the axis of symmetry.

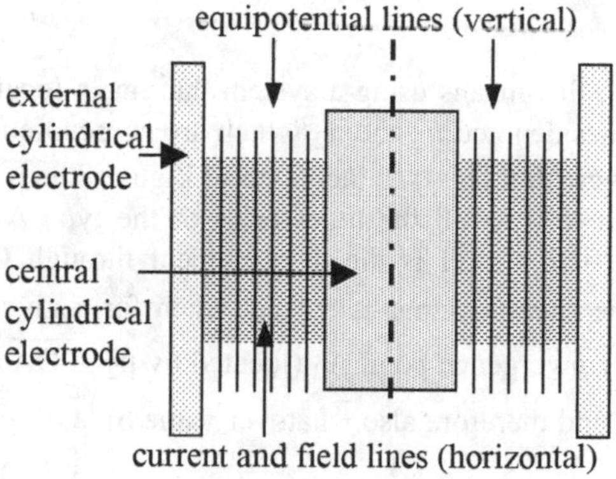

Figure 8.7. *Schematic view in the plane of symmetry of a paper-based conductor where the equipotentials are generated by a system of coaxial electrodes.*

8.2.3. Focusing trajectories with an electrostatic lens of axial symmetry (generating a radial field)

8.2.3.1. Principle of an electrostatic lens

The lens focuses the various trajectories that pass through A_1 onto a given point A_2. This happens whatever the angle of incidence (α_1) of the trajectories at A_1 (although in order to remain within the Gauss or "small angle" approximation, α_1 should be relatively small). The result is that A_2 thus appears as the image of A_1 through the lens.

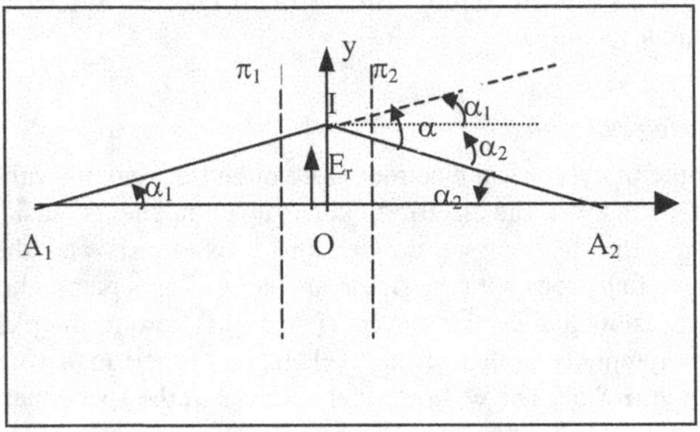

Figure 8.8. *Representation of a convergent electrostatic lens.*

In practical terms, this means using a system that has a localized electrostatic field between two planes denoted π_1 and π_2 which are perpendicular to the axis Oz and close to one another (thin lens), as indicated in Figure 8.8. A ray coming from A_1 is focused at A_2 as long as all the trajectories of the type A_1I are bent easily enough by the radial field (E_r); I is the displacement through Oy which can be located by $\overline{OI} = r$. The curvature thus is correct when for a given A_1 (or rather a given $p_1 = \overline{A_1O}$) the convergence point A_2 (located by $p_2 = \overline{OA_2}$) is the same for whatever value of α_1, and therefore also whatever value of $\overline{OI} = r$. In other words, for a given p_1, p_2 must remain independent of r as well as $\left(\frac{1}{p_1}+\frac{1}{p_2}\right)$.

In the Gauss approximation, the angle of deviation is such that:

$$\alpha = \alpha_1 + \alpha_2 \approx \mathrm{tg}\,\alpha_1 + \mathrm{tg}\,\alpha_2 = \frac{r}{p_1} + \frac{r}{p_2} = r\left(\frac{1}{p_1}+\frac{1}{p_2}\right).$$

With $\left(\frac{1}{p_1}+\frac{1}{p_2}\right)$ independent of r, the deviation (α) produced by the radial field of the electrostatic lenses must be proportional to r.

By making

$$\frac{1}{f}=\left(\frac{1}{p_1}+\frac{1}{p_2}\right), \qquad (2)$$

we therefore must have

$$\alpha=\frac{r}{f}. \qquad (2')$$

The $\frac{1}{f}$ is given by Eq. (2) and represents the convergence of the electrostatic lens. In effect we have an equation bringing together the terms analogous to those found in classic optics.

8.2.3.2. Form of the radial field to ensure the convergence condition $\alpha=\frac{r}{f}$

To obtain a convergence at A_2, we have seen that E_r must give a deviation (α) of the trajectories that is proportional to r (as the convergence condition, given by $[1/p_1 + 1/p_2]$, is independent of r). We now will look for a satisfactory form of E_r.

For a beam of particles each of mass m_0 coming from infinity ($A_1 \to \infty$) to be converged by the lens on its focal point at A_2 under the effect of E_r, the radial component of the speed v_r will be modified at the level of the thin lens.

Along the radial direction, the law governing the movement of an electron is given by:

$$m_0 \frac{dv_r}{dt} = -e\,E_r\,.$$

In the thickness (ε) traveled during an interval of time (Δt) the radial speed (v_r) undergoes a variation (δv_r) which is such that:

$$\delta v_r = \int_{\Delta t} \frac{dv_r}{dt} dt = -\int_{\Delta t} \frac{e}{m_0} E_r dt = v_r(L_2) - v_r(L_1) = v_r(L_2),$$

where it is supposed that $v_r(L_1) = 0$ as the speed v of the particle presents a single component (v_{0z}) along Oz before going into the lens system (so that in the space $z'L_1$, v_r is zero as indicated in Figure 8.9).

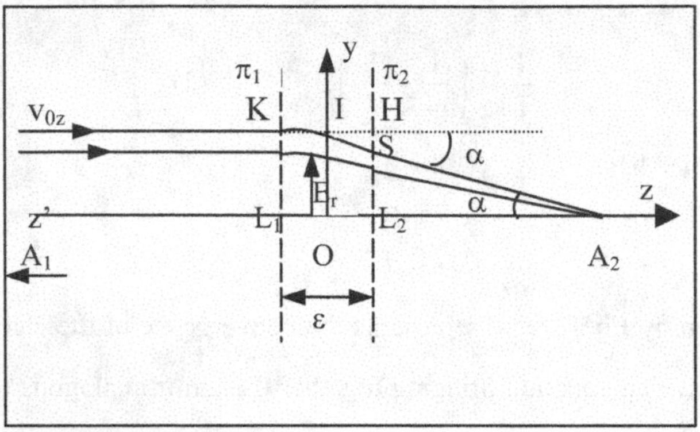

Figure 8.9. *Representation of focusing a parallel beam of particles using a radial field (E_r).*

The axial speed with respect to Oz, $v_z = \dfrac{dz}{dt}$, as a first approximation can be assumed to be constant to a high degree and equal to an average value v_{0z}. The latter is such that $\dfrac{1}{2}mv_{0z}^2 = eV$, where V represents the potential at which the particle beam is extracted.

With A_1 having been moved to infinity and within Gauss's approximation we have (Figure 8.9):

$$\alpha \approx \operatorname{tg} \alpha = \frac{HS}{KH} = \frac{|\delta v_r|\,\Delta t}{v_{0z}\,\Delta t} = \frac{|\delta v_r|}{v_{0z}} = \frac{e}{m_0}\frac{1}{v_{0z}}\int_{\Delta t} E_r dt\,.$$

With $v_z = \dfrac{dz}{dt}$ (instantaneous speed with respect to Oz between π_1 and π_2), we can definitively write:

$$\alpha \approx \frac{e}{m_0}\frac{1}{v_{0z}}\int_{L_1}^{L_2}\frac{E_r}{v_z}dz\,. \qquad (3)$$

So that α is proportional to r, E_r needs to be of the form $E_r = \mathcal{E}.\ r$; that is to say that E_r is itself also proportional to r. From this we obtain:

$$\alpha \approx \frac{e}{m}\frac{r}{v_{0z}}\int_{L_1}^{L_2}\frac{\mathcal{E}}{v_z}dz. \qquad (4)$$

So that the focalization condition is true, the intensity of the radial field therefore must increase with the distance r from the axis. This would allow a greater folding of the more external trajectories so that all particles would converge on the same point A_2.

8.2.4. Electrostatic lens with a rotational symmetry (generating an electrical field consisting of radial and longitudinal components)

8.2.4.1. Effect of a rotational electrostatic field around the z'Oz axis

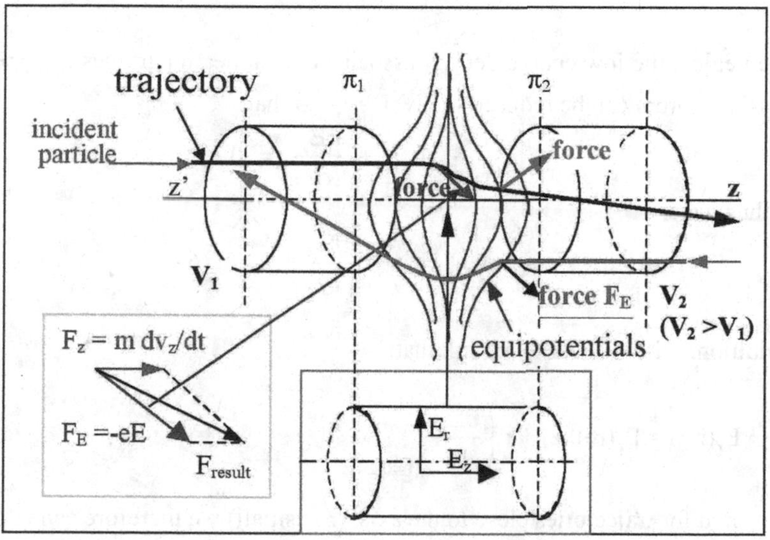

Figure 8.10. *Trajectory of an electron in a system of coaxial electrodes, with schematization of the electric force (F_E) at various points.*

An electrostatic lens can be made up of two disks each pierced in the middle, or rather two cylinders hollowed out along the same axis (for the same or different rays) as shown in Figure 8.10. Once again we look at the trajectory of a charged particle, initially rectilinear along z'z, penetrating the internal space of the electrodes (defined by $\pi_1\pi_2$).

At each point along its trajectory, a particle is submitted to the action of a force that is perpendicular to the equipotentials (collinear with the lines of the field, as $\vec{F}_E = -e\vec{E}$ for electrons with charge –e). As in Figure 8.10, if $V_2 > V_1$, the component along Oz for the V potential gradient is orientated toward the right and the component E_z for the electric field (E = – grad V) is directed toward the left so that the electrical force (F_E) exerted on the electrons has a component Oz directed toward the right.

It is assumed that the electric field between the electrodes has a cylindrical symmetry (shown in the central box of Figure 8.10). This is to say that E is independent of θ, and therefore, E_θ = constant. This component is taken to be zero to simplify the calculations.

The only components of the electric field with cylindrical coordinates are thus given by: $E_r = E_r(r,z)$ and $E_z = E_z(r,z)$, and as a consequence:

$$\vec{E} = E_r(r,z)\,\vec{e_r} + E_z(r,z)\,\vec{e_z}\,. \qquad (5)$$

If we neglect the low charge density associated with beam particles ($\rho_{real} \approx 0$), then Gauss's theorem can be reduced to div $\vec{E} = 0$, so that:

$$\frac{1}{r}\frac{\partial}{\partial r}(rE_r) + \frac{\partial E_z}{\partial z} = 0$$

We thus have:

$$\frac{\partial}{\partial r}(rE_r) = -r\,\frac{\partial E_z}{\partial z}. \qquad (6)$$

In addition, it is possible to remark that:

- $E_z(r,z) = E_z(r{=}0,z) + r\left(\frac{\partial E_z}{\partial r}\right)_{(r=0,z)}$

  and for trajectories close to the axis, (r is small) we therefore can take:

$$E_z(r,z) \approx E_z(r{=}0,z) = E_z(z). \qquad (7)$$

- For reasons of symmetry, along the axis Oz the radial field is zero and $E_r(r{=}0,z) = 0$, so that in the neighborhood of the axis, for a small r we have:

$$E_r(r,z) = E_r(r{=}0,z) + r\left(\frac{\partial E_r}{\partial r}\right)_{(r=0,z)} = r\left(\frac{\partial E_r}{\partial r}\right)_{(r=0,z)}. \qquad (8)$$

In Eq. (6), $\frac{\partial E_z}{\partial z}$ does not depend on r [according to Eq. (7)], so that the integration of Eq. (6) gives:

$$r\,E_r = -\frac{r^2}{2}\frac{dE_z(z)}{dz} + \varphi(z).$$

From this we can see that when r = 0, then $\varphi(z) = 0$, and therefore we finally have:

$$E_r(r,z) = -\frac{r}{2}\left[\frac{dE_z(z)}{dz}\right]_{(r=0,\,z)}. \qquad (9)$$

In this configuration based on cylinders, the radial field (E_r), which is proportional to r, fulfills the convergence conditions; $\mathcal{E}$, introduced in Section 5.2.3.2, takes on the form:

$$\mathcal{E} = -\frac{1}{2}\frac{dE_z(z)}{dz}, \qquad (10)$$

which makes $\mathcal{E}$ a function of z. In addition, E_z is not zero in this configuration where the two components E_r and E_z are not independent from each other, but tied by Eq. (9).

As

$$E_z(r{=}0,z) = -\left(\frac{\partial V}{\partial z}\right)_{r=0} = -\,V'$$

and that

$$\left(\frac{\partial E_z}{\partial z}\right)_{r=0} = -\left(\frac{\partial^2 V}{\partial z^2}\right)_{r=0} = -\,V''$$

we also can write Eq. (9) in the form:

$$E_r(r,z) = \frac{r\,V''}{2}. \qquad (11)$$

This equality shows again that the radial component of the electric field, and therefore the force that moves the particle to the axis, increases linearly with distance from the axis. So that the radial force is well directed toward the axis (as a focalizing action) then V" must be positive for electrons (if $V'' > 0$, $f_r = -\,e\,E_r < 0$ is thus directed toward the axis).

In can be seen in Figure 8.10 that during the first half of the incident particle's pathway (from left to right), the action of the radial component of the electric force is to focus the beam toward the axis. The particle moves progressively closer to the axis. However, in the second part of the trajectory, the radial force tends to move the particle away from the axis and has a defocalizing effect.

Nevertheless, in the first part of the trajectory the particle moves more slowly than in the second. The upshot is that the focusing of the particle is the stronger of the two effects as it lasts for the longer period of time.

8.2.4.2. Expression for the focal length as a function of potentials

By substituting Eq. (10) into Eq. (4), we obtain:

$$\alpha = \frac{e}{m}\frac{r}{v_{0z}}\int_{L_1}^{L_2}\frac{E}{v_z}dz = -\frac{1}{2}\frac{e}{m}\frac{r}{v_{0z}}\int_{L_1}^{L_2}\frac{1}{v_z}\frac{dE_z}{dz}dz,$$

so that

$$\frac{1}{f} = \frac{\alpha}{r} = -\frac{1}{2}\frac{e}{m}\frac{1}{v_{0z}}\int_{L_1}^{L_2}\frac{1}{v_z}\frac{dE_z}{dz}dz.$$

If V_0 represents the extraction potential of a particle, then we have $\frac{1}{2}m_0 v_{0z}^2 = eV_0$, so that

$$v_{0z} = \sqrt{\left(2\frac{e}{m_0}V_0\right)}.$$

Similarly, if V(z) represents the potential on the axis of the lens where the radial field is produced, then we have:

$$v_z(z) = \sqrt{\left(2\frac{e}{m_0}V(z)\right)}.$$

Using Eq. (11) makes it possible to definitively state that

$$\frac{1}{f} = \frac{1}{4\sqrt{V_0}}\int_{L_1}^{L_2}\frac{V''(z)}{\sqrt{V(z)}}dz. \qquad (12)$$

We therefore can remark that if V_0 increases, then f increases as well.

8.2.5. Equation for the trajectory in the electrostatic lens

The equation for the movement of a particle (of mass m_0 and charge $-e$) is given by the fundamental dynamic equation, as in $m_0 \vec{\gamma} = -e\vec{E}$, so that in terms of cylindrical coordinates:

$$\begin{cases} \gamma_r = \ddot{r} - r\dot{\theta}^2 = -\dfrac{e}{m_0}E_r(r,z) & (13) \\[2ex] \gamma_\theta = \dfrac{1}{r}\dfrac{d}{dt}\left(r^2\dot{\theta}\right) = 0 \\[2ex] \gamma_z = -\dfrac{e}{m_0}E_z(r,z)\,. \end{cases}$$

As E_θ is zero, there is no drive to move the particle along $\overrightarrow{e_\theta}$, and we can neglect the movement associated with a variation in θ (we can note mathematically that the second equation leads to $r^2\dot{\theta} =$ constant, which is zero when there is no initial rotation, as assumed).

Therefore, when $\dot{\theta} \approx 0$, Eq. (13) can be reduced to $\ddot{r} \approx -\dfrac{e}{m_0}E_r(r,z)$, so that with Eq. (11) we have:

$$\ddot{r} \approx -\frac{e}{2m_0}rV''. \qquad (14)$$

For trajectories at small inclinations, and using the small-angle approximation for *para*-axial rays where $\alpha \approx 0$ and $\cos\alpha \approx 1$, the axial component v_z of the speed can be approximated to the total speed (v) of the electron, as in

$$v_z = \frac{dz}{dt} = v\cos\alpha \approx v. \qquad (15)$$

We also can say that $\vec{v} = \overrightarrow{v_r} + \overrightarrow{v_z} \approx \overrightarrow{v_z}$ as $v_r = \dfrac{dr}{dt} = \dfrac{dr}{dz}\dfrac{dz}{dt} \approx 0$, because $\dfrac{dr}{dz} = \tan\alpha \approx \alpha \approx 0$ from the approximation made for small angles.

If the particles are subject to a potential given by $V(r,z)$, which brings the particles into a small angle approximation, then it is possible to state that $V(r,z) \approx V(r = 0,z)$. If the extraction is performed with an initial speed equal to zero (which is the case for ions extracted from a source), then we have $\frac{1}{2}mv^2 = e\,V(r,z)$, so that:

$$v = \sqrt{2\frac{e}{m_0}V(r,z)} \approx \sqrt{2\frac{e}{m_0}V(r=0,z)}\ . \qquad (16)$$

From Eq. (15) $v \approx v_z$, and therefore:

$$v_z = \frac{dz}{dt} \approx \sqrt{2\frac{e}{m_0}V(z)}\ . \qquad (17)$$

In addition, we can state that:

$$\ddot{r} = \frac{d}{dt}\left(\frac{dr}{dt}\right) = \frac{d}{dz}\left(\frac{dr}{dt}\right)\frac{dz}{dt} = \frac{d}{dz}\left(\frac{dr}{dt}\right).v_z = v_z\frac{d}{dz}\left(\frac{dr}{dz}\frac{dz}{dt}\right) = v_z\frac{d}{dz}\left(\frac{dr}{dz}v_z\right),$$

so that with Eq. (17)

$$\ddot{r} \approx 2\,\frac{e}{m_0}\,\sqrt{V(z)}\,\frac{d}{dz}\left(\sqrt{V(z)}\frac{dr}{dz}\right). \qquad (18)$$

By substituting Eq. (18) into Eq. (14), then the Eq. (13) finally gives:

$$\sqrt{V(z)}\,\frac{d}{dz}\left(\sqrt{V(z)}\frac{dr}{dz}\right) = -\frac{r}{4}V''. \qquad (19)$$

This last equation also can be written in another form:

$$\frac{d}{dz}\left(\sqrt{V(z)}\frac{dr}{dz}\right) = -\frac{r}{4}\frac{V''}{\sqrt{V(z)}}, \qquad (20)$$

and hence

$$\frac{d^2r}{dz^2} + \frac{V'}{2V}\frac{dr}{dz} + \frac{V''}{4V}r = 0\ . \qquad (21)$$

Equation (21) is the differential equation for the trajectory. By integrating between two points on the axis s'Oz (points A_1 and A_2, for example) where the form of V(z) is known, we can obtain the trajectory of a particle within a given field.

The classic case study of the three electrode lens—which corresponds to the generation of a field containing uniquely radial and longitudinal components ($E_\theta = 0$)—is described in the following section.

8.2.6. Focal length of a three- electrode lens

Figure 8.11 shows a lens made of two identical electrodes placed at the same potential (V_0) and symmetrically surrounding another electrode at a potential V_L. The electric field is most effective in the neighborhood of the central electrode, as at the exterior of the lens the potential is practically constant.

As outside the lens the field is zero, the trajectory of an emitted particle (without initial speed) from A_1 onward is rectilinear. The speed is given by Eq. (16) and can be written: $v = \sqrt{2\frac{e}{m_0}V_0}$.

This particle meets the lens at a distance r_0 from the axis, is deviated by the lens, and exits through a new rectilinear trajectory that cuts the z'Oz axis at A_2.

Integrating the differential equation [Eq. (20)] between A_1 and A_2 with the potentials distributed as shown in Figure 8.11 gives Eq. (22) shown after the figure.

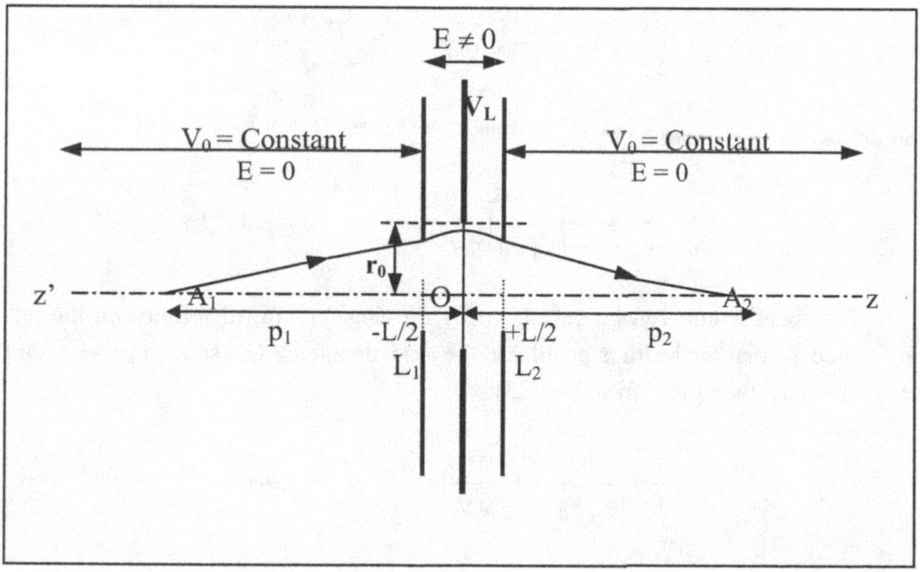

Figure 8.11. *Schematization of a three- electrode symmetrical lens, with the corresponding potential distribution.*

$$\left[\sqrt{V}\frac{dr}{dz}\right]_{A_2} - \left[\sqrt{V}\frac{dr}{dz}\right]_{A_1} = -\frac{1}{4}\int_{A_1}^{A_2} r\frac{V''}{\sqrt{V}}dz\,. \qquad (22)$$

Once again using the small-angle approximation, we can assume that in the central zone around the lens the distance of the particle from the axis varies negligibly, and therefore:

$$\left[\frac{dr}{dz}\right]_{A_2} = -\frac{r_0}{p_2} \text{ and } \left[\frac{dr}{dz}\right]_{A_1} = \frac{r_0}{p_1}.$$

Thus Eq. (22) gives:

$$-\frac{r_0}{p_2}\sqrt{V_0} - \frac{r_0}{p_1}\sqrt{V_0} \approx -\frac{1}{4}\int_{A_1}^{A_2} r\frac{V''}{\sqrt{V}}dz\,.$$

As $V = V_0$ = constant outside of the lens, then V'' is nonzero only in the zone L_1L_2 of the lens where $r \approx r_0$. This gives

$$-\frac{r_0}{p_2}\sqrt{V_0} - \frac{r_0}{p_1}\sqrt{V_0} \approx -\frac{r_0}{4}\int_{-L/2}^{+L/2} \frac{V''}{\sqrt{V}}dz\,,$$

so that:

$$\frac{1}{p_1} + \frac{1}{p_2} \approx \frac{1}{4\sqrt{V_0}}\int_{-L/2}^{+L/2} \frac{V''}{\sqrt{V}}dz\,. \qquad (23)$$

The focal length can be calculated by moving A_1 toward infinity on the left hand side, so that the beam is parallel to the axis denoted z'Oz, so then $p_1 \to \infty$ and $p_2 \to f$, which then gives from Eq. (23):

$$\frac{1}{f} = \frac{1}{4\sqrt{V_0}}\int_{-L/2}^{+L/2} \frac{V''}{\sqrt{V}}dz\,. \qquad (24)$$

This is in a form identical to that seen for Eq. (12) above.

Equations (23) and (24) then can be used to find the more well-known Eq. (2), as in:

$$\frac{1}{p_1}+\frac{1}{p_2}=\frac{1}{f}.$$

Equation (24) shows that if V_0 or V (governed by the tension V_L of the central lens) increases, then the f also increases. The variation in V", however, is harder to define.

8.3. Problems

8.3.1. Problem 1. Mathematical study of a cycloid

By definition, the trajectory of a point M is a cycloid if the movement of the point is that of a fixed point in a circle with center C which itself is rolling along a given straight line D. If after a roll along the line, I denotes the point at which the circle is in contact with the straight line D and we can make $\theta = \left(\overrightarrow{CM}, \overrightarrow{CI}\right)$.

In this problem, establish the parametric equations for the cycloid so that you have the x and y coordinates for the point M following a roll of the circle given in terms of θ and the radius of the circle.

Answer

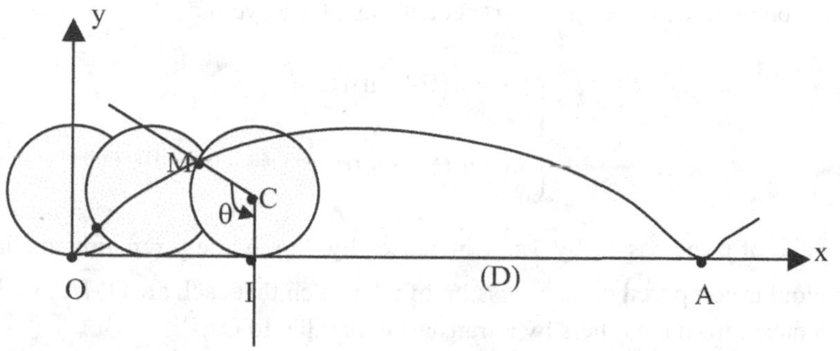

The figure shown above can be traced from the details of the question. The condition of rolling without slipping can be mathematically expressed as:

$$\overline{OI} = \widehat{MI},$$

(to have a simple demonstration of this, it would suffice to roll a circle along a line as shown in the figure).

With $\theta = \left(\overrightarrow{CM}, \overrightarrow{CI}\right)$, IC = a, we have $\overline{OI} = \widehat{MI} = a\theta$.

To determine the x and y coordinates of M as a function of θ, we can use the relation: $\overrightarrow{OM} = \overrightarrow{OI} + \overrightarrow{IC} + \overrightarrow{CM}$.

By projection on Ox, we have:

$$x = a\theta + 0 + a\cos\left(\overrightarrow{Ox}, \overrightarrow{CM}\right).$$

However, $\left(\overrightarrow{Ox}, \overrightarrow{CM}\right) = \left(\overrightarrow{Ox}, \overrightarrow{CI}\right) + \left(\overrightarrow{CI}, \overrightarrow{CM}\right) = 3\frac{\pi}{2} - \theta$

and as $\cos\left(3\frac{\pi}{2} - \theta\right) = -\sin\theta$, we finally obtain:

$$x = a(\theta - \sin\theta).$$

Similarly we obtain:

$$y = a + a\cos\left(\overrightarrow{Oy}, \overrightarrow{CM}\right),$$

from which with $\left(\overrightarrow{Oy}, \overrightarrow{CM}\right) = \left(\overrightarrow{Oy}, \overrightarrow{CI}\right) + \left(\overrightarrow{CI}, \overrightarrow{CM}\right) = \pi - \theta$, we find

$$y = a(1 - \cos\theta).$$

We conclude with the parametric equations of the cycloid:

$$\begin{cases} x = a(\theta - \sin\theta) \\ y = a(1 - \cos\theta). \end{cases}$$

For a value of θ increasing by 2π, x increases by 2πa, while y remains unchanged. The cycloid is composed of a succession of arcs, such that each arc $\widehat{OMA}$, with each being deduced from the others by a translation parallel to Ox.

8.3.2. Problem 2. The effect of a crossed field $\vec{E} \times \vec{B}$ ***on a charged particle q***

A charged particle q is initially placed at O, the defining origin, and has a velocity v with respect to the fields $\vec{E}$ and $\vec{B}$ which have the components given by:

$$\vec{E}\begin{cases} E_x = E \\ E_y = 0 \\ E_z = 0 \end{cases} \quad \text{and} \quad \vec{B}\begin{cases} B_x = 0 \\ E_y = 0 \\ B_z = B \end{cases}$$

for which the velocity components are : $\vec{v}\begin{cases}\dfrac{dx}{dt}\\ \dfrac{dy}{dt}\\ \dfrac{dz}{dt}\end{cases}$

1. From the dynamic fundamental equation projected on the axes, deduce the equation for the movement of the particle with respect to the axes x, y, and z (where the initial velocity is assumed in the first instance to be zero).

2. Show that for a system of axes to be defined, we can once again for the equation for a cycloid described by a circle of radius $a = \dfrac{qE}{m\omega^2}$, where $\omega = \dfrac{qB}{m}$.

3. Study in more detailed terms the movement as a function of the sign of q.

4. Study the speed of the particle on its trajectory. At what point is the speed at its greatest.

5. How does the particle move if it has an initial velocity v_0 which is parallel to the axis Oz?

Answer

1. The dynamic fundamental equation can be written:

$$\vec{F} = m\vec{\gamma} = q\left[\vec{E} + \vec{v}\times\vec{B}\right].$$

Projected onto the axes, it gives:

$$\begin{cases} m\dfrac{d^2x}{dt^2} = qE + qB\dfrac{dy}{dt} & (1) \\ m\dfrac{d^2y}{dt^2} = 0 - qB\dfrac{dx}{dt} & (2) \\ m\dfrac{d^2z}{dt^2} = 0 & (3) \end{cases}$$

The integration of Eq. (3) with respect to time gives rise to:

$$\frac{dz}{dt} = \text{constant}$$

The initial speed along Oz being zero, this constant is also equal to zero. From this we deduce that z is constant, and with the particle initially placed at O ($z_0 = 0$), then the new constant is equal to zero and $z = 0$.

The integration of Eq. (2) gives: $m\frac{dy}{dt} = -qBx + \text{constant}$.

When t = 0, then $x = x_0 = 0$ and $v_y = v_{y0} = 0$, from which is deduced that the constant is zero, and from which:

$$m\frac{dy}{dt} = -qBx\,.$$

Making $\omega = \frac{qB}{m}$, then we can write that $\frac{dy}{dt} = -\omega x$. By taking this expression into Eq. (1), we find that

$$\frac{d^2x}{dt^2} + \omega^2 x = \frac{qE}{m}\,.$$

The characteristic equation of this differential equation is $r^2 + \omega^2 = 0$, and the solutions of the equation without a second member are $x = \lambda\cos\omega t + \mu\sin\omega t$.

Evidently, a particular solution to the equation with a second member is given by $x = \frac{qE}{m\omega^2}$, so that the general solution is of the form:

$$x = \lambda\cos\omega t + \mu\sin\omega t + \frac{qE}{m\omega^2}\,.$$

The initial condition, that t = 0, $x = x_0 = 0$, gives $0 = \lambda + \frac{qE}{m\omega^2}$, from which $\lambda = -\frac{qE}{m\omega^2}$.

Similarly, the initial condition $t = 0$, $\frac{dx}{dt} = \frac{dx_0}{dt} = 0$, gives $\mu = 0$.

Finally

$$x = \frac{qE}{m\omega^2}\left(1 - \cos\omega t\right).$$

By moving this expression into $\frac{dy}{dt} = -\omega x$, we obtain:

$$\frac{dy}{dt} = -\frac{qE}{m\omega}\left(1 - \cos\omega t\right).$$

With y = 0 when t = 0, we find that following integration, we have

$$y = \frac{qE}{m\omega^2}\left(\sin\omega t - \omega t\right).$$

2. Ultimately, by making $a = \dfrac{qE}{m\omega^2}$ we obtain:

$$\begin{cases} x = \dfrac{qE}{m\omega^2}(1-\cos\omega t) = a(1-\cos\omega t) \\ y = -\dfrac{qE}{m\omega^2}(\omega t - \sin\omega t) = -a(\omega t - \sin\omega t) \\ z = 0. \end{cases}$$

Within the system of axes (−y, x) we once again find the equation for a cycloid generated by a circle of radius $a = \dfrac{qE}{m\omega^2}$. The length of the arc is given by $\ell = 2\pi a$ and is situated in the plane y0x.

3.

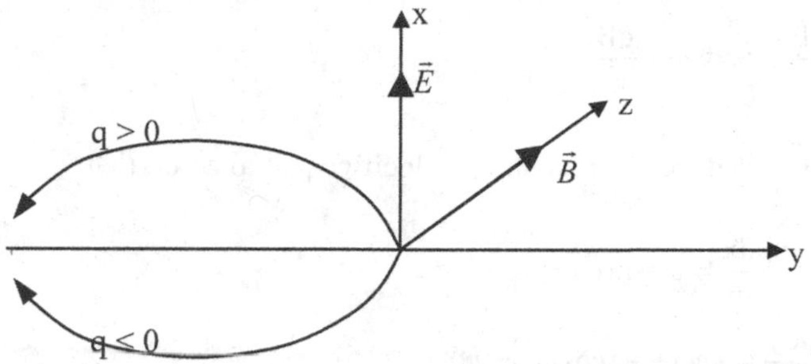

Noting that *y* has the same sign as $-\dfrac{q}{\omega}$, we can go on to write that in effect,

$$y = -\frac{qE}{m\omega^2}(\omega t - \sin\omega t) = -\frac{q}{\omega}\frac{E}{m}t\left(1 - \frac{\sin\omega t}{\omega t}\right).$$

As $\dfrac{\sin\theta}{\theta} \le 1$, the bracketed term is always positive, and with $\dfrac{E}{m}t$ also being positive (E is positive and the time t increases from t = 0 onward where t > 0), y is also the same sign as $-\dfrac{q}{\omega}$. As $\omega = \dfrac{qB}{m}$, we have $-\dfrac{q}{\omega} = -\dfrac{m}{B}$. Thus, for a given positive value of B, the latter term is always negative, along with y.

As $(1-\cos\omega t) \ge 0$, the side $x = a(1-\cos\omega t)$ takes on the sign of $a = \dfrac{qE}{m\omega^2}$, which means the sign of the charge q under consideration. Finally, we therefore have:

- when $q > 0$, we have $x > 0$ and $y < 0$; and
- when $q < 0$, we have $x < 0$ and $y < 0$.

From this we have the behavior schematized in the preceding figure.

In practical terms, the movement along Ox is that indicated by the contribution from the electric force, namely $\vec{F}_E = q\vec{E}$. With $E > 0$, $x > 0$ when $q > 0$ and $x < 0$ when $q < 0$.

4. The equations, which describe the coordinates of the particle, thus are given by:

$$\begin{cases} x = a(1 - \cos\omega t) \\ y = -a(\omega t - \sin\omega t) \\ z = 0 \end{cases}$$

where $a = \dfrac{qE}{m\omega^2}$, and $\omega = \dfrac{qB}{m}$.

From this can be deduced (for an initial velocity equal to zero), that:

$$\begin{cases} \dfrac{dx}{dt} = a\omega \sin\omega t = v_x \\ \dfrac{dy}{dt} = -a\omega + a\omega\cos\omega t = v_y \\ v_z = 0 \end{cases}$$

from which

$v^2 = v_x{}^2 + v_y{}^2 = (a\omega)^2\left[\sin^2\omega t + 1 + \cos^2\omega t - 2\cos\omega t\right] = 2(a\omega)^2\left[1 - \cos\omega t\right]$, so that

$$v = 2(a\omega)\sin\frac{\omega t}{2}.$$

With $a\omega = \dfrac{qE}{m\omega} = \dfrac{qE}{m}\dfrac{m}{qB} = \dfrac{E}{B}$, we finally reach:

$$v = 2\frac{E}{B}\sin\frac{\omega t}{2}.$$

The speed is at a maximum when $\sin\frac{\omega t}{2} = 1$, *i.e.* $\omega t = \pi$.

This plugged into v gives:

$$v = v_{Max} = 2\frac{E}{B},$$

as a maximum velocity independent of $\frac{q}{m}$.

The complete arc of the cycloid corresponds to $\omega t = 2\pi$, and the value of $\omega t = \pi$ (associated with the maximum speed) corresponds to half the arc. Thus, the summit of the arc is the point at which the speed is at a maximum.

5. The integration of the starting Eq. (3) thus gives us $v_z = v_{0z} = v_0$. In addition, keeping the same limiting conditions ($z = z_0 = 0$ when $t = 0$), we obtain:

$$z = v_0 t.$$

With the other limiting conditions being retained (zero velocity with respect to Ox and Oy, for a particle at O at an origin instant t = 0), the x and y coordinates remain unchanged.

Finally, the projection of the movement in the plane Oxy is the same as the cycloid defined above. The only additional movement (derived) is that along Oz defined by $z = v_0 t$.

8.3.3. Problem 3. Movement of a particle in a uniform B field

A particle of charge q and mass m placed at an instant t = 0 at the origin O of a reference trihedral penetrates at the instant t = 0 a magnetic field B directed along Oz, as in $\vec{B} = B\vec{e}_z$. Here we set $\omega = -\frac{qB}{m}$.

1. The initial speed of the particle is assumed to be of the form: $\vec{v}_0 = v_{0x}\vec{e}_x$.

Determine the components of the speed and the position of the particle at any given instant t. Show that the trajectory is a circle, for which the center and radius are denoted C and r, respectively.

2. Calculate the kinetic ($\vec{L}$) and the magnetic ($\vec{M}$) moments associated with the movement of a particle of charge q in the field B. Give the relations that exist between $\vec{M}$ and the kinetic energy (E_C) and between $\vec{M}$ and $\vec{L}$. Also show that the normal of the magnetic moment is proportional to the flux of the magnetic field that traverses the surface of the orbit of the particle.

3. The initial speed of the particle is assumed to be of the form: $\vec{v}_0 = v_{0x}\vec{e}_x + v_{0z}\vec{e}_z$.

Velocities perpendicular and parallel to the field ($\vec{B}$) are, respectively, denoted $\vec{v}_\perp$ and $\vec{v}_\parallel$. Show that the normals to the velocity components are conserved.

Give the expressions for $\vec{v}_\perp$ and $\vec{v}_\parallel$ and show that the projection on the plane Oxy of the trajectory of the particle is a circle of radius R described for an angular velocity $\dot{\theta}$.

Answers

1.

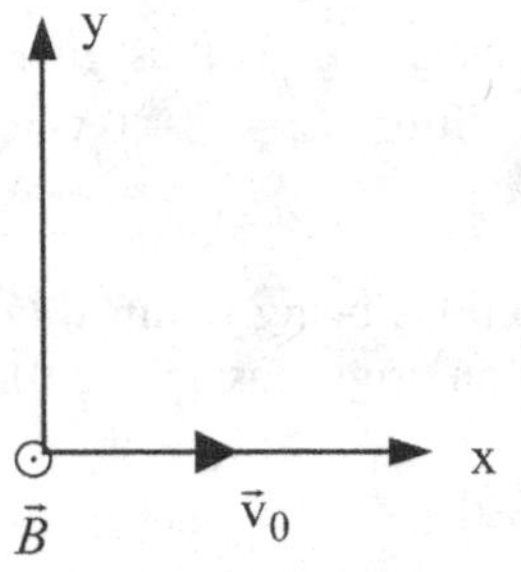

The fundamental dynamic equation makes it possible to state that:

$$\frac{d\vec{v}}{dt} = \frac{q}{m}\vec{v}\times\vec{B},$$

of which the projection on the axes gives:

$$\begin{cases} \dfrac{dv_x}{dt} = \dfrac{q}{m} v_y B = \dfrac{q}{m} B \dfrac{dy}{dt} & (1) \\ \dfrac{dv_y}{dt} = -\dfrac{q}{m} v_x B = -\dfrac{q}{m} B \dfrac{dx}{dt} & (2) \\ \dfrac{dv_z}{dt} = 0. & (3) \end{cases}$$

The first integration of Eq. (3) gives, with $v_{0z} = 0$, that $v_z = 0$. The second integration gives $z = 0$ (as $z_0 = 0$), and thus the movement is in the plane Oxy.

For its part, the first integration of Eq. (2) gives $v_y = -\frac{q}{m}Bx + cte$. The initial condition, $v_y = v_{0y} = 0$ when $x = 0$ gives $cte = 0$, from which by making $\omega = -\frac{qB}{m}$ we have

$$v_y = -\frac{q}{m}Bx = \omega x. \qquad (4)$$

By plugging this into Eq. (1), we obtain:

$$\frac{dv_x}{dt} = -\left(\frac{q}{m}B\right)^2 x, \text{ so that } \frac{d^2x}{dt^2} + \left(\frac{q}{m}B\right)^2 x = 0.$$

With $\omega = -\frac{qB}{m}$, we have $\frac{d^2x}{dt^2} + \omega^2 x = 0$, from which $x = \lambda \cos \omega t + \mu \sin \omega t$.

The initial condition, $x = x_0 = 0$ when $t = 0$ gives $\lambda = 0$, from which: $x = \mu \sin \omega t$. From this can be deduced that:

$$v_x = \omega\mu \cos \omega t.$$

The initial condition $v_x = v_{0x}$ when $t = 0$ give: $v_{0x} = \omega\mu \cos 0$, from which we have $\mu = \frac{v_{0x}}{\omega}$. Finally, we reach:

$$v_x = v_{0x} \cos \omega t \text{ and } x = \frac{v_{0x}}{\omega} \sin \omega t.$$

By substituting this into Eq. (4), we obtain:

$$\frac{dy}{dt} = v_{0x} \sin \omega t,$$

from which: $y = -\frac{v_{0x}}{\omega} \cos \omega t + \text{constant}$.

When $t = 0$, $y = 0$, from which constant $= \frac{v_{0x}}{\omega}$ and then,

$$y = \frac{v_{0x}}{\omega}(1 - \cos \omega t).$$

To summarize, we have:

$$\begin{cases} x = \dfrac{v_{0x}}{\omega}\sin\omega t \\ v_x = v_{0x}\cos\omega t \end{cases} \qquad \begin{cases} y = \dfrac{v_{0x}}{\omega}(1-\cos\omega t) \\ \text{and } v_y = \omega x = v_{0x}\sin\omega t. \end{cases}$$

The equation for the trajectory is obtained through:

$$x^2 + \left(y - \frac{v_{0x}}{\omega}\right)^2 = \left(\frac{v_{0x}}{\omega}\right)^2,$$

which is the equation for a circle:

- with a center C $(0, \dfrac{v_{0x}}{\omega}, 0)$
- and a radius R: $R = \left|\dfrac{v_{0x}}{\omega}\right| = \left|\dfrac{mv_{0x}}{qB}\right|$.

$\dfrac{v_{0x}}{\omega} = -\dfrac{mv_{0x}}{qB}$ is positive if $q < 0$ (C above Ox) and negative if $q > 0$ (C below Ox).

The movement of the particle is a rotation defined by the rotation vector ($\vec{\Omega}$), which is such that:

$$\frac{d\vec{v}}{dt} = \vec{\Omega}\times\vec{v}.$$

This vector can be directly deduced from the fundamental dynamic equation which gives:

$$\frac{d\vec{v}}{dt} = \frac{q}{m}\vec{v}\times\vec{B} = -\frac{q}{m}\vec{B}\times\vec{v},$$

from which, by identification:

$$\vec{\Omega} = -\frac{q}{m}\vec{B} = \omega\vec{e}_z$$

and here $\omega = -\dfrac{qB}{m}$.

Depending on the sign of q, the rotation vector $\vec{\Omega}$ is parallel to $\vec{B}$ $(q < 0)$ or antiparallel to $\vec{B}$ $(q > 0)$ (see the figure below).

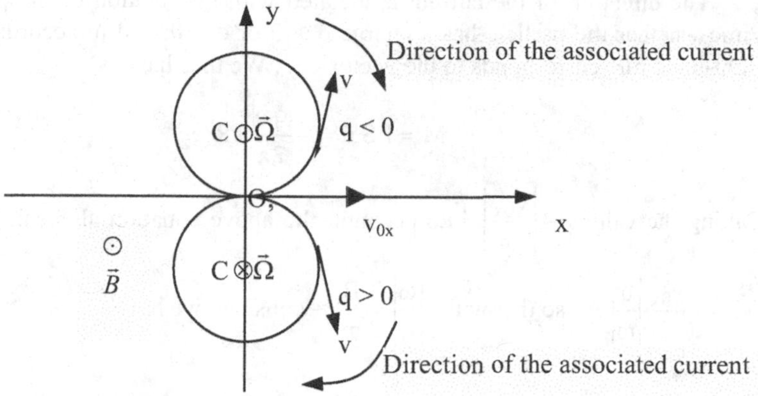

2. The kinetic moment with respect to C of a particle is, by definition, $\vec{L} = \overrightarrow{CM} \times m\vec{v} = \vec{R} \times m\vec{v}$. It is such that (kinetic moment theory):

$$\frac{d\vec{L}}{dt} = \vec{R} \times \vec{F} = \vec{\Gamma} \text{ (moments of the forces).}$$

As $\vec{F} = \vec{F}_n$ is parallel to $\vec{r}$, then $\frac{d\vec{L}}{dt} = 0$ and $\vec{L} = \text{constant}$ (independently of the position of the point M on the trajectory).

We can take $M \equiv O$, and we therefore have:

$$\vec{L} = \overrightarrow{CM} \times m\vec{v} = \overrightarrow{CO} \times m\vec{v} = \text{constant.}$$

With $\overrightarrow{CO} = -\frac{v_{0x}}{\omega}\vec{e}_y$, we obtain

$$\vec{L} = -\frac{v_{0x}}{\omega}\vec{e}_y \times mv_{0x}\vec{e}_x = \frac{mv_{0x}^2}{\omega}\vec{e}_z,$$

so that with $\omega = -\frac{qB}{m}$ we have:

$$\vec{L} = -\frac{m^2 v_{0x}^2}{qB}\vec{e}_z$$

For its part, the magnetic moment is given by $\vec{M} = I\iint \overrightarrow{dS} = I\,S\,\vec{n}$ where the intensity I is of the form $I = q\,\nu = q\frac{|\omega|}{2\pi}$ (ν being the rotational frequency of the particle).

The direction of the current associated with the rotation of the particle is the same whether the particle has a charge q > 0 or q < 0, and in accordance with the right-hand rule corresponds to the vector $-\vec{e}_z$. We thus have:

$$\vec{M} = I\,S\,\vec{n} = -\frac{|q\omega|}{2\pi}\pi R^2 \vec{e}_z$$

Taking the value $R=\left|\frac{v_{0x}}{\omega}\right|$ into account, the above equation also can be written as

$\vec{M} = -\frac{v_{0x}^2}{2}\left|\frac{q}{\omega}\right|\vec{e}_z$, so that with $\left|\frac{\omega}{q}\right| = \frac{B}{m}$ = constant, we have

$$\vec{M} = -\frac{mv_{0x}^2}{2B}\vec{e}_z .$$

As $v^2 = v_x^2 + v_y^2 + v_z^2 = v_{0x}^2\left(\cos^2\omega t + \sin^2\omega t\right) = v_{0x}^2$ = constant, we have:

$$E_C = \frac{1}{2}mv^2 = \frac{1}{2}mv_{0x}^2 .$$

By plugging E_C into $\vec{M}$, we can write that:

$$\vec{M} = -\frac{mv_{0x}^2}{2B}\vec{e}_z = -\frac{E_C}{B}\vec{e}_z$$

Given the expression of $\vec{L} = -\frac{m^2 v_{0x}^2}{qB}\vec{e}_z$, we also have:

$$\vec{M} = \frac{q}{2m}\vec{L} .$$

The preceding expression $\vec{M} = -\frac{|q\omega|}{2\pi}\pi R^2 \vec{e}_z$ makes it equally possible to write:

$$\left|\vec{M}\right| = \frac{|q\omega|}{2\pi}\pi R^2 = \frac{q^2 B}{2\pi m}\pi R^2 .$$

By introducing the term for the flux (Φ) of the magnetic field which traverses the orbit of the particle, given by $|\Phi| = \pi R^2 B$, we also have:

$$\left|\vec{M}\right| = \frac{q^2}{2\pi m}|\Phi| .$$

3. We now have $\vec{v}_0 = v_{0x}\vec{e}_x + v_{0z}\vec{e}_z$. For their part, the velocities $\vec{v}_\perp$ and $\vec{v}_{||}$ are such that:

$$\begin{cases} \vec{v}_{||} = v_z\vec{e}_z \\ \vec{v}_\perp = v_x\vec{e}_x + v_y\vec{e}_y \end{cases}$$

According to the fundamental dynamic equation, we still have $\frac{dv_z}{dt} = 0$ [Eq. (3) from question 1] but here, given the limiting conditions, we have:

$$\vec{v}_z = v_{0z}\vec{e}_z = \vec{v}_{||} = \text{constant}.$$

With respect to $\vec{e}_x$ and $\vec{e}_y$ the conditions do not change and the movement therefore still is described by the same values of $\vec{v}_x$, $\vec{v}_y$, x, and y.

We therefore have:

$$\vec{v}_\perp = v_x\vec{e}_x + v_y\vec{e}_y = v_{0x}\cos\omega t\ \vec{e}_x + v_{0x}\sin\omega t\ \vec{e}_y .$$

The equation of the trajectory therefore is

$$\begin{cases} x = \dfrac{v_{0x}}{\omega}\sin\omega t \\[2ex] y = \dfrac{v_{0x}}{\omega}(1 - \cos\omega t) \\[2ex] z = v_{0z}t . \end{cases}$$

The trajectory is helical over a circular base. Its projection on the plane Oxy is a circle with a radius given by

$$R = \left|\frac{v_{0x}}{\omega}\right| = \left|\frac{mv_{0x}}{qB}\right|,$$

and covered at an angular velocity given by $\dot{\theta} = \omega = -\frac{qB}{m}$.

Chapter 9

Electromagnetic Processes Applied to a Large-Scale Apparatus: The Ion Accelerator

9.1. Introduction: general principles and overall design of a machine for implanting ions

On designing such a machine, the first question asked is what type of ions we would like to deposit and what energies they will need to carry out the physical study in question. The energy of the ions should be of the order of a GeV to study the elementary particles in a material. To introduce ions (which will end up in the atomic state) into a material, to obtain electronic or optical doping, the energy necessitated for the ions is of the order of keV or MeV. In effect, to minimize the faults that can be formed during the process of ion implantation, lower energies are solicited. These can be around several tens of keV and are now widely used. Lower energy ions also resolve the problem of X-ray protection, which otherwise is necessary when using energies higher than 100 keV. If surface treatments are planned, such as cleaning by ionic pulverization or densification by ion-assisted deposition, then the energy levels used are more of the order of 100 eV to tens of keV.

Therefore, machines that operate over a range of 100 eV to 100 keV (while benefiting from the use of multicharged ions) now can be used for a wide range of physical treatments. It is this apparatus that this chapter describes.

The nature of ions and the required density of their ion current determines the choice of the source. In practical terms, electronic cyclotron resonance (ECR) sources make it possible to reach high current densities (10 μA cm^{-2}) while delivering large numbers of multicharged ions. As at a given acceleration potential triple-charged ions exhibit thrice the energy of monocharged ions, the operating energy range can be increased considerably. In addition, ECR sources permit continuous, long durations of irradiation with reactive ions. This is not often found with other sources, which often use filaments. This chapter therefore will be limited also to an ECR source.

The primary ion beam, direct from the ECR ion source, needs to be well defined with respect to the mass, charge states, energy, and the dimensions of the beam spot. The first three conditions can be fulfilled by using a mass filter, a suppression of neutral particles resulting from collisions between the ions and any residual gases, and choosing the potential between the source and the target, respectively. These techniques are schematized in Figure 9.1. The last parameter—the spot size—is controlled mainly by the ionic optics of the machine, using lenses and diaphragms detailed in Chapter 8.

In addition, the quality of a beam of low energy ions depends heavily on any space charge effects that may result in a dispersion of the low-energy beam. The ions move through a region of high energy that must include a large number of elements that shape the beam. This is done with the help of a tube in which the ions have a kinetic energy equal to that of their extraction. The tube contains a series of focalizing lenses, diaphragms, a mass filter (circular section filter or rectilinear filter otherwise known as a Wien filter), a neutral trap, and a sweeping zone that permits the equal distribution of ions on the beam. This zone is followed by one containing the acceleration and deceleration lenses, which is adjacent to the implantation chamber. The latter is earthed for experimental simplicity and security (the link between the chamber and the targets has a resistance of the order of 4 ohms).

Finally, experiments involving the treatment of materials are extremely sensitive to residual gas pressures. A good vacuum therefore is obligatory. At the source, a pressure associated with plasma conditions ($\leq 10^{-5}$ mbar) is necessary. At the level of the target chamber, a pressure between 10^{-6} and 10^{-5} mbar is common. In order to reach these levels, the apparatus should be capable of delivering a pressure of the order of 10^{-7} mbar when there is no inlet gas.

9.2. Setup of an ion beam

9.2.1. Overall description

The above-described accelerator has three sections, each set out in Figure 9.1.

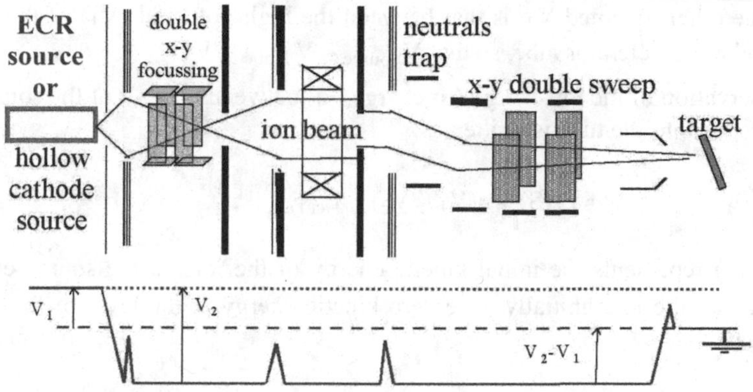

Figure 9.1. *Schematization of the different stages required in an ion accelerator (low energies from 100 eV to 70 keV), where qV_1 is the energy of the ions, qV_2 is the energy of the ions following a selection by mass and the shape of the beam, and V_2-V_1 is the deceleration voltage.*

The first stage corresponds to that of a cyclotronic resonance source working with gases under classic conditions. A turbomolecular pump directly connected at this level gives a pressure of the order of 10^{-5} mbar during plasma formation and 10^{-7} mbar at rest. The resolution of the various electromagnetic problems encountered during the design of such a device is detailed in Section 9.3.

The second stage consists of the acceleration–deceleration tube which has one end placed at the source exit. At this point, there also is a single lens—a so-called Einzel lens—and the beam thus is introduced into the mass filter. The latter in this case is of the Wien type (by rectilinear mass filtering). Neutral and as yet unfiltered particles are removed by a double deviation, of around 3 ° of the beam. The ions are then doubly swept in the x and y directions before reaching the postacceleration or postdeceleration zone.

The third stage consists of an earthed irradiation chamber. The spot of the beam thus is "swept" across the target (a surface typically around 5 x 5 cm) and a Faraday cage connected to a current integrator makes it possible to control the density of the ion current deposited on the target.

9.2.2. Use and distribution of high tensions within the apparatus

Two high tensions are applied:

- one (denoted the potential V_1) is between the high potential of the ion source and the true mass (in the irradiation chamber). This can be denoted as $V_{Source} - 0 = V_1$;

- the other, denoted V_2, is that between the high potential (V_1) of the source and the accelerator tube, so that $V_{Source} - V_{Tube} = V_2$.

The conservation of the ion energy (of charge +q) between the exit of the source and their entrance into the tube is written as:

$$[qV_1 + E_c(S) = q(V_1 - V_2) + E_{c(Tube)}], \qquad (1)$$

where $E_c(S)$ represents the initial kinetic energy of the ions at the source exit. On assuming that the ions initially have zero kinetic energy at the level of the source, we can state that

$$\frac{1}{2}mv_2^2 = E_{c(Tube)} = qV_2, \qquad (2)$$

where v_2 is the velocity of the ions in the tube.

With the exception of the variations in the electric potential energies associated with effective zones for beam control (lenses at potentials given by the general terms VL and applied between the central electrode and the symmetrical electrodes at the floating tension of the tube), the ions display the same velocity along the tube, such that

$$v_2 = \sqrt{\frac{2q}{m}V_2}. \qquad (3)$$

This velocity will be increased or decreased only at the lenses (situated post-acceleration of postdeceleration) placed in the neighborhood of the implantation chamber.

Writing down the conservation of energy between two extremities of the machine,

$$[qV_1 + E_c(S) = qV_{(chamber)} + E_{c(target)}]$$

makes it possible to write [with $E_c(S) = 0$ and $V_{(chamber)} = 0$] that:

$$E_{c(target)} = qV_1. \qquad (4)$$

With the potential energy of ions arriving at the target being equal to zero by construction (as the implantation chamber is earthed) then the total energy of the implanted ions thus is also equal to qV_1.

To conclude, given the scheme of potentials involved, it is the potential difference applied between the source and the implantation chamber (V_1) that determines the implantation energy of the ions (equal to qV_1). It is by the use of V_2 that the transport energy of the ions down the tube is controlled [which are therefore at a floating potential $(V_1 - V_2)$ along with the lenses, the Wien filter, the system for the deviation of neutral species, and the sweep].

For an accelerator that delivers monocharged ions going from 100 eV or so up to 30 keV (which can be used to reach 90 keV with triply charged ions), the potentials used are for $V_1 = 0.40$ keV and for $V_2 = 10$ or 20 keV (the kinetic energy for ion transport in the tube, which makes it possible to guide the beam without risking its dispersion by space charges).

9.2.3. *The Wien filter*

9.2.3.1. How it works

This type of filter has the advantage of selecting along the axis of the accelerator, thus limiting its geometric dimensions. Only ions selected from the given values of $\vec{E}$ and $\vec{B}$, which insure an equilibrium between electrical and magnetic forces (Figure 9.2), will follow their initial trajectory without deviation. However, this type of filter has the disadvantage of creating localized strong space charges associated with nonselected ions. These are deviated toward a zone where their confinement can result in perturbation fields (of the applied field $\vec{E}$) which then can influence the required effect of separation.

These filters are in principle limited to machines dealing with quite low currents, or the order of less than 100 μA, so as to limit the aforementioned space charge. Nevertheless, by locally generating electrons (by the thermoelectronic effect) in the space charge zone, the positive ions can be neutralized. In practical terms, this means placing a filament that generates electrons that will go toward an electrode connected to a plate of the Wien filter, on which the nonselected positive ions accumulate. The magnetic force generated by permanent magnets in the Wien filter dominates (at the beginning where $\vec{E} \approx 0$) over the weak electrostatic force applied, and the ions thus accumulate on the positive plate (weakly so at the start). A manual adjustment of the intensity of the current in the filament controls the supply to the filter plates so that a permanent flow is not brought about. In effect, an equilibrium is brought about between the current of ions associated with the space charge and the electronic current produced by the heated filaments (Figure 9.2).

9.2.3.2. Velocity filter and its configuration

Given the following experimental conditions (and when $q > 0$):

- velocity of the ions in the tube along Oz is denoted v_2,

- there is a magnetic field (B) directed along –Ox, of a fixed value, and produced by permanent magnets of several hundred Gauss, and
- electric field directed along Oy,

then the magnetic forces [$\vec{F}_m = q(\vec{v}_2 \times \vec{B})$] and the electric forces [$\vec{F}_e = q\vec{E}$] are anti-parallel. Their effect cancels out (so that the initial trajectory of the ion is not changed) when either $|F_e| = |F_m|$, or when $qE = qv_2B$, which is the same as when

$$v_2 = \frac{E}{B}. \qquad (5)$$

This is the condition for a rectilinear trajectory.

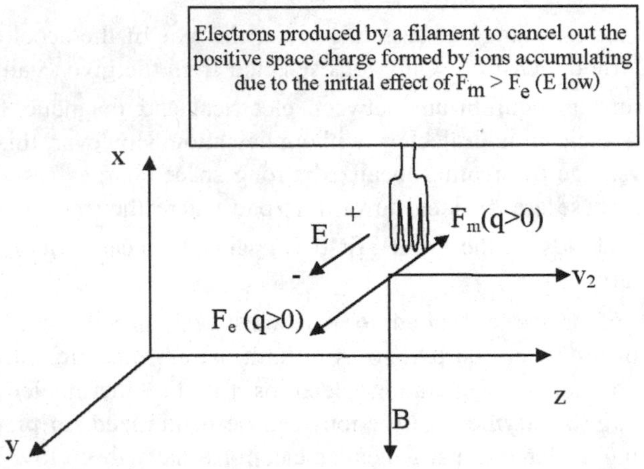

Figure 9.2. *Direction of electric (F_e) and magnetic (F_m) forces in a Wien filter.*

At this level of reasoning, we can consider that the filter acts like a velocity filter. In effect, the velocity (v_2) of the ion at the point of the Wien filter is determined by the extraction potential (V_2), which has a value given by Eq. (3).

Numerically speaking, for example, using potassium where m = A uma = 38/N, gives $m = 6.31 \times 10^{-23}$ g, we find that $v_2 = 2.25\ 10^3$ m s^{-1}.

9.2.3.2. Selection by mass

For a given extraction tension (V_2), the velocity indicated by Eq. (3) thus is of the form:

$$v_2 \overset{(3)}{=} \sqrt{2V_2}\sqrt{\frac{q}{m}} = K\left(\frac{q}{m}\right)^{1/2}, \qquad (6)$$

where $K = \sqrt{2V_2}$ is a constant at the given extraction tension.

It thus appears that for ions of the same charge, the velocity filter acts as a mass filter. For multicharged ions (with n charges), nq is substituted for q into Eq. (6). The corresponding particle is selected as if it had a mass given by m/n. The doubly charged ions appear as if they have half their mass, and similarly triply charged ions appear to have one third of their mass.

By writing the equality between the squared Eqs. (3) and (5), we obtain: $(E/B)^2 = 2qV_2/m$, so that

$$m = 2qV_2\left(\frac{B}{E}\right)^2. \qquad (7)$$

For given vales of B, E, and V_2, Eq. (7) gives the value of the selected mass (in a rectilinear trajectory). As B and V_2 are determined by the initial experimental conditions, the control of m is made simply by varying the tension applied to the Wien filter which in turn forms the field E.

In practical terms, if the experimental controls remain unchanged (i.e., V_2 and B remain untouched), we can state that for ions of the same charge that

$$m\,E^2 = \text{constant.}$$

On using this equation for two elements of given masses m_1 and m_2, we can determine the field (E_2) necessary to select m_2 when the standard field (E_1) determined for the standard mass (m_1) is known. This is done using:

$$E_2 = E_1\left(\frac{m_1}{m_2}\right)^{1/2}. \qquad (8)$$

9.2.3.3. Separating power

A simple calculation (using the results from Section 8.1.2.1) shows that for a distance ℓ from the middle of a filter of length a, the deviation denoted by Y of the particles of mass $m + \Delta m$ with respect to an axis along which nondeviated particles of mass m travel is equal to:

$$Y = \frac{a\,\ell\,E\,\Delta m}{4\,mV_2}. \qquad (9)$$

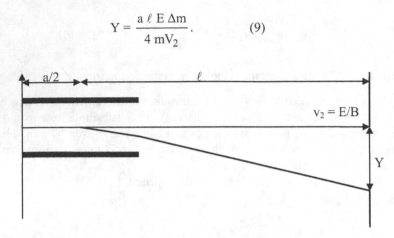

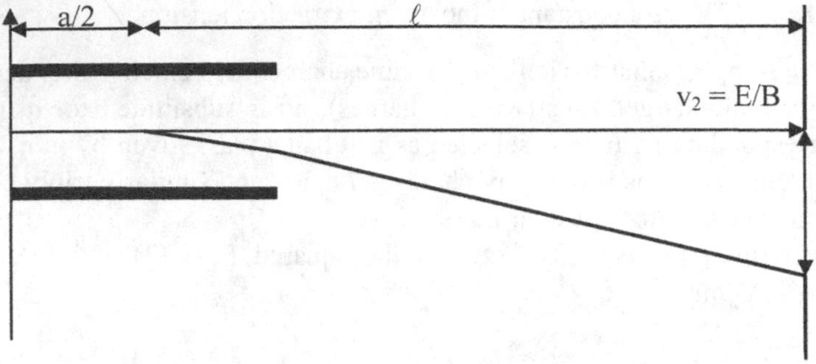

Figure 9.3. *Deviation (D) of ions of mass m + Δm with respect to the trajectory of non-deviated ions with mass m and velocity v_2 = E/B (Wien filter).*

Therefore, the separation and hence the resolution power of the filter can be modified by modifying the electric (E) or the magnetic (B) field (as $V_2 = mv_2^2/2q = mE^2/2qB^2$).

We can note that Y decreases if m increases (from which the difficulty of separating high-mass ions).

We also can state that two ion beams will be resolved if the separation of the beams is greater than or equal to the sum of the diameter of the beams (2r) and the slit width (F), i.e., when $Y \geq F + 2r$. In concrete terms, F is controlled by the diaphragm diameter at the exit point of the Wien filter. In $Y = F+2r$, the equal sign is obtained at the separation limit where Y is given by $Y = C\ \Delta m/m$ [C is the constant determined in Eq. (9) preceding that for Y].

9.2.4. The neutrals' trap

Any neutral particles need to be eliminated as they are not compatible with the Faraday cage (which counts the number of charged ions), and because they create implantation inhomogeneities as they cannot be spread about the target by the sweep system.

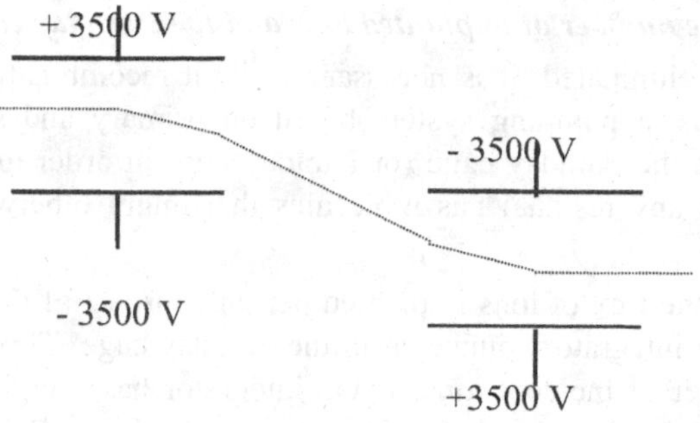

Figure 9.4. *Electrostatic double deflection making up the neutral trap.*

The trap thus works by two electrostatic deflections of the ion beam–which retains its parallel nature, as shown in Figure 9.4.

9.2.5. Sweeping

With the help of diaphragms and electrostatic lenses we can produce a focused implantation "spot" (image of the source). It is this spot that is swept across the target describing its outline. If the spot is out of focus, then the spot is larger and less able to define the target, but can yield a more homogeneous implantation.

The sweep is controlled through electrostatic deviations. For a given tension (V_d) applied between the two parallel sweep plates of length ℓ and separated by a distance h, then the deviation (Y) at a distance (D) from the middle of each plate (Section 8.1.2.1) can be given by:

$$Y = \frac{\ell D}{2h} \frac{V_d}{V_2}. \qquad (10)$$

To obtain a sweep in X and Y directions, two parallel pairs of plates are used, one operating in one direction, the other in a perpendicular direction (Figure 9.1). The two systems are set up in parallel with the potentials of the plates being crossed so that the deviations each compensate one another (as for the neutral deviation plates).

The tensions used in the deflection process are typically of a triangular form, with a frequency of between 10 and 200 Hz. Lissajous curves (where the frequencies are in a simple integer ratio) should not be generated as they would otherwise lead to an inhomogeneous implantation, and means that the frequencies used are near (for example 10 and 11 Hz) or far apart (for example 10 and 200 Hz).

9.2.6. Determination of the number of implanted ions and the Faraday cage

With the neutral particles eliminated, it is necessary to limit recombination of the remaining ions. To do this, a pumping system based on primary and secondary pumps is installed close to the Faraday cage (or Faraday cup) in order to limit the number of collisions with any residual gas molecules that might otherwise favor recombination processes.

The measurement of the flux of ions implanted per unit surface of the target is carried out using a current integrator connected to the Faraday cage. This acts as a collector and compatibilizer of incident ions. If the integrator has a high entrance resistance (which opposes the flow of charges), then the cage becomes positively charged and attracts secondary electrons that result in a current that tends to diminish the real current of positive ions. This is the reason why integrators with low entrance resistances are generally used.

In practical terms, a current density of 1 A cm^{-2} on the cage is equivalent to it receiving 6.26 x 10^{18} monocharged ion cm^2 s^{-1} (as 1 A = 1C/1s and 1C = Ne, where $N = 1/e =$ 6.25 x 10^{18} when e = 1.6 x 10^{-19} C). Finally, if:

- I is the current (in amps) arriving on the Faraday cage;
- t is the irradiation time in seconds;
- S is the surface of the Faraday cage in cm^2; and
- n is the state of the implanted ions,

then the Faraday cage which receives a charge given by Q = It actually has per cm^2 a charge of Q/S. This quantity of charge corresponds to a number of monocharged ions N (Q/S). If the ions carry n charges the flux (D) that by definition represents the number of received ions per cm^2 thus is given by:

$$D = \frac{N}{n}\frac{Q}{S}, \qquad (11)$$

so that numerically speaking (with S in cm^2):

$$D\ (\text{ions/cm}^2) = (6.2\ \text{x}\ 10^{18}\ \text{I t})/(\text{S n}).$$

In practical terms, the Faraday cage is exchanged for the target to be treated with the ion beam. Any measure of the flux thus is performed prior to and following the ion treatment so as to insure a stable ion beam. A more stable and elegant solution is found in using a Faraday cage with a hole in the middle for the target. This insures a stable beam during the exposure of the target to the beam.

9.2.7. General remarks on the mechanism used to produce ions

9.2.7.1. Ion formation

Positive ions generally are formed from electric discharges from a chamber containing, either wholly or partly, the gas to be ionized. Classically, the ions are produced following a bombardment of the gas by electrons accelerated between the source cathode and anode, with the level of ionization being determined by several factors.

9.2.7.1.1. Ionization potential of the atoms present as a gas or vapor

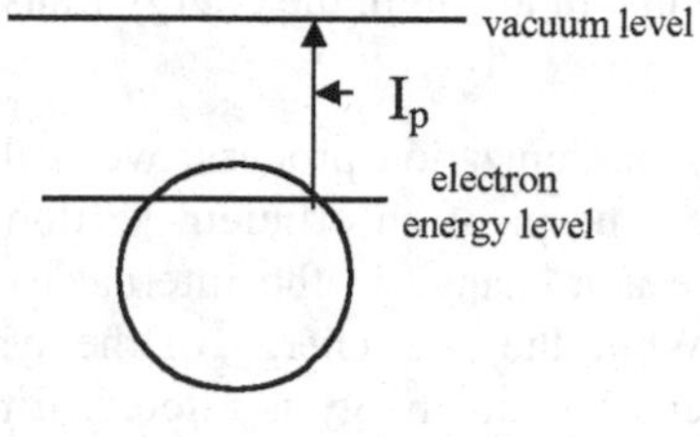

Figure 9.5. *Atom ionization.*

The ionization potential of an electron is equal to its bonding energy (Ip), which is the difference between its energy at the level of a vacuum (taken as the origin for energies) and the original level of the electron. Metals with a single valence (column 1A of the periodic table) typically exhibit low ionization potentials (Ip = –5.4, –5,1 and –3.9 eV for Li, Na, and Cs, respectively) and are ionized by low tensions. Atoms with more saturated outer layers have high bonding energies (24.6 eV for He, – 21.6 eV for Ne, – 13.6 eV for O, and – 17.4 eV for F).

9.2.7.1.2. The electronic current

The current of electrons controls the ionization of the atoms or molecules with which it comes into contact. At different pressures, the gases or vaporized atoms are ionized on impact with electrons. The latter are generated by cold or hot cathodes or directly by an inductive or capacitive coupling between the gas itself and a radiofrequency field due to the high number of collisions in the volume caused by the high-frequency electrons generated by the field at a frequency $\nu \approx 10$ MHz. The density of the ions depends, in the absence of a discharge, on the electronic current between the cathode and the anode.

Emission of electrons from electrodes can be obtained by thermoionic emission, photoelectric emission, electronic impact (electrons at 100 eV to 1 keV being the most efficient to yield secondary electrons), or by ionic emission using a field effect (due to a tunneling effect with fields of the order of 10^6 V cm^{-1}).

9.2.7.1.3. Collision efficiency and elastic and inelastic collisions

Rather than considering an individual electron–atom collision, it is the statistical result of the collisions that should be studied. In order to do this, we have to look at the effective surface (or efficient surface) for the collisions, which corresponds to an efficient section. The latter can be used just as well for elastic or inelastic collisions.

The efficient elastic collision section for an atom (with a diameter denoted as d) with an electron (which is assumed to be a point with a practically negligible diameter) is given by $\sigma = \pi\, d^2/4$. With an atomic density given in terms of n atoms cm^{-3}, the total surface is given by $n\sigma$. Taking into account the low mass of an electron with respect to that of an atom, the energy transferred in elastic collisions is nevertheless small.

In order to study the ionization process, we really need to look at inelastic collisions to which the concept of an efficient section also can be applied. In an inelastic collision, there is a change in the internal excitation of the atom, which follows the collision. While the total energy of the system is conserved, the ratio between the potential and kinetic energy is altered. Either totally or partially, the kinetic energy of the incident electron is used to modify the potential energy of the valence electron of the atom.

Two types of collisions thus can come about:

- the first type where the kinetic energy of the incident electron is transferred to the valence electrons and increases their potential energy. The atom thus carries an excited state for which the limiting state is the ionized state; and
- the second type where the atom is already in an excited state and on meeting an electron the atom transfers its excitation energy (thus reducing its potential energy) to the incident electron, which sees its own kinetic energy increase. This type of collision also can come about between atoms or ions. A collision at an atom (A) with an atom (B) in a metastable state gives: if eV meta B > eVi (where Vi is the ionization potential of A) then A is ionized. This process is that of the Penning effect which is applied to the ionization of mixtures of rare gases. Molecular gases with ionization potentials of the order of 15 eV are easily ionized by this method, notably through an intermediate such as neon which exhibits a metastable state with a long lifetime at 16.5 eV. In addition, this type of collision involves a large efficiency section, and as a consequence shows a high ionization yield, greater than that of electron–atom collisions (for which the efficiency section is small given the relatively small size of the electrons with respect to atoms).

9.2.7.1.4. Effects caused by the pressure and geometry of the discharge chamber on the ionization

The breakdown voltage for a gas depends on a large number of parameters, such as the type of gas, its purity, the space between the electrodes, and the nature of the materials making up the electrodes and the enclosure.

The Paschen law takes into account the combined effects of the breakdown voltage (V_c), the pressure (P) of the gas, and the distance (d) between the electrodes. The law indicates that V_c depends on the product of P and d for a given electrode system, so that $V_c \approx P\ d$. A plot of V_c as a function of the product (P d) shows a minimum which signifies that there is for a certain value $(P\ d) = (P\ d)_{min}$ a minimum breakdown voltage. Finally, for a given separation d between the electrodes, this law implies that the breakdown is least easy at both low and high pressures. The minimum is due to the frequency at which the collisions occur and the effect of the energy acquired on each collision.

9.2.7.1.5. Effect of the form of the magnetic field lines on the pathway and confinement of the electrons: increasing the ionization yield

This effect, and most notably that of a magnetic bottle, is dealt with in a problem at the end of this chapter.

9.2.8. Nature of the electric discharge

Naturally occurring ionization, due to cosmic rays, UV radiation, and radioactivity and under normal condition of temperature and pressure, can be found in the atmosphere at levels of around 1000 positive and negative charge per cm^3. At equilibrium, the generation and recombination of charges is equal.

Under such conditions, two electrodes carrying a low electric field can circulate a small current with a density given by $J_{sat} = e\ d\ (dn/dt)$. Here, the equilibrium conditions remain unmodified so that the total number of charges which arrive at each electrode as dn/dt per cm^3 and per second is equal to the number of charges produced. The value of the current density, under normal conditions, is $J_{sat} \approx 10^{-9}$ A cm^{-2} and thus insufficient to produce a visible discharge.

If, however, the tension between the electrodes is increased, then the current increases as a function of the gas pressure. A rapid increase in current with tension is due to:

- a volume effect where collision with rapid primary electrons with atoms in the volume generate ions and hence an internal ionization of the gas. The electrons are those that are accelerated by the electric field applied between the cathode and the anode. Following this ionization there is a multiplication of electron–ion pairs, resulting in an avalanche toward an eventual discharge;

- a surface effect where the collision of ions with the cathode surface generates secondary electrons which result in a high increase in the intensity. A steady luminescent discharge thus appears when there is a sufficient number of ions to generate enough electrons at the cathode.

Electric discharges thus can be obtained by:

- application of a continuous tension with the pressure playing an essential role for the luminescent discharge which appears at low pressures. At equilibrium, the discharge is maintained by electrons coming from the cathode, bombarded with positive ions, and absorbed by the anode. At sufficiently high pressures (≈ 10 Torr), a very high current discharge arc can appear (of a density reaching 10^6 A cm^{-2} at the cathode) making the positive column appear very luminescent;
- by using tensions at various frequencies:
  - o up to 10 MHz, while the frequencies remain low (less than 1 kHz) the period associated with a cycle (≈ 1 ms) is generally higher than the transit time of the ions and the breakdown tension is of the same order as the continuous current. When the frequency increase, the value of the breakdown voltage (Vc) can increase and then decrease for the following reasons:
    - at low frequencies and with an increase in the frequency an increasing number of ions see their direction inversed, and the actual number of ions reaching the cathode diminishes just as does the number of secondary electrons emitted by the cathode. Thus the breakdown voltage increases;
    - at higher frequencies, the efficiency of oscillating electrons in ionizing collisions increases. The length of the oscillations at these high frequencies become very small, the electrodes are no longer reached, and phenomena at the electrodes become negligible with respect to the high number of collisions in the volume. Finally, the important contribution of ionizing collisions makes it possible to decrease the breakdown voltage for frequencies higher than several MHz.

  - o At frequencies greater than 10 MHz, when discharges occur at these frequencies, the current is maintained without resorting to electrodes. In effect, a high number of collisions occur in the volume due to the high-frequency, and the electrodes become unnecessary to produce electrons. Generally, a high-frequency field is produced using a simple capacitive coupling (with electrodes being outside the enclosure). The devices are termed radiofrequency (RF) sources.

- Above 1 GHz and with a low gas pressure (from 10^{-4} to 10^{-5} Torr), the application of a high-frequency field is capable of generating several oscillations for an electron between two consecutive collisions with gas atoms. At what is a cyclotronic frequency, the electrons continuously gain energy from the electric field and develop a high electronic temperature.

One of the benefits of using such sources is the high production of multicharged ions. Multi-ions can be produced from single-impact events with energetic electrons and also through step-by-step ionizations (equivalent to multiple impacts) with lower-energy electrons. High current densities of monocharged ions can be equally obtained. In general, the system is used outside of resonance conditions and the discharges thus are produced using a high-pressure medium which increases the energy of the ionizing electrons and results essentially in the production of monocharged ions. Microwaves at 10 GHz often are used, but nowadays the ready availability of magnetron oscillators at 2.45 Ghz (microwave ovens) has permitted the development of sources close to this high frequency. We now will present in more detail these types of sources, commonly referred to as ECR sources.

9.3. ECR-type source of ions ("cyclotronic resonance")

9.3.1. Principles of an ECR source

An ECR source (schematized in Figure 9.6) functions by the ionization of a gas (or vapors) through the collisions of gas atoms and highly energetic electrons (high ionization yield). A magnetic field (B), such that its field lines confine the electrons, produces a helical trajectory of the electrons, with a rotational frequency corresponding to the cyclotronic frequency given by $\omega_c = eB/m$. The application of an electromagnetic field (EM) makes it possible to transfer from the EM field the energy of the electrons. The resonance condition for this energy transfer is that the pulsation of the EM field is equal to the cyclotronic pulsation (ω_c). For an electric field at a frequency of 2.45 GHz, the necessary magnetic field is 875 Gauss. The radius of the trajectory is given by $r_c = u/\omega_c$ where u is the component of the electron velocity perpendicular to the direction of B. For a frequency of 2.45 GHz and electrons with energies ($E = \hbar\omega_c$) typically between 5 and 10 eV, the circular orbitals have a radius of around 0.01 cm, which is considerably smaller than most chambers associated with ion sources.

The density of the plasma is given by the equation $N_p = (m\varepsilon_0\omega_p^2/e^2)$, and therefore the frequency of the microwave field is higher than the plasma frequency (so that the field propagates). Figure 9.7 recalls the propagation condition detailed in Chapter 8, Volume 1, *Basic Electromagnetism.* When ω ($\approx \omega_c$) is too small or when ω_p is too large (meaning that in practical terms N_p is too large), the EM wave cannot penetrate the plasma. For a frequency of 2.45 GHz, the critical density (N_p) is thus 7 x 10^{10} ions cm^{-3}.

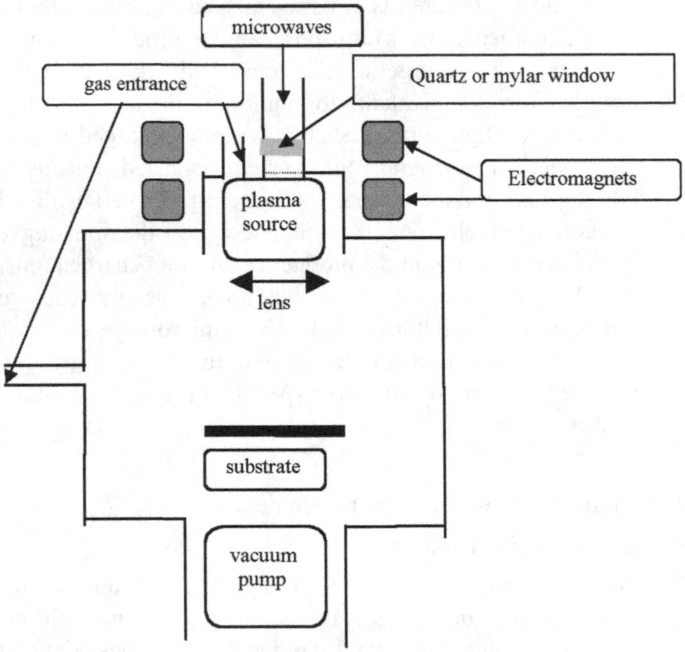

Figure 9.6. *Schematization of the workings of an ECR source.*

As the external flux of ions coming from the plasma is directly proportional to the plasma density, a dense plasma is necessary to obtain an elevated ion current (beam current). Using the preceding value for N_p, a saturation current of the order of 10 mA cm^{-2} can be obtained.

A powerful EM field is injected into the source through a quartz, ceramic, or mylar window, as shown in Figure 9.6.

B can be produced from two electromagnetic reels that surround the external sides of the source so that the field is axial.

The gas is directly introduced into the upper part of the source at the level of the sample. The source unit is cooled using a circulation of water, as the level of power used to drive the source develops a considerable amount of heat.

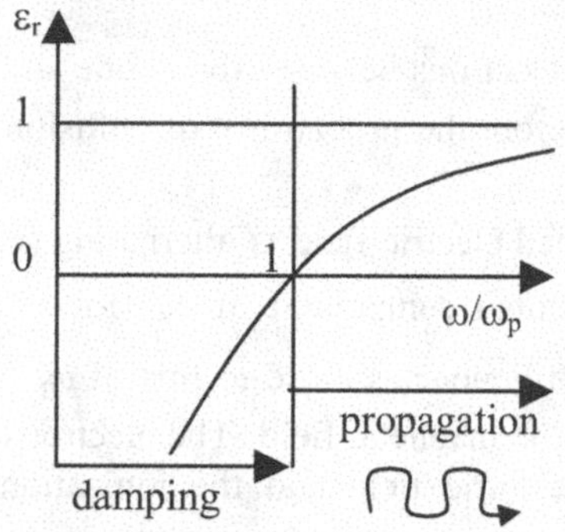

Figure 9.7. *Propagation of an EM wave in a plasma.*

9.3.2. Magnetic field effects on the pathway and confinement of electrons: increasing the ionization yield

9.3.2.1. Trajectory in the presence of a uniform magnetic field (B)

A particle of a given charge (q) and velocity ($\vec{v}$) placed in a magnetic field ($\vec{B}$) undergoes a force given by $\vec{F} = q\left(\vec{v} \times \vec{B}\right)$ which results in a circular movement if $\vec{v}$ has a component (u) perpendicular to $\vec{B}$. The corresponding angular velocity is:

$$\vec{\omega}_c = -\frac{q\vec{B}}{m}$$

for a cyclotronic angular frequency for a particle of mass m (if $q < 0$, $\vec{\omega}_c \,//\, \vec{B}$). The curvature radius (R) of the trajectory is given by $R = u/\omega_c$, so that $R = mu/qB$ (we also can say that $F = q\ u\ B$, as $\vec{u} \perp \vec{B}$, so that $q\ u\ B = m\ v^2/R$, from which $R = mu/qB$).

If in addition the particle is subject to a velocity that has a component $\vec{v}_{//}$ parallel to $\vec{B}$ [and therefore the magnetic force is zero as $q\left(\vec{v}_{//} \times \vec{B}\right) = 0$], then the translation effect in the direction of this component superimposes itself on the preceding circular rotation. The resultant movement therefore is that of a helical movement with a constant helical step parallel to $\vec{B}$ (see problem 3).

If a continuous electric field ($\vec{E}$) is applied perpendicularly to $\vec{B}$, the resulting movement is that of a cycloid directed in the direction normal to $\vec{E}$ and $\vec{B}$. When $\vec{E}$ exhibits an additional component in the direction of $\vec{B}$, then there is a deviation in this direction.

The movement of the electrons between the anode and cathode is increased on application of $\vec{B}$, and therefore the probability of collision and hence ionization is increased.

If in addition the applied electric field is alternating (rotating field as shown in Figure 9.8), then the rotational component of $\vec{E}$ (for example, $\vec{E}^+$ denoting an angular frequency $+\omega$) for frequencies close to that of ω_c ($+\omega = \omega_c$) will produce a resonance effect driven by the magnetic field. The electrons quickly will take on the energy coming from the electric field and the ionization yield will be strongly increased.

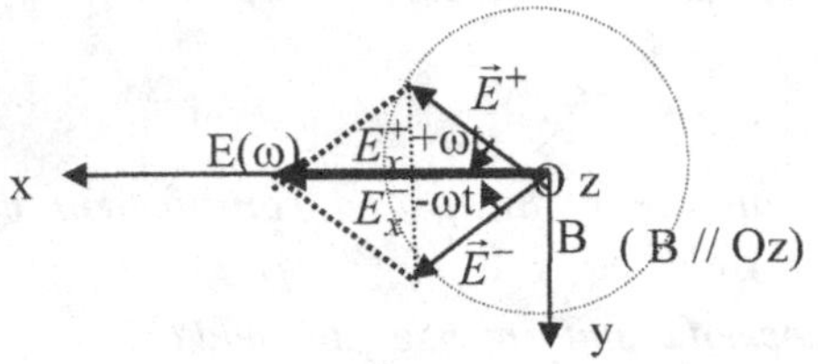

Figure 9.8. *Rotating field ($\vec{E}^+$) such that*

$$E_x^+ = \text{proj}_{Ox}\vec{E}^+ = E\cos\omega t,\ E_x^- = \text{proj}_{Ox}\vec{E}^- = E\cos(-\omega t),$$

from which

$$\vec{E}(\omega) = \vec{E}^+ + \vec{E}^- = \left(E_x^+ + E_x^-\right)\vec{e}_x = E(\omega)\vec{e}_x \text{ with } E(\omega) = E_x^+ + E_x^- = 2E\cos\omega t.$$

9.3.2.2. Confinement of electrons with the help of a magnetic bottle

The discharge electronic current density between an anode and a cathode depends on the respective values of the creation and annihilation of electrons. The electrons can be localized by using a magnetic mirror formed by an *ad hoc* distribution of magnetic field lines. To further understand this, it is of interest to study the helical trajectory of a charged particle placed in a convergent field (such that B crosses from point P_1 toward another point P_2), and such that $\vec{B}$ has a cylindrical symmetry (that is to say that B is independent of θ), so that $B_\theta = 0$, and thus also:

$$\vec{B} = B_r\vec{e}_r + B_z\vec{e}_z.$$

As $\mathrm{div}\vec{B} = 0$, so that $\oiint \vec{B}.\vec{dS} = 0$, the flux of $\vec{B}$ is constant through a tubular section of the field. As the section (S) decreases on going from P_1 toward P_2, $\vec{B}$ increases in the same direction and $B(P_2) > B(P_1)$.

From $\mathrm{div}\vec{B} = 0$, we can deduce that:

$$\frac{1}{r}\frac{\partial}{\partial r}(rB_r) + \frac{dB_z}{dz} = 0\text{, so that } rB_r = -\frac{r^2}{2}\frac{dB_z}{dz} + \varphi(z).$$

As when r = 0 we have $r\,B_r = 0$ and a result that gives

$$\varphi(z) = 0\text{, from which } B_r(r,z) = -\frac{r}{2}\frac{dB(z)}{dz}.$$

The radial component (B_r) of B is directed toward the center of the helice. We denote $v_\theta = u$ as the component of the velocity normal to B (and therefore to B_r). The corresponding magnetic field $q\left(\vec{v}_\theta \times \vec{B}_r\right)$ works as a decelerating force that reduces the derived velocity in the direction P_1P_2.

In effect, we can show that not only the movement of the particle is slowed, but even zeroed and inversed in direction. This is as if the system is working as a magnetic mirror, which can be used to form a magnetic bottle of charged particles (Figure 9.9).

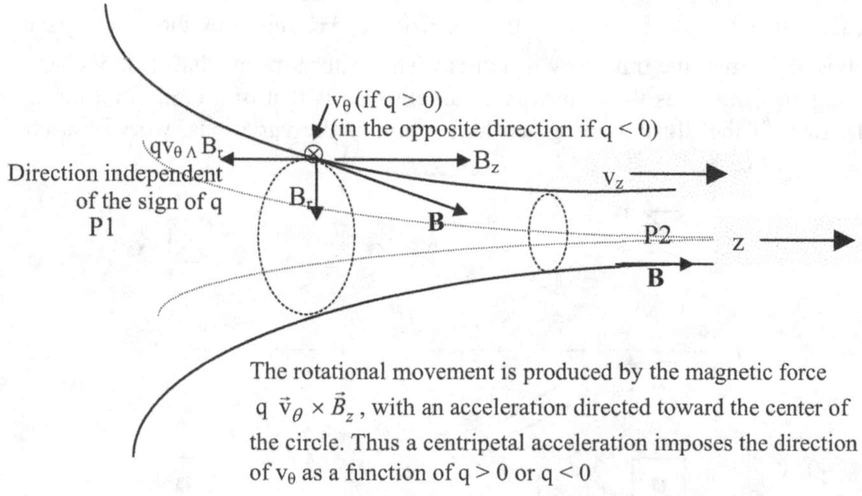
vθ (if q > 0)
(in the opposite direction if q < 0)
qvθ ∧ Br
Direction independent
of the sign of q
P1
Bz
Br
B
vz
P2
z
B
The rotational movement is produced by the magnetic force
q vθ × Bz, with an acceleration directed toward the center of
the circle. Thus a centripetal acceleration imposes the direction
of vθ as a function of q > 0 or q < 0

Figure 9.9. *Confinement of charged particles.*

We can show, by projection of the dynamic fundamental equation on the co-ordinate axes of the cylinder that, as demonstrated in the problem at the end of this chapter:

$$\frac{dv_z}{dt} = -r\frac{q}{m}\dot{\theta}B_r\,,$$

so that by taking into account the preceding expression for B_r:

$$\frac{dv_z}{dt} = +\frac{q}{m}r\dot{\theta}\frac{r}{2}\frac{dB_z}{dz} = -\frac{v_\theta^{\,2}}{2B_z}\frac{dB_z}{dz}\,.$$

This equation shows that v_z decreases when in a region where B_z increases with z. Similarly, the analytical expressions in cylindrical coordinates (r,θ,z) show that v_θ increases in modulus while B_z increases with z.

Comment. This latter result can be confirmed by the following method. Denoting the magnetic force as $\vec{F} = q\left(\vec{v} \times \vec{B}\right)$ normal to the velocity of the particle, this force does not produce any work. Thus, $dW = 0 = d(mv^2/2) = mvdv$, so that $dv = 0$ and finally, $v = Cte$. The total velocity remains constant, that is to say that the velocity with respect to P_1P_2 (v_z) being diminished; the velocity normal to this direction is increased. If we assume that $v_r \approx 0$, then with $v_r = r'$ supposes that $r \approx$ constant, which is to say that the trajectory is a circle. This is turn means that B only changes very slightly with z (as the trajectory of an electron is that of a circle in a uniform field). Even if the illustrations given here magnify the variations, we will assume that this is the case.

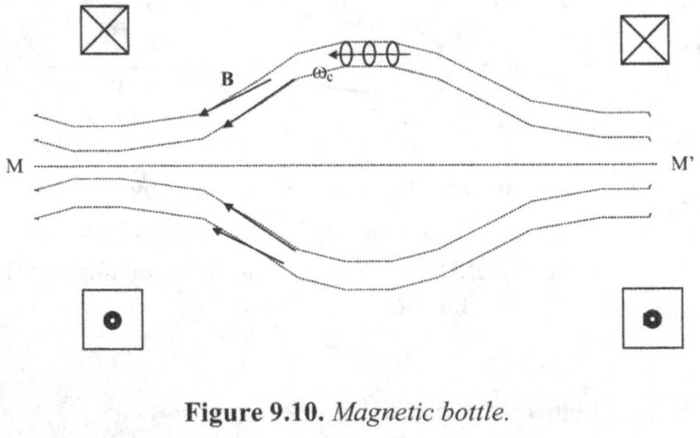

Figure 9.10. *Magnetic bottle.*

With the formation of these field lines in the ion source, we can make considerable gains in the ionization efficiency. In fact, for a magnetic bottle, there is a double convergence of field lines in space. The field tubes exhibit the form described in Figure 9.10, and the trajectories follow circles as the orbits of the particles are rolled up on the tubes of the field. On moving toward the points M or M', the intensity of B reaches a maximum, and the decelerating force $q\left(\vec{v}_\theta \times \vec{B}_r\right)$ increases up to a point where movement toward M and M' is cancelled. The electronic charges thus are confined between these points.

Such a magnetic field configuration can be produced by two reels carrying current in the same sense. As indicated above, the angular frequency associated with the rotation produced by B gives $|\omega_c| = eB/m$. If an alternating field (rotating field) at a frequency ω also is applied perpendicularly to $\vec{B}$, then a resonance rotation of the electrons is produced when ω_c takes on a value (ω_{c0}) such that $\omega = \omega_{c0} = eB_0/m$. For an electric field with a given frequency ω, the resonance occurs at a point in space where the field (B) takes on the precise value B_0 (which insures that $\omega = \omega_{c0}$). In the case of the magnetic bottle, the two points P and P' correspond to this value and it is in the neighborhood of these points that the ionization yield is at a maximum (the rotation of the electrons is in phase with the rotating electric field which concedes its energy to the electrons, which in turn use the energy to increase the degree of ionization).

9.4. Problem

Charged particle (q) in a slightly nonuniform magnetic field: the magnetic mirror

Here a charged particle (q) is subject to a magnetic field ($\vec{B}$) of which the component (B_z) along Oz varies only slightly with z. In addition, it is supposed that B exhibits a cylindrical symmetry, that is to say that B has the same value at whatever value of θ. As a consequence, B(θ) = constant, and to simplify, the component B_θ of B with respect to θ is taken to be equal to zero. As B_θ is zero, in cylindrical coordinates, we thus suppose that $\vec{B}$ has the form:

$$\vec{B} = B_r(r,z)\vec{e}_r + B_z(z)\vec{e}_z .$$

1. Calculate B_r that is expressed as a function of r and of $\dfrac{\partial B_z}{\partial z}$.

2. Show that the kinetic energy of the particle is retained.

3. Here we introduce: $\omega = -\dfrac{qB_z}{m}$.

From the components v_r, v_θ, and v_z of the velocity, give in cylindrical coordinates the differential equations associated with the movement of the particle (expressions for $\frac{dv_r}{dt}$, $\frac{dv_\theta}{dt}$, and $\frac{dv_z}{dt}$ as a function of v_r, v_θ, v_z, ω, $\dot{\theta}$, B_z, and $\frac{dB_z}{dz}$).

4. It also is supposed that B varies slightly with z, so that it can be assumed that locally B is uniform. Locally, the trajectory is considered a circle, for which:

$$v_r = \dot{r} = 0\text{, and } \frac{dv_r}{dt} = 0.$$

Give the angular velocity ($\dot{\theta}$) and give an expression for the radius of gyration (ρ) as a function of ω and of v_θ.

Express $\frac{dv_\theta}{dt}$ and $\frac{dv_z}{dt}$ as a function of v_θ, v_z, B_z and $\frac{dB_z}{dz}$.

5. Show that M_z is a constant of the movement.

6. Calculate the magnetic field flux through the orbit described by the particle. It is supposed that the particle moves in a tubular field.

7. Study the directions in the variations of the components v_z and v_θ of the velocity of the particles when B_z is crossed with z. Show that a reflection effect (magnetic mirror) can result.

8. It is supposed that the initial velocity ($\vec{v}_0$) of the particles is at an angle α to the Oz axis and that the particle moves in a region where the field B_z goes from a value B_1 to a value B_2 (where $B_2 > B_1$). Establish the condition that must be verified by α so that the particles are reflected on traversing this region.

Give the size of R for the reflected particles when they are emitted isotropically in a half-space.

Carry out a numerical appreciation of the calculation using the values of $B_1 = 10^{-4}$ T and $B_2 = 5 \times 10^{-4}$ T.

Answers

1. As $\vec{B}$ verifies $\operatorname{div}\vec{B} = 0$, we have in terms of cylindrical coordinates:

$$\operatorname{div}\vec{B} = \frac{1}{r}\frac{\partial}{\partial r}(rB_r) + \frac{\partial B_z}{\partial z} = 0.$$

from which we have

$$\frac{\partial}{\partial r}(rB_r) = -r\frac{\partial B_z}{\partial z},$$

so that

$$rB_r = -\frac{r^2}{2}\frac{\partial B_z}{\partial z} + K(z).$$

When r = 0, we have $rB_r = 0$ from which K(z) = 0, and we thus can deduce, that:

$$B_r = -\frac{r}{2}\frac{\partial B_z}{\partial z}.$$

2. The theory for kinetic energy is written as $dE_C = \vec{F}.\vec{dl}$, from which

$$dE_C = = q(\vec{v} \times \vec{B})\vec{v}\, dt = 0,$$

so that

$$E_C = \frac{1}{2} m\, v^2 = \text{constant},$$

and also v = constant.

3. In cylindrical coordinates, we have:

$$\begin{cases} \overrightarrow{OM} = r\vec{e}_r + z\vec{e}_z \\ \vec{v} = \dfrac{d\overrightarrow{OM}}{dt} = \dot{r}\vec{e}_r + r\dot{\theta}\vec{e}_\theta + \dot{z}\vec{e}_z == v_r\vec{e}_r + v_\theta\vec{e}_\theta + v_z\vec{e}_z \\ \vec{\gamma} = \dfrac{d\vec{v}}{dt} = \left(\ddot{r} - r\dot{\theta}^2\right)\vec{e}_r + \left(2\dot{r}\dot{\theta} + r\ddot{\theta}\right)\vec{e}_\theta + \ddot{z}\vec{e}_z. \end{cases}$$

The relation $\vec{\gamma} = \frac{q}{m}\vec{v} \times \vec{B}$ (coming from the fundamental dynamic equation) thus gives:

$$\begin{bmatrix} \ddot{r} - r\dot{\theta}^2 \\ 2\dot{r}\dot{\theta} + r\ddot{\theta} \\ \ddot{z} \end{bmatrix} = \frac{q}{m}\begin{bmatrix} \dot{r} \\ r\dot{\theta} \\ \dot{z} \end{bmatrix} \times \begin{bmatrix} B_r \\ 0 \\ B_z \end{bmatrix} = \frac{q}{m}\begin{bmatrix} r\dot{\theta}B_z \\ \dot{z}B_r - \dot{r}B_z \\ -r\dot{\theta}B_r \end{bmatrix}.$$

From this can be deduced that:

$$\begin{cases} \dfrac{dv_r}{dt} = \ddot{r} = r\dot{\theta}^2 + \dfrac{q}{m} r\dot{\theta}B_z \\ \dfrac{dv_\theta}{dt} = \dfrac{d\left(r\dot{\theta}\right)}{dt} = \dot{r}\dot{\theta} + r\ddot{\theta} = -\dot{r}\dot{\theta} - \dfrac{q}{m}\dot{z}\dfrac{r}{2}\dfrac{dB_z}{dz} - \dfrac{q}{m}\dot{r}B_z \\ \dfrac{dv_z}{dt} = \dfrac{q}{m} r\dot{\theta}\dfrac{r}{2}\dfrac{dB_z}{dz}. \end{cases}$$

By making $\omega = -\dfrac{qB_z}{m}$, we finally obtain:

$$\begin{cases} \dfrac{dv_r}{dt} = \left(\dot{\theta} - \omega\right)v_\theta & (1) \\ \dfrac{dv_\theta}{dt} = \left(\omega - \dot{\theta}\right)v_r + \dfrac{\omega}{2}\dfrac{v_\theta v_z}{\dot{\theta}B_z}\dfrac{dB_z}{dz} & (2) \\ \dfrac{dv_z}{dt} = -\dfrac{\omega v_\theta^2}{2\dot{\theta}^2 B_z}\dfrac{dB_z}{dz}. & (3) \end{cases}$$

4.

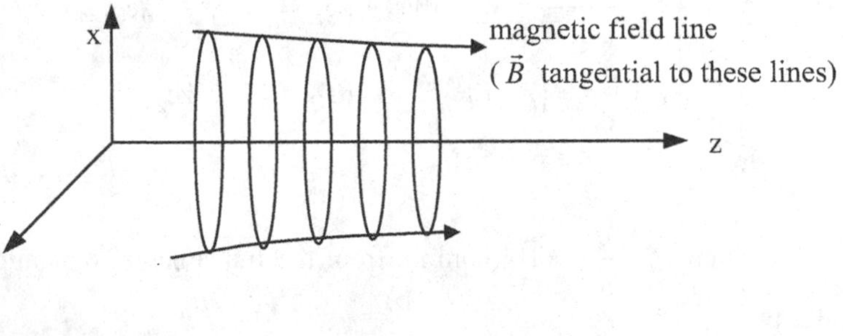

As $r \approx$ constant (circle), we have $v_r = \dot{r} = 0$.

Therefore, by assuming that

$$\frac{dv_r}{dt} \approx 0$$

we thus have from Eq. (1) $\dot{\theta} \approx \omega$ and $v_\theta = r\dot{\theta} \approx r\omega$. This equation gives a good estimation of the radius of gyration (ρ) (which becomes the radius r of the circle when B is uniform) as a function of ω and of v_θ, so that

$$\rho \approx \frac{v_\theta}{\omega}.$$

With $\dot{\theta} \approx \omega$, the preceding Eqs. (2) and (3) become:

$$\begin{cases} \dfrac{dv_\theta}{dt} = \dfrac{v_\theta v_z}{2B_z}\dfrac{dB_z}{dz} & (2') \\ \dfrac{dv_z}{dt} = -\dfrac{v_\theta^2}{2B_z}\dfrac{dB_z}{dz} & (3') \end{cases}$$

5. With (see question 2, from problem 3 of Chapter 8, this volume):

$$M_z = -\frac{|q\omega|}{2\pi}\pi\rho^2 = -\frac{q^2}{2m}\rho^2 B_z,$$

and $\rho \approx \dfrac{v_\theta}{\omega}$, we have:

$$M_z = -\frac{q^2}{2m}\frac{v_\theta^2}{\omega^2}B_z,$$

so that in addition (using $\omega = -\dfrac{qB_z}{m}$) we have :

$$M_z = -\frac{mv_\theta^2}{2B_z}.$$

For the calculation of $\dfrac{dM_z}{dt} = -\dfrac{m}{2B_z}\dfrac{dv_\theta^2}{dt} - \dfrac{mv_\theta^2}{2}\dfrac{d}{dt}\left(\dfrac{1}{B_z}\right)$, by writing

$$\frac{d}{dt}\left(\frac{1}{B_z}\right) = \frac{d}{dz}\left(\frac{1}{B_z}\right)\frac{dz}{dt} = \frac{d}{dz}\left(\frac{1}{B_z}\right)v_z,$$

we have :

$$\frac{dM_z}{dt} = -\frac{m}{2}\left\{\frac{2v_\theta}{B_z}\frac{dv_\theta}{dt} + v_z v_\theta^2 \frac{d}{dz}\left(\frac{1}{B_z}\right)\right\} = -\frac{m}{2}\left\{\frac{2v_\theta}{B_z}\frac{dv_\theta}{dt} - v_z v_\theta^2 \frac{1}{B_z^2}\frac{dB_z}{dz}\right\}.$$

By plugging Eq. (2') into this last equation, it now becomes:

$$\frac{dM_z}{dt} = -\frac{m}{2}\left\{\frac{v_z v_\theta^2}{B_z^2}\frac{dB_z}{dz} - \frac{v_z v_\theta^2}{B_z^2}\frac{dB_z}{dz}\right\} = 0$$

from which $M_z = \text{constant}$; M_z is the first integral of the movement.

6. The flux of the magnetic field through the orbit of the particle is given by $\Phi = \pi\rho^2 B_z$. With (see question 5)

$$M_z = -\frac{q^2}{2m}\rho^2 B_z$$

and hence $\rho^2 B_z = -\frac{2m}{q^2}M_z$, we have (as $M_z = \text{constant}$), then:

$$\Phi = \pi\rho^2 B_z = -\frac{2\pi m}{q^2}M_z = \text{constant}$$

The flux of the B_z field being constant through the section of a tube of the field (general property of vectors of which the divergence is zero, for example, that given in Section 1.3.3.2 of Volume 1), the section $\pi\rho^2$ of radius ρ through which the flux of B_z remains exactly constant must correspond to the section of the tube. The result is that the orbit (circle of radius ρ) encloses (rolls around) the tube of the field.

It is worth noting that if B_z increases, we should have a value of ρ which decreases (the section of the field tube will thus decrease). See also the preceding figure where B_z increases toward the right-hand side while the section simultaneously decreases.

7. According to Eq. (3'), we have

$$\frac{dv_z}{dt} = -\frac{v_\theta^2}{2B_z}\frac{dB_z}{dz}$$

so that with $\frac{dB_z}{dz} > 0$ and $B_z > 0$ (positive B_z field increasing with z)

$$\frac{dv_z}{dt} < 0 ,$$

and v_z decreases when z is increasing in a region where B_z increases with z.

In order to study the variation of v_θ, we can write that:

$$\frac{dv_\theta^2}{dt} = 2v_\theta \frac{dv_\theta}{dt} .$$

With (2'), we thus have:

$$\frac{dv_\theta^2}{dt} = 2\frac{v_\theta^2 v_z}{2B_z}\frac{dB_z}{dz}$$

from which when $\frac{dB_z}{dz} > 0$ and $B_z > 0$, we have

$$\frac{d\left|v_\theta\right|}{dt} > 0 ;$$

$\left|v_\theta\right|$ increases with z in a region where B_z increases with z (a variation of this demonstration is given in Section 9.3.2.2.

To conclude, the velocity v_z of the particles of charge q decreases as the field B_z increases. If the field B_z increases sufficiently, then a moment can arrive when the velocity v_z cancels out and then changes sign. The particles return in the inverse direction, undergoing the effect of a "magnetic mirror".

8. As we assume that $v_r \approx 0$, the velocity essentially has two components, one denoted v_z and the other v_θ (normal to v_z). At the initial time t = 0 where $\vec{v} = \vec{v}_0$, we thus have $v_{0z} = v_0 \cos\alpha$ and $v_{0\theta} = v_0 \sin\alpha$.

We therefore have $v^2_{0z} + v^2_{0\theta} = v^2_0$. The conservation of kinetic energy, which can be written as

$$E_C = \frac{1}{2}mv^2 = \frac{1}{2}mv_0^2$$

shows that $v^2 = v^2_0 = v^2_{0z} + v^2_{0\theta} = \text{constant}$.

For its part, the invariance of $M_z = -\frac{mv_\theta^2}{2B_z}$ implies (as $B_z = B_1$ at $t = 0$ and $z = 0$ where $v_\theta = v_{0\theta}$) that

$$\frac{v_\theta^2}{B_z} = \text{constant} = \frac{v_{0\theta}^2}{B_1}.$$

With $v_{0\theta} = v_0 \sin\alpha$, this law of invariance can be written as

$$\frac{v_\theta^2}{B_z} = \frac{v_0^2 \sin^2\alpha}{B_1}.$$

For particles to be reflected in the zone where B_z varies from B_1 to B_2, the velocity v_z needs to be cancelled out before the particles reach the maximum of the field at B_2. When v_z is cancelled out by a field B_d (of value $B_1 < B_d \le B_2$) we simply have $v^2 = v^2_\theta$ (= v^2_0 by conservation of kinetic energy). This condition, $v^2_\theta = v^2_0$ substituted into the law for the invariance of M_z results in there being a condition of the angle α which must have a value α_d such that:

$$\frac{\sin^2\alpha_d}{B_1} = \frac{1}{B_d}$$

so that

$$\alpha_d = \arcsin\sqrt{\frac{B_1}{B_d}}.$$

As we must have $B_d \le B_2$, we also can state that for particles to be reflected between B_1 and B_2, we should find that the following relation is true

$$B_d = \frac{B_1}{\sin^2\alpha_d} \le B_2$$

so that $\sin^2\alpha_d \geq \frac{B_1}{B_2} = \sin^2\alpha_\ell$. This condition is written as

$$\alpha_d \geq \alpha_\ell = \arcsin\sqrt{\frac{B_1}{B_2}}.$$

All the particles emitted with an angle $\alpha_d \geq \alpha_\ell = \arcsin\sqrt{\frac{B_1}{B_2}}$ can be reflected within the field between B_1 and B_2.

Thus, particles emitted isotropically within the solid angle $\Omega = 2\pi(1-\cos\alpha_\ell)$ cannot be reflected. If we consider that the particles are only emitted within the half-space given by the solid angle of 2π steradians, the corresponding NR of non-reflected (unreflected) particles is given by

$$\mathrm{NR} = \frac{\Omega}{2\pi}.$$

The level R of particles reflected being such that R + NR = 1, we have

$$\mathrm{R} = 1-\frac{\Omega}{2\pi} = 1-(1-\cos\alpha_\ell) = \cos\alpha_\ell.$$

With $\cos^2\alpha_\ell = 1-\sin^2\alpha_\ell = \frac{B_2-B_1}{B_2}$, we finally have:

$$\mathrm{R} = \sqrt{\frac{B_2-B_1}{B_2}}.$$

In numerical terms, with $B_1 = 10^{-4}$ T and $B_2 = 5 \times 10^{-4}$ T, we obtain:

$$\alpha_\ell \approx 26° \text{ and } \mathrm{R} = \sqrt{\frac{4}{5}} \approx 90\ \%.$$

Chapter 10

Electromagnetic Ion–Material Interactions

This chapter describes the interactions of a charged particle (an ionized atom) with a target material. For the most part, the application of these interactions is in the field of ion implantation, widely used in material physics to obtain electric doping of semiconductors. This chapter is limited to the basic theories, with some results being established by way of problems and exercises and others being given directly without going into the details—which can be found in specialist books such as, for example, M. Nastasi, J.W. Mayer, J.K. Hirvonen, *Ion–Solid Interactions*, Cambridge University Press (1996). Material surface treatments using ion beams and the various configurations applied also are detailed. Finally, it is shown how basic electromagnetism can be used when realizing ion sources.

10.1. High-energy collisions between atoms and ions: the nature of the interaction potentials

10.1.1. General form of interaction potentials

Here we consider the collisions of high-energy atoms or ions with a target (generally a solid material). It is assumed that the incident particles have energies and velocities well beyond those acquired by thermal excitation (thermal vibration of atoms in equilibrium in a solid). The interatomic forces now are exerted over a distance less than the equilibrium distance (r_0) in the solid. The interaction distance (r) during a collision thus depends on the relative energies of the particles during the collision. As a consequence, there arises an interpenetration of the electronic orbitals that eventually overlap to generate a modification of the wave functions of the systems.

In order to resolve such a problem (ion–solid interactions and radiation damage of solids), the interatomic potential should be known to the highest possible degree of accuracy. This is because the potential permits a determination of the energy loss of ions as they penetrate the solid.

10.1.1.1. General analytical form of the interatomic potential: Lennard–Jones potentials

The schematic form of this potential that presides between two atoms is represented in Figure 10.1. It describes the potential, or more precisely the Lennard–Jones potential energy, given by $W = W_r + W_a$, where W_r is the repulsive potential that varies with A / r^{12}, while W_a is the attractive potential that follows the law of proportionality of $(- B / r^6)$.

Taking the analytical form into account, we can state that the attractive energy has a radius of action greater than that of the repulsive force. When $r >> r_0$, only the attractive force intervenes. However, when $r << r_0$, the repulsive force dominates.

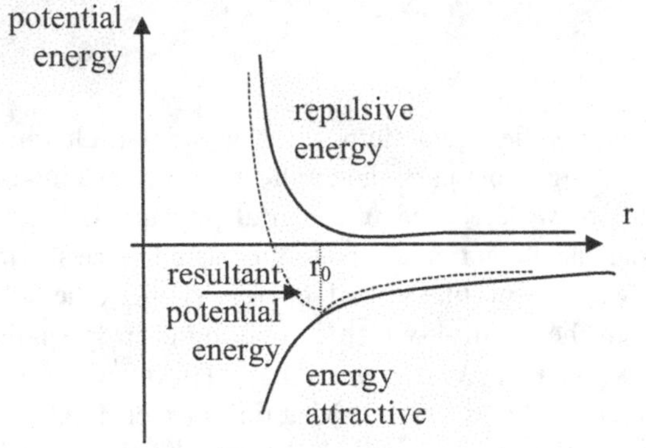

Figure 10.1. *Form of the Lennard–Jones potentials.*

10.1.1.2. Ion–atom interactions: study of the limiting case

To model the interaction of an energetic ion with the atoms in a solid, we can study the limiting scenario for a collision between two atoms separated by a distance r of masses M_1 and M_2 and atomic numbers Z_1 and Z_2, respectively. The force is best described by a potential energy [V(r)] resulting from the interaction of several bodies combining electrons and nuclei.

Two constants can be introduced to give an idea of the scale of dealing with separations between particles:

- the Bohr radius (a_0) for a hydrogen atom which gives an indication of the extension of hydrogen atoms ($a_0 = 0.0529$ nm $= \hbar^2/me^2$); and

- the lattice constant (r_0) which defines the distance between two nearest atoms in a crystal (typically of the order of 0.2 to 0.3 nm).

10.1.1.2.1. Under extreme conditions ($r >> r_0$), the electrons are distributed throughout the levels of individual atoms in accordance with Pauli's principle. Here, the lowest levels correspond to the internal layers which are totally occupied, while the external levels are orbitals which give rise to relatively weak *attractive interactions*, namely Van de Walls forces.

10.1.1.2.2. For the opposing extreme case, when $r << a_0$, the two nuclei become the closest particles of the system and their *repulsive coulombic interaction* dominates any other possible interactions. The corresponding (positive) potential energy thus is given by (to within the MKS coefficient, often omitted in the physics of interactions)

$$V(r) = Z_1 Z_2 e^2 / r. \quad (1)$$

10.1.1.2.3. For intermediate distance, $a_0 < r \leq r_0$, a positive interaction *potential energy* comes from a *repulsive force* between the two atoms. It is worth noting that in general terms, a repulsive force always is associated with a positive potential energy.

The two principal contributions to this potential are:

- electrostatic repulsion ($V_{nn} > 0$) between positively charged nuclei; and
- an increase in the energy necessary to maintain the electrons of the closest atoms in the same spatial region while respecting the Pauli principal. As two electrons cannot occupy the same position, the orbital overlapping of the two atoms (which generates a positive repulsion between the electrons of the two atoms, $V_{ee} > 0$) is accompanied by the displacement of a certain number of electrons to empty or even higher-energy atomic levels. The latter can correspond to antibonding functions with an increase in the potential energy of these electrons to $V_a > 0$ along with a modification of their kinetic energy). The energy necessary for these processes increases with the closeness of the atoms, as the number of electronic orbitals involved also increases.

For these intermediate distances, the coulombic potential generated by the nuclei is reduced by a screening effect due to internal layer electrons. The effective and screened potential thus is written in the form:

$$V(r) = (Z_1 Z_2 e^2 / r) \, \chi(r), \qquad (2)$$

where $\chi(r)$ is the screening coefficient defined as being the ratio between the effective potential at a distance r and the coulombic potential of the unscreened nuclei.

Under ideal conditions, $\chi(r)$ cooperatively modulates the coulombic potential to give a description of the ion–atom interaction over all distances. For long distances (positively charged nuclei totally screened by negative electronic charges) $\chi(r)$ should tend to zero, while for short distances $\chi(r)$ should go toward unity (no more screening of the nuclei by the electrons). Hence, a single potential energy function is introduced and Eq. (2) describes the totality of collision processes:

$$V(r) = (Z_1 Z_2 e^2 / r)\ \chi(r).$$

The estimation of the function $\chi(r)$ in effect depends on the hypotheses used to model the interactions.

10.1.2. The Thomas–Fermi model

In this case we consider a regime of collisions at relatively low velocities (generally used for implantation and surface treatments). The closest distance used respects the general condition $a_0 < r < r_0$. The nuclear charge is in effect screened by electrons. All the electrons follow Fermi-Dirac statistics and behave as if an ideal gas of particles filling potential wells distributed around a nucleus. In turn, the electronic filling of levels is characterized by the function denoted $N(E_F)$. This gives the total number of electrons (given by the number of spin-orbitals—meaning the number of orbitals, spins included) per unit volume with an energy lower than E_F, as in

$$N(E_F) = \frac{1}{2\pi}\left[\frac{2mE_F}{\hbar^2}\right]^{3/2}. \qquad (3)$$

This equation is established from the ideas developed in Chapter 5, where simply writing that $N(E_F) = \int_0^{E_F} Z(E).F(E)dE = \int_0^{E_F} Z(E)dE$ (as at T = 0 K the Fermi-Dirac function verifies that F(E) = 1 when $E \le E_F$).

10.1.2.1. For a single atom

We introduce the screened atomic potential (see, for example, Nastasi *et al.*, *Ion–solid interactions*, Cambridge University Press (1996) in the form:

$$\frac{V(r)}{-e} = \frac{Ze}{r}\chi(x), \qquad (4)$$

where $x = \frac{r}{a_{TF}}$ and $a_{TF} = \frac{0{,}885\, a_0}{Z^{1/3}}$ (a_{TF} is termed the Thomas–Fermi screening length). We can note that this potential is of a form resembling that coming from Eq. (2).

The Poisson equation which must verify this potential, $\Delta[V(r)/(-e)] + \rho/\varepsilon_0 = 0$, with $\rho = -\ N(E_F)\, e$, makes it possible to establish an equation without dimensions and which the function χ should also verify. Thus the Thomas–Fermi (TF) equation is obtained:

$$x^{1/2}\frac{d^2\chi}{dx^2} = \chi^{3/2}. \qquad (4)$$

Various approximate forms of the function χ have been determined, notably that called Moliere's form:

$$\chi = 0.35\exp(-0.3x) + 0.55\exp(-1.2x) + 0.10\exp(-6.0x). \qquad (5)$$

10.1.2.2. Approximation for the screening function using inverse powers

It has been mathematically shown that an inverse power law is acceptable as a solution to the Thomas–Fermi equation. It is written as:

$$\chi(x) = \chi(\frac{r}{a_{TF}}) = \frac{k_s}{s}\left(\frac{a_{TF}}{r}\right)^{s-1}$$, where $s = 1, 2, \ldots$ and k_s is a constant.

It is this form of the equation that is most often used to calculate the nuclear stopping power (S_N) as it provides the most effective route to obtaining S_N.

10.1.2.3 For an interaction between two atoms with atomic numbers Z_1 and Z_2

For this interaction we have

$$\chi(x) = \frac{V(r)}{Z_1 Z_2 e^2/r}, \qquad (6)$$

and here $a_{TF} = \frac{0.885\, a_0}{Z_{eff}^{1/3}}$, where Z_{eff} is the number of effective charges and has been approximated in a number of different forms (the simplest being the average value given by $Z_{eff} = (Z_1^{1/2} + Z_2^{1/2})^2$.

It is this form of $\chi(x)$ which is used to model the electronic stopping power under a regime of not too high velocities (outside of the Bethe region where the energies of the electrons are high).

10.1.3. The universal interatomic potential

The preceding potentials come from a simple statistical distribution of charges (the Fermi–Dirac distribution) and do not include any information about the internal distribution of the charges, which notably can occur when the two electronic clouds of the atoms interpenetrate (for nuclear atomic collisions which are predominant at low energies, as will be detailed below).

Ziegler, Biersack, and Littmark (ZBL) performed a more precise determination of the interatomic potential in their model. They supposed that each atom presents a spherical charge distribution, and the total energy of the interaction can be written in the form:

$$V = V_{nm} + V_{en} + V_{ee} + V_k + V_a\,, \qquad (7)$$

where:

- V_{nm} is the electrostatic potential energy between nuclei (repulsive so $V_{nm} > 0$);
- V_{en} is the energy of interaction between each nucleus and other electrons in the distribution (attractive and thus negative);
- V_{ee} is the electrostatic potential energy between the two electron distributions (so is > 0 as is repulsive);
- V_k is the increase in the kinetic energy of the electrons following their movement to the excited orbitals (in accordance with the Pauli principle); and
- V_a is the increase in the potential energy of these electrons.

The resulting coulombic interaction thus is given by

$V_C(r) = V_{nm} + V_{en} + V_{ee}$.

Ziegler's calculation [J.F. Ziegler *et al.*, *The Stopping and Range of Ions in Solids,* Pergamon Press (1985)] to determine the universal screen function uses the total interaction potential (V) and is based on an approximation for localized densities. Each volume element and the corresponding electron densities of the two atoms colliding are considered. It is supposed that the electronic density of the volume elements of the two overlapping atoms does not change. In this case, the mixture is treated as a gas of free electrons. For volume elements where there is an overlap, the attractive potential diminishes as electrons "hop" to higher energy states, in accordance with the Pauli principle. Thus the screening function is obtained as a potential that is the sum of two terms:

- a term for the coulombic interaction between electrons and nuclei (V_{en}) which is attractive and therefore negative; and
- a term for the exchange of electrons between the two atoms (V_{ee}, V_a, V_k) which tend to increase the repulsion between the atoms (hence is positive).

A universal screen potential thus can be calculated:

$$\chi_u(x) = 0.1818e^{-3.2x} + 0.5099e^{-0.9423x} + 0.2802e^{-0.4028x} + 0.02817e^{-0.2016x}, \quad (8)$$

with $x = \dfrac{r}{a_u}$, where $a_u = \dfrac{0.8854a_0}{Z_1^{2/3} + Z_2^{2/3}}$ is the universal screening length; χ_u is the screen function, such that the interatomic potential (V) can be given by

$$V(r) = \frac{Z_1 Z_2 e^2}{r}\chi_u\left(\frac{r}{a_u}\right). \quad (9)$$

10.2. Hypotheses for the dynamics of inelastic and elastic collisions between two bodies and various energy losses and electron and nuclear stopping powers

10.2.1. Introduction

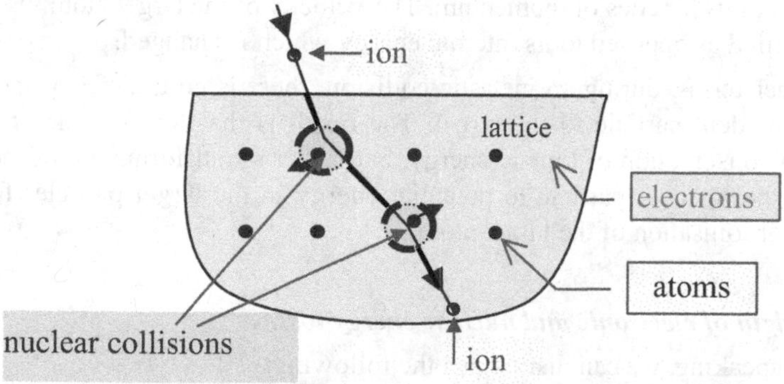

Figure 10.2. *Schematization of ion–material interactions.*

When an ion penetrates a solid, it undergoes collisions with the atoms of the target and its initial trajectory is modified. The ion also collides with electrons in the solid and loses energy during these collisions. Nevertheless, the major changes in the trajectory are due to collisions with the atoms met in the target. These collisions are termed “two-body collisions”.

In general terms, the simplest collision is that between a charged particle and a resting atom. The collision then can be inelastic or elastic, as detailed below.

10.2.2. Various hypotheses and assumptions concerning classic (Rutherford) diffusion theory

1. Only two bodies are considered. This situation arises when the collisions are sufficiently violent so that the two particles are very close to one another, to the point of excluding interactions with other particles. It is only at lower energies (below 1 keV) that collective effects can be introduced (problems involving three or more bodies).

2. The excitation or ionization of electrons simply results in a loss of energy, but without a modification in the collision dynamics.

3. It is assumed that the target atom is at rest (this approximation comes under doubt when particularly dense cascades of collisions are generated).

10.2.3. Elastic and inelastic collisions

Elastic collisions are those where there is a conservation of kinetic energy and of momentum in the system. In contrast, an inelastic collision does not retain the same momentum; for example, when overlapping of K orbitals occurs, the electrons are displaced toward higher energy levels. The energy used to promote an electron to a higher level is energy that is not used in the postcollision kinetics of the particle–atom pair. The energy is simply absorbed by the target atom and no longer participates in its kinetics of momentum. The velocity of the target atom, its kinetics, is not modified as opposed to its internal energy which is changed.

In other terms, during an inelastic collision, there is no transfer of momentum from the incident particle to the target. The result is that for the pair of particles there is no conservation of kinetic energy, but rather a transformation of the kinetic energy of the incident particle to potential energy in the target particle (following excitation or ionisation of the target atom).

10.2.4. Origin of electronic and nuclear energy losses

Generally speaking, we can distinguish the following:

- the *loss of electronic energy* due to *inelastic* collisions by *interactions* between an incident ion and electrons bound to or free of target atoms.

The incident ion transfer energy to the target so that the atoms are excited and electrons move through the effect of, for example, V_{en}, or by an overlapping of K orbitals (from the most external layers); and

- the *loss of nuclear energy* due to elastic coulombic interactions between screened nuclear charges (by escorting electrons) of the incident ion particle and atoms in the target. The energy is transferred by the incident ion to the target nuclei.

The energy losses due to nuclear and electronic interactions are correlated, as both are present in short-distance collisions. In effect, and in general terms, we tend to ignore these correlations and assume that the electronic breaking acts as a continuous phenomenon. For their part, nuclear collisions are quite rare, being more related to short distance collisions that produce a high number of recoiling nuclei. The exchange of electrons between incident ions and target atoms at their closest points of interception can also contribute to energy losses.

Finally, it is the velocity of the incident ions and the target atoms that controls the relative importance of each of the different interaction processes between the participants.

10.2.5. Electronic stopping power

10.2.5.1. High-energy collisions (incident ion at high velocity)

When an incident ion moves at a velocity (v) higher than $v_1 = Z_1e^2/\hbar$, that is to say at a velocity greater than that of the electrons in the K layer, then there is a very high probability that all its electrons will be "stripped". Here, Z_1 is the atomic number of the ion. The Bohr velocity, denoted v_0 and relative to the hydrogen atom (for which $Z_1 = 1$), is such that $v_1 = Z_1v_0$, where $v_0 = \hbar/ma_0 = e^2/\hbar \cong 2 \times 10^8$ cm s^{-1}.

In this case, the influence of an incident particle on the target atom can be thought of as a very abrupt and weak external perturbation. The collision results in a sudden transfer of energy of the projectile toward a stationary particle. The loss in energy from a rapidly moving particle to a stationary target particle (here the electron exhibits an electronic stopping power) can be calculated from the diffusion in a central force field. The effective stopping section (σ) decreases with the increasing velocity of the incident particle that spends less and less time in the neighborhoods of the target atom. We can show that in this case the electronic stopping power is proportional to $(Z_1 / v)^2$. With M_1 and M_2 being the masses of the incident and target particles, and Z_1 and Z_2 being their atomic numbers, respectively, then with v denoting the velocity of the incident particle, we obtain (in the MKS system)

$$\sigma(\theta) = \left[\frac{Z_1 Z_2 e^2}{8\pi\varepsilon_0 M_1 v^2}\right]^2 \frac{1}{\sin^4\theta}, \qquad (10)$$

and the law of variation in $(Z_1/v)^2$ clearly can be seen.

The expression for energy loss was obtained by Bohr (1913) via a classical calculation. In this hypothesis, the incident ion is thought of as a "naked" nucleus so that its interactions with the target electrons are purely coulombic. This interaction is between the incident particle (here a charge Z_1 e of nucleus of the ion from which the electrons have been completely removed at the high velocity) and the Z_2 electrons of the target atom by which the incident particle passes. We thus have:

$$-\frac{dE}{dx} = \frac{4\pi Z_1^{\,2} Z_2 e^4}{mv^2} \ln\left(\frac{b_{max}}{b_{min}}\right),$$

where b_{min} and b_{max} are the inferior and superior limits of the approach distance between the ion and the electron (b is an impact parameter).

This equation (then modified by Bethe to introduce a relativistic correction) shows that the loss in energy varies as $\frac{1}{v^2}$, which is to say with $\frac{1}{E}$, as in

$$-\frac{dE}{dx} = kE^{-1}. \qquad (11)$$

This loss in energy, as low as the energy (or velocity) is high, corresponds to velocities (v) of incident particles such that $v \gg v_0$.

10.2.5.2. Incident ions with low or medium velocities

Three models have been developed up until now for this regime. Each leads to a proportionality between the effective stopping section and the velocity of the projectile.

For projectiles with velocities lower than $v_1 = v_0 Z_1$, the majority of the electrons in the target move faster than the incident ions. For ions to move under this velocity regime, the target electrons cannot take energy from direct collisions with the ions, as otherwise found when using high velocity ions (greater than $v_0 Z_1$).

Firsov introduced in 1959 (Sov. Phys. JETP 36, 1076) a model for which each ion–atom collision is thought of as leading to an overlapping of electronic orbitals. The transfers of electrons, be they temporary or permanent from the ion to the atom or *vice versa*, give rise to a transfer of momentum that leads to a slowing down of the ion. The transfer of momentum happens when the target electrons are captured by the incident ion, as the electrons must be accelerated up to a velocity c (which is a vector) of the incident ion. This results in the ion losing a small part of its momentum proportional to $m_e v$. Inversely, when it is the electrons of the projectile that are transferred to the target atom, then momentum is transferred to the target atom but does not result in a slowing down in the projectile.

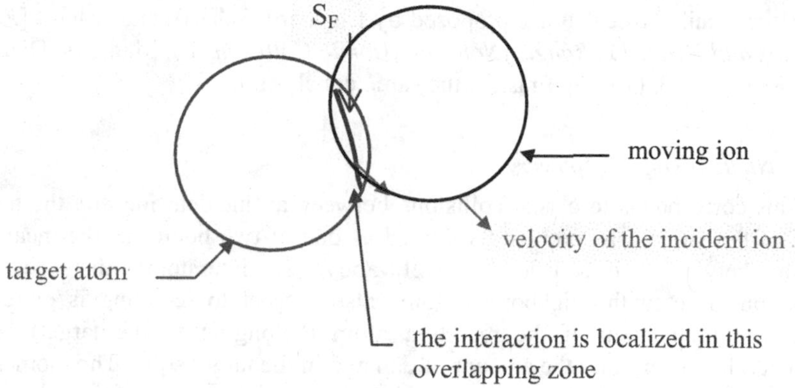

Figure 10.3. *Schematization of the interaction of slow or medium velocity incident ion with the electrons surrounding the target atom, according to Firsov.*

To understand the problem, Firsov proposed a geometric model. He supposed that the incident ion and the target atom formed a quasimolecule during the collision, given the low velocity of the ion. He assumed that the electronic spheres of the two elements were to all practical purposes not deformed during the collision. The energy loss from the incident particle, during the formation of the quasimolecule, is via the transfer of velocity and energy to the electrons of the atoms in the target. The energy lost by the ion finally results in the ionization of target atoms at the end of the collision.

This is why in this model we are first of all interested in the relative velocity (v) of the ion with respect to the electrons bound to the solid target (the Fermi velocity v_F is taken as a reasonable approximation for the value of v). An arbitrary plane is defined between the nuclei of the ion and the target atom. Hence the loss of electrons and energy is defined as a flux through this plane (Figure 10.3). So that an electron of the target atom is captured by the incident ion (or *vice versa*) and traverses the plane, the electron must take on a velocity higher than v. This exchange of electrons thus is characteristic of the energy loss of an ion in an electronic collision. The relative velocity is required knowledge and we can intuitively understand the conclusion that Firsov made in stating that the energy loss of a slow ion is proportional to its relative velocity (v). This is itself proportional to the square root of the energy, of kinetic origin, of the incident ion. We thus have:

$$S_e(E) = KE^{1/2} . \qquad (12)$$

More detailed calculations, proposed by Lindhard, Scharff, and Schiott [*Range Concepts and Heavy Ion Range, (Notes on Atomic Collisions II),* Mat. Fys. Dan Vid. Selsk. 33 (n°14), 3, (1963)], finish at the same conclusion.

10.2.6. Nuclear stopping power

This corresponds to elastic collisions between an incident ion and the atomic nuclei of the target. The problem is treated as one of two bodies as the mean free pathway between two collisions is well above the interatomic distances. The correlation effect with neighboring atoms also subject to recoiling is extremely small. The momentum of the recoiling atom (belonging to the target) is the parameter that determines the amount of damage in the target solid. The momentum transferred to the recoiling atom also is responsible for a large part of the energy loss of the incident ion.

It is the observation of this type of behavior with incident α-particles that gives rise to a recoil phenomenon at large angles, which led Rutherford to an atomic model based on nuclei. However, the value of the field created by the nucleus is limited for heavy ions by the existence of an important number of electrons that form a partial screen. The “screening” comes from the fact that the positive charge of the nucleus is surrounded by a cloud of electrons in their orbits. As the incident ion penetrates this cloud, it senses an increasing *electrostatic repulsion* that deviates its trajectory.

In the following Section 10.3, we will look at the principal steps in calculating the stopping powers.

10.3. The principal stages in calculating stopping powers

In this section the main ideas for the theoretical calculations of stopping powers are described. These analyses are classically used to theoretically describe the distribution of ions implanted in a material.

10.3.1. Rutherford-type diffusion for a particle of charge + $Z_1 e$ and mass M_1, by a particle of charge + $Z_2 e$ and mass M_2

Here M_2 is assumed to be immobile (application of the power of nuclear stopping) through a sudden interaction (velocity of the particle is very high and is the application of electronic stopping power in a region of very high velocities). The interaction potential is strictly coulombic, without screening effects, and the results can be applied only in the high-velocity regime (theorems presented in Section 10.2.5.1) for the stopping powers or nuclei or electrons.

10.3.1.1. Experimental localizations

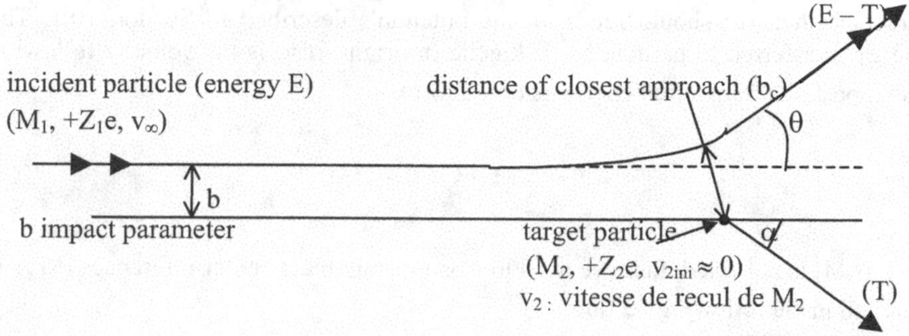

Figure 10.4. *Schematization of a collision at experimentally located positions.*

Figure 10.4 specifies the parameters of the impact against laboratory reference marks.

In the particular case of a frontal shock, as shown in Figure 10.5, the closest distance of approach is denoted b_c, and is such that if E denotes the energy of the incident particle, then (by the law of conservation of energy):

$$E = \frac{1}{2} M_1 V_\infty^2 = \frac{Z_1 Z_2 e^2}{b_c} \quad \text{(cgs system).} \qquad (13)$$

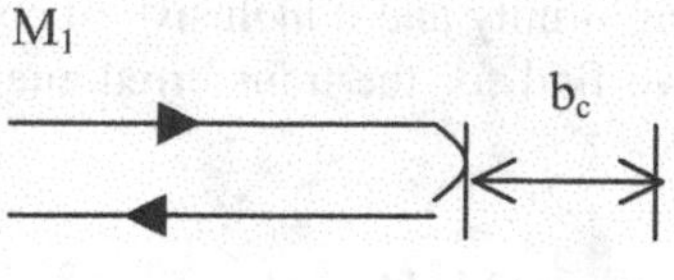

Figure 10.5. *head-on collision with an indicated closest distance b_c.*

The law of the conservation of energy written for diffusion in the general case shown in Figure 10.4 makes it possible to relate b, b_C, and the diffusion angle θ, as in:

$$b = \frac{b_c}{2} \operatorname{cotg} \frac{\theta}{2}. \qquad (14)$$

This formula is uniquely variable when the incident particle is very rapid and deviated by θ through coulombic effects alone [as in Eq. (13)]. If the potential is screened, then we should return to the potentials described in Section 10.1. The energy transferred to particle M_2 is kinetic in origin. If v_2 is the velocity following the impact, the transferred energy is of the form:

$$T = \frac{1}{2} M_2 v_2^2. \qquad (15)$$

If M_2 is not fixed, then we need to reason using mass center references (R_G) as detailed in the following section.

Equation (13) finally gives

$$b_C = \frac{Z_1 Z_2 e^2}{E}, \qquad (16)$$

while from Eq. (14) it can be deduced that:

$$\operatorname{tg} \frac{\theta}{2} = \frac{b_c}{2b}. \qquad (17)$$

10.3.1.2. Locating the center of mass

The rigorous calculation (see problem at end of chapter) necessitates the use of a located mass center with which the trajectories can be schematized, as shown in Figure 10.6.

For values of b between infinity and 0 inclusive (0 corresponding to the frontal impact described above), we find for the transferred energy (T) values between 0 and T_{Max}, which are such that:

$$T_{Max} = 4 \frac{M_1 M_2}{(M_1 + M_2)^2} E. \qquad (18)$$

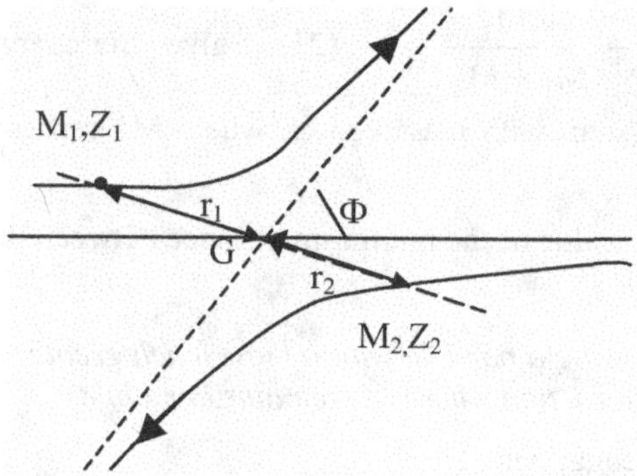

Figure 10.6. *Schematization of the trajectories with respect to located mass centers.*

The general expression for T is obtained by writing the conservation of momentum and kinetic energy. With $\alpha = \frac{\pi - \phi}{2}$ (relationship between laboratory and barycentric coordinates) we obtain (see also problem):

$$T = \frac{4\, M_1\, M_2}{(M_1 + M_2)^2}\, E \cos^2\alpha - \frac{4\, M_1\, M_2}{(M_1 + M_2)^2}\, E \sin^2\frac{\phi}{2}, \tag{19}$$

where ϕ is the angle of reflection in the reference given by the center of mass. Its expression is:

$$\phi = \pi - 2\,b \int_0^{u_{max}} \frac{du}{\left[1 - \frac{V(u)}{E_r} - b^2 u^2\right]^{1/2}}\,. \tag{20}$$

where

- $u = \frac{1}{r}$, $r = r_1 + r_2$ is the distance between the particles in the center of mass reference;
- V(u) is the interaction potential;

- $E_r = \frac{1}{2} M^* V_\infty^2 = \frac{E M_2}{M_1 + M_2}$ (21), gives the energy of the incident particle in the system with mass center, where M* represents the effective mass; and
- u_{max} is the inverse value of the minimum distance between the particles.

10.3.1.3. Simplification using a particle with a mass much greater than the other (or for a sudden interaction where M_2 remains immobile)

10.3.1.3.1. Form of the equations

In these calculations we assume that $M_2 >> M_1$, so that

$$M^* = \frac{M_1 M_2}{M_1 + M_2} \sim \frac{M_1 M_2}{M_2} = M_1 \text{ and}$$

$$OG = \frac{M_1 r_1 + M_2 r_2}{M_1 + M_2} \sim \frac{M_2 r_2}{M_2} \sim r_2 \Rightarrow G \equiv A_2.$$

The result is that the reference given by the center of mass is identical to that of the particle M_2 situated at A_2.

Within this reference, corresponding to the equations of Section 10.3.1.1, we have

- from Eqs. (18) and (19) that $T = T_{Max} \sin^2 \frac{\phi}{2}$, so that with our hypotheses (location G ≈ point tied to the M_2) we have $\Phi \approx \theta \Rightarrow$. $T = T_{Max} \sin^2 \frac{\phi}{2}$.

  (We can demonstrate that in this case, $T = T_{Max}\cos^2\alpha$, where α is the angle of deviation of M_2 *cf.* Figure 10.3.)

- $\frac{1}{\sin^2 \frac{\theta}{2}} = 1 + \mathrm{cotg}^2 \frac{\theta}{2} = 1 + \left(\frac{2b}{b_c}\right)^2$ [via Eq. (14)],

from which finally:

$$T = \frac{T_{MAX}}{1 + \left(\frac{2b}{b_c}\right)^2}. \quad (22)$$

10.3.1.3.2. Calculation of energy loss

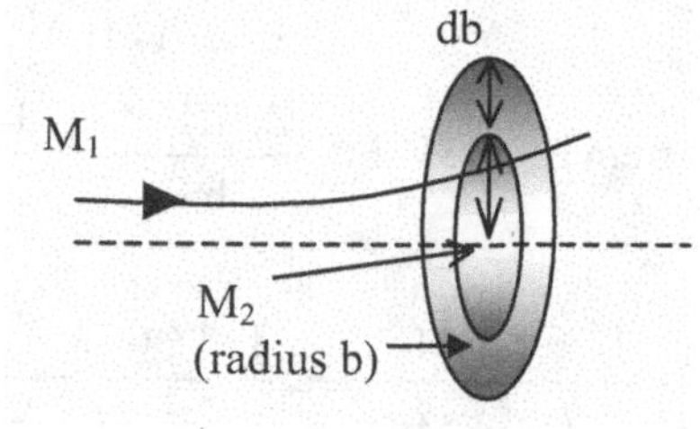

Figure 10.7. *Schematization of the interaction of an incident particle with the surface* $s = \int ds = \int 2\pi\, b\, db$.

Figure 10.7 shows a particle (M_1) interacting with a surface (s) that takes up the form $s = \int ds = \int 2\pi\, b\, db$. The energy loss of M_1 is given by $S = \int T ds$, so that:

$$S = \int T\, ds = 2\pi\, T_{Max} \int_{b_{min}}^{b_{max}} \frac{b\, db}{1 + \left(\frac{2b}{b_c}\right)^2}. \qquad (23)$$

By differentiating Eq. (22) under the form $1 + \left(\frac{2b}{b_c}\right) b^2 = \frac{T_{Max}}{T}$, we obtain

$$2\left(\frac{2b}{b_c}\right)\frac{2}{b_c}\, db = T_M \frac{dT}{T^2} \Rightarrow \frac{4}{b_c^2} 2b\, db = \frac{T_{Max}}{T}\frac{dT}{T}, \text{ from which}$$

$$b\, db\, T = \frac{b_c^2}{4}\frac{1}{2} T_{Max} \frac{dT}{T}.$$

By replacing T by its expression given in the left-hand side of Eq. (22), we then have for S:

$$S = T_{Max}\left(\frac{b_c}{2}\right)^2 \pi \int_{T_{min}}^{T_{Max}} \frac{dT}{T} = T_{Max}\left(\frac{b_c}{2}\right)^2 \pi \ln\left(\frac{T_{Max}}{T_{min}}\right).$$

$$\left.\begin{array}{ll}\text{From Eq. (18)} & T_{Max} = 4\dfrac{M_1 M_2}{(M_1 M_2)^2} E \\ \text{and from Eq. (13)} & E = \dfrac{Z_1 Z_2 e^2}{b_c} = \dfrac{1}{2} M_1 v_\infty^2 \end{array}\right\} \Rightarrow$$

$$T_{Max}\left(\frac{b_c}{2}\right)^2 = 4\frac{M_1 M_2}{(M_1+M_2)^2} E \frac{1}{4}\frac{\left(Z_1 Z_2 e^2\right)^2}{E^2} = \frac{M_1 M_2}{(M_1+M_2)^2}\frac{Z_1^2 Z_2^2 e^4}{\frac{1}{2} M_1 V_\infty^2}.$$

From this can be deduced that

$$S = \frac{M_1 M_2}{(M_1+M_2)^2}\frac{Z_1^2 Z_2^2 e^4}{E}\ln\left(\frac{T_{Max}}{T_{min}}\right). \qquad (24)$$

If $M_2 >> M_1$ (M_2 is immobile or the incident atom is light), we have:

$$\frac{M_2}{(M_1+M_2)^2} \approx \frac{M_2}{M_2^2} = \frac{1}{M_2}$$, and finally:

$$S \approx \frac{2\pi Z_1^2 Z_2^2 e^4}{M_2 v_\infty^2}\ln\frac{T_{Max}}{T_{min}}, \qquad (25)$$

which gives the form of the stopping power for two interacting nuclei Z_1 and Z_2 where the potential is strictly coulombic, with $M_2 >> M_1$.

10.3.1.4. One of the interacting particles is an electron ($M_2 = m_{electron}$): electronic stopping power in a regime of high energies (high velocities)

Here the trajectory takes on the form shown in Figure 10.8. The mass M_2 must be replaced by the mass m of the electron and Z_2e by e. The stopping power for an electron is thus

$$S_e\ (1\ \text{electron}) = \frac{2\pi Z_1^2 e^4}{m v_\infty^2}\ln\frac{T_{Max}}{T_{min}}.$$

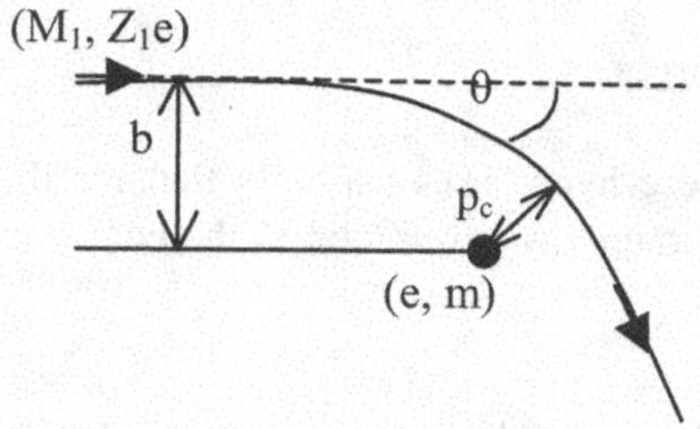

Figure 10.8. *Interaction between an incident ion and an electron target*

As there are Z_2 electrons to each target atom, the loss in energy (said to be of electronic origin as the incident particle interacts with the electrons of the target) is thus:

$$S_e = \frac{2\pi Z_1^2 Z_2 e^4}{m v_\infty^2} \ln \frac{T_{max}}{T_{min}}.$$

10.3.1.5. Order of scale of the stopping powers

- If we compare the equations in Sections 10.3.1.3 and 10.3.1.4 concerning stopping powers, those in Section 10.3.1.3 correspond to interactions between nuclei, i.e., $S = S_N$, which in fact is not a nuclear stopping power as the coulombic potential for the nuclear stopping power is screened (see Section 10.1). As remarked above, Section 10.3.1.4 corresponds to the losses termed electronic (S_e) for projectiles at very high velocities (under the Bethe regime when relativity also is brought into play, while here it remains strictly coulombic).

  With $M_2 \approx Z_2 \times$ (proton mass + neutron mass) $\approx 2\,Z_2 \times$ proton mass, the ratio $\frac{S_e}{S_N} = \frac{M_2}{Z_2\, m} \sim 2\,\frac{\text{proton mass}}{\text{electron mass}} \approx 4000$.

- The equation for the electronic stopping power under a regime of high velocity are such that:

  $$S \propto \frac{1}{v_\infty^2} \frac{1}{\frac{1}{2} M_1 v_\infty^2} = \frac{1}{E_{inc}} = \frac{1}{E}.$$

From this can be deduced that $-\frac{dE}{dx} = S\,N_i \propto kE^{-1}$, where N_i is the concentration of i electron particles in the target.

To conclude, with these hypotheses (notably that for the high velocity of the incident particle), the loss in energy is reduced as the energy of the incident particle is increased.

10.3.2. Low-velocity incident particle (inferior to the velocity of electrons in the K layer): expression for the electronic stopping power S_e

In the scenario given in Section 10.3.1.1, Eq. (20) for Φ brings in the interaction potential [V(u)] which we took in the form

$$V\left(u=\frac{1}{r}\right)=\frac{Z_1\,Z_2\,e^2}{r}. \text{ [see Eq. (13)]}$$

The velocity of the projectile is sufficiently high to assume that it has been "stripped" and that the Rutherford theorem developed for the particular case of α particles can be applied.

However, if the projectile has a velocity that is sufficiently low so that it is not completely "stripped" of electrons (see the nuclear state given in Section 10.1), then the strictly coulombic potential must be modified so as to bring in the screening effect of electrons that are in the occupied electronic levels of the incident ion. This representation of a unstripped ion, with practically all of its surrounding electrons, brings into play an interaction potential (between the surrounded ion and the target electrons) of the form:

$$V(r)=\frac{Z_1\,Z_2\,q^2}{r}\,\chi\left(\frac{r}{a_{TF}}\right),$$

where $\chi\left(\frac{r}{a_{TF}}\right)$ is the screening function which has the form detailed in Section 10.1.2.3 to determine S_e.

This electronic stopping power (in a regime of low enough velocities) was established with the help of an equation in the form $S = \int T ds$, as already used in Eq. (23). The transferred energy $T = T_e$ (lost through collision) can be obtained using Firsov's model where it is supposed that the force exerted on the projectile (incident ion) is due to the transfer of momentum ($m_e v$) by movement of an electron from the target to the incident ion, which has a velocity v_1. This gives

$$dF = (m_e v) d\Phi_F ,$$

where $d\Phi_F$ is the electron flux (number of electrons per unit time traversing the Firsov surface element; see Figure 10.2). This flux is expressed as a function of the electronic density, which in turn is expressed in the Thomas–Fermi theory as a function of the screening potential. A similar form to that of Section 10.1.1.3 has been proposed by Lindhard and Scharff, and it brings in a term $(Z_1^{2/3} + Z_2^{2/3})^{1/2}$ as the authors suppose that the average coulombic field of interaction of the quasi-molecule is well represented by twice the average geometry of the individual coulombic fields. From this, they deduced the electronic stopping power (S_e) given in the following expression:

$$S_e(E) = \frac{8\pi e^2 a_0 Z_1^{7/6} Z_2}{\left(Z_1^{2/3} + Z_2^{2/3}\right)^{3/2}} \frac{v_1}{v_0},$$

where v_1 is the velocity of the incident particle which is such that $E = \frac{1}{2} M_1 v_1^2$, so that $v_1 \propto E^{1/2}$.

Finally, we obtain the law also observed by Firsov, as in $S_e(E) = k\,E^{1/2}$.

It is worth understanding that this mechanism corresponds to a loss of energy of the incident ion following a transfer of velocity and energy to the electrons of the target atoms. An ionization (or excitation) of the target atoms thus is produced by an incident particle. This transfer of velocity that occurs during the formation of the quasi-molecule (which corresponds to the fusion of the incident ion and the target atom during which time their electrons may be exchanged) is equivalent to loss of momentum ($m_e v_1$) by the incident ion, and thus contributes to the process of breaking the ion's path.

The interaction potential is given by the electrostatic interaction between the electrons of the target atoms and the screened nuclei of the incident particle. The screened nucleus gives rise to a (positive) potential $V = \frac{Z_1 e}{r} \chi\left(\frac{r}{a_{TF}}\right)$ where $\chi\left(\frac{r}{a_{TF}}\right)$ represents the screening effect. The resultant potential energy is given by $W = -\,e\,V$ which is negative (due to the attractive effect of the target electrons to the incident particle). The final result is that the target electrons either can be stripped (and captured by the incident particle), thus giving rise to an ionization effect, or carried by the upper electronic layers in an electronic excitation of the target.

10.3.3. Nuclear stopping power

In this interaction, the electronic "displacements" are no longer of importance (attractive potential between the incident ion and the electron). What is of concern, however, is the repulsion between the incident ion and the target atom which undergoes a recoil (repulsive potential). When the incident ion, screened by its own electron cloud, penetrates the target, a Rutherford-type electrostatic (repulsive) diffusion between the nuclei of the incident particle and the target atoms is modified by the intervention of screen potentials. These modify the interaction potential energy [V(u)], which intervenes in Eq. (20), which gives the deviation angle for the two particles (incident and target) within the barycentric reference coordinates.

The analytical solution for the screening function can be taken in the form already described above in Section 10.1.1.2, as in:

$$\chi\left(\frac{r}{a_{TF}}\right) = \frac{k_s}{s}\left(\frac{a_{TF}}{r}\right)^{s-1},$$

where the function is such that k_s is a constant and s varies:

s = 1 is the collisions involve the transfer of high amounts of energy; and

s = 2 if the collisions involve transfer of low amounts of energy.

We thus can write:

$$V(r) = C_s r^{-s} \text{ where } C_s = \frac{Z_1 Z_2 e^2 k_s}{s\, a_{TF}^{1-s}}.$$

The determination of the energy transferred in an elastic collision (which uses an approximation of impulsion) makes it possible to determine in turn the nuclear stopping power through the same method previously described, i.e., the equation $S_N = \int T d\sigma$ is used. See also Figure 10.7 and Eq. (23).

The use of reduced coordinates (by way of the Lindhart formalism) gives the equation: $\left(\frac{d\varepsilon}{d\rho}\right)_N = f(\varepsilon)$ where ε and ρ are specified in the following section.

The shape of this curve is independent of the incident particle and the target. Only ε depends on these parameters.

10.3.4. Total energy loss

If N denotes the atomic density of the target, the energy loss by nuclear collisions is given by $\left[\frac{dE}{dx}\right]_N = NS_N(E)$, while the energy loss through electronic collisions is $\left[\frac{dE}{dx}\right]_e = NS_e(E)$. The processes of energy loss by nuclei and electrons are independent and additive. The total loss in energy thus is of the form:

$$\left[\frac{dE}{dx}\right]_{total} = \left[\frac{dE}{dx}\right]_N + \left[\frac{dE}{dx}\right]_e = N\left[S_N(E) + S_e(E)\right].$$

By using Lindhart, Scharff, and Schiott (LSS)-type reduced coordinates, we then make:

$$\varepsilon = C_E E = \frac{M_2}{M_1 + M_2}\,\frac{a}{Z_1\,Z_2\,e^2}\,E \qquad (\varepsilon \text{ is the reduced energy})$$

$$\rho = xN4\pi a^2 \frac{M_1\,M_2}{(M_1 + M_2)^2} = C_R\,x \qquad (\rho \text{ is the reduced length})$$

$$a = \frac{0.88\,a_0}{\left(Z_1^{2/3} + Z_2^{2/3}\right)^{1/2}} \qquad (\Rightarrow a \approx 0.1 \text{ to } 0.2\text{ Å}).$$

We thus obtain:

$$\left(\frac{d\varepsilon}{d\rho}\right)_N \approx \frac{1.7\,\varepsilon^{1/2}\,\ln(\varepsilon + e)}{1 + 6.8\,\varepsilon + 3.4\,\varepsilon^{3/2}}$$

$$\left(\frac{d\varepsilon}{d\rho}\right)_e = k\,\varepsilon^{1/2}$$

with $k = \zeta_e \dfrac{0.08\, Z_1^{1/2}\, Z_2^{1/2}\, (M_1 + M_2)^{3/2}}{\left(Z_1^{2/3} + Z_2^{2/3}\right)^{3/4}\, M_1^{3/2}\, M_2^{1/2}}$, where ζ_e is a dimensionless constant of amplitude $Z_1^{1/6}$.

Finally:

$$\left(\frac{d\varepsilon}{d\rho}\right)_{total} = \left(\frac{d\varepsilon}{d\rho}\right)_e + \left(\frac{d\varepsilon}{d\rho}\right)_N$$

A graphic representation, covering a wide energy range up to more than several MeV, is shown below in Figures 10.9 and 10.10.

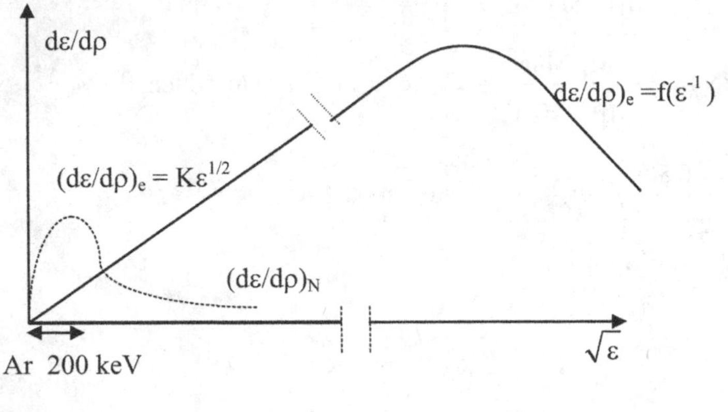

Figure 10.9. *Representation of dε/dρ over a wide energy range.*

In the region of the lowest energies up to 100 keV, the evolution of stopping powers can be schematized as in Figure 10.10.

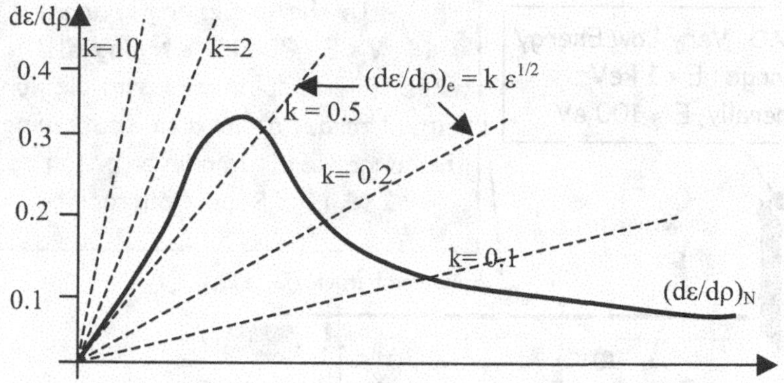

Figure 10.10. *Representation of dε/dρ in a range of energies [0 – ≈ 100 keV].*

10.4. The various phenomena of ion–material interactions and their applications

10.4.1. The various phenomena

Depending on the energy range of the incident ion, the result of an ion–material interaction will be (see also Figure 10.11):

1. for ions with energies typically above 20 or so keV, the ion implantation is poor;
2. for ions with energies of the order of ten to several tens of keV, the result can be an amorphism of the target and an eventual displacement (insertion) in the volume of atoms deposited on the surface of the target;
3. when the energies of the ions are low, or the order of several keV, a pulverization of the target can give rise to graving of the target or a deposition of a thick layer of a targeted substrate; and
4. when the incident ions have a very low energy, then there can be a growth of the target using an assisted deposition of atoms through their densification.

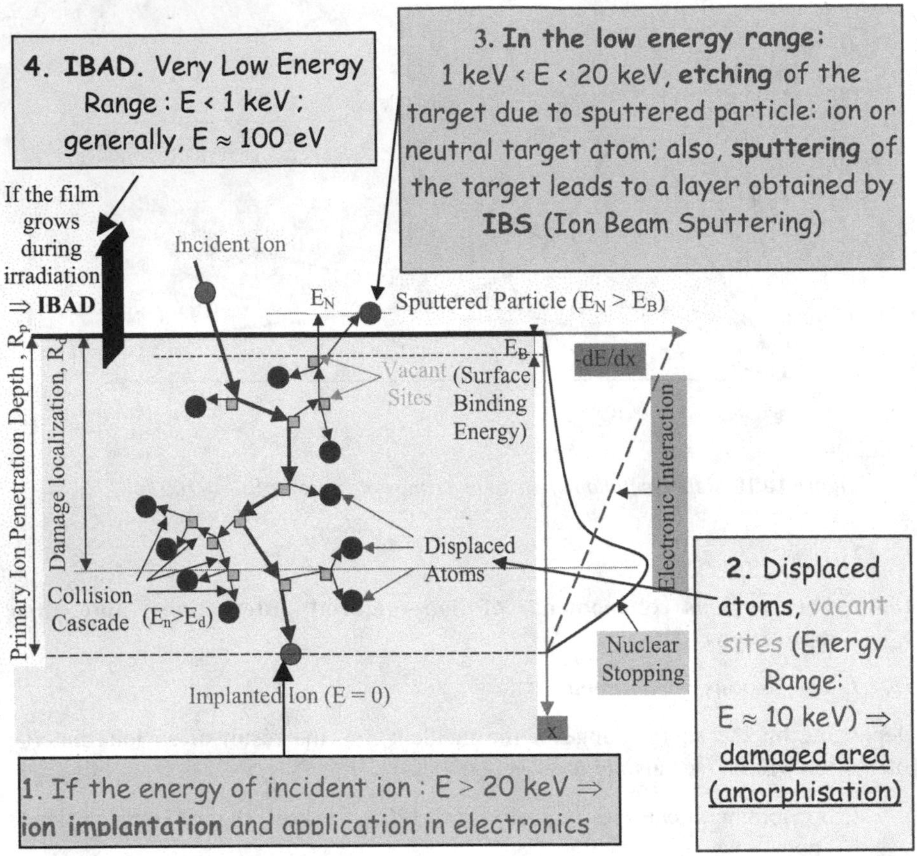

Figure 10.11. *Various mechanisms in ion–material interactions.*

10.4.2. Ion implantation

Ion implantation is used in a large number of electronic applications where it is necessary to have ions that are stopped at an atomic state in a target. These include the use of atoms from column III (B or P, for example) or from column V (As) that are used to dope target atoms from column IV (such as Si). The doping occurs because the projected ions are arrested in an atomic state by their trapping an electron liberating a hole in the semiconductor latter, hence, the term n doping.

The profile of the doping corresponds to that of the implantation, obtained by numerical simulation with the help of the equation of the electronic and nuclear stopping powers established in the above text. See also Figure 10.12 which describes implantation in a polyimide. It shows that the nuclear stopping power predominates at lower energies (E < 10 to 20 keV), while at higher energies it is the electronic stopping power that is the dominant mechanism. The average depth of the implantation corresponds to a value R_P which indicates the average projectile pathway; from which the difference, ΔR_p, gives the average thickness of the implanted layer. In order to attain sufficiently deep implantations of an acceptable thickness, the projected ions generally have energies of the order of 100 keV.

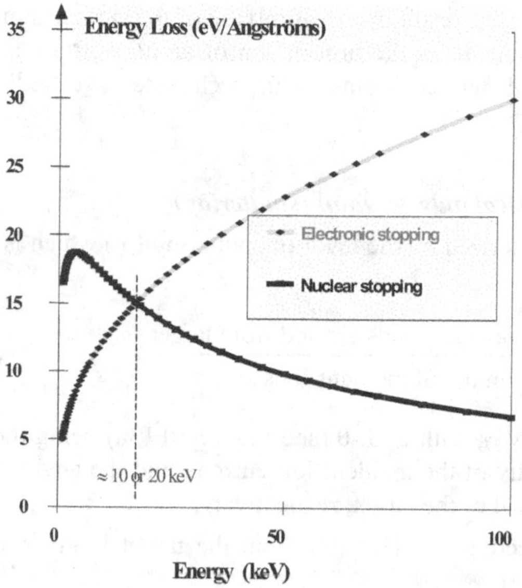

Figure 10.12. *Example of a theoretical implantation profile for a polyimide.*

An experimental determination of these profiles can be obtained by secondary ion mass spectroscopy (SIMS). Following the pulverization of atoms implanted by assistance from an ion beam (typically Cs^+ or O^+), the characterization of the mass of the pulverized atoms as a function of the erosion depth makes it possible to determine the distribution by depth of the atoms contained in the target.

10.4.3. Target amorphism and "mixing" in the volume of initially surface-deposited atoms ("ion beam mixing")

In order to obtain an important degree of amorphization of a layer, the target atoms need to be displaced. This is obtained by incident ions colliding with the atoms in the target. These collisions mostly occur in the energy range where the nuclear stopping power dominates, i.e., around 10 keV. At this level:

- either the incident ions arrive at such a value at the end of their pathway and have an initial energy of the order of 100 or more keV (see Figure 10.11); or
- they have such an energy at the beginning of their interaction with the target if their energy is of the order of 10 keV.

In the latter case, if we have already deposited on the target a layer of atoms that we wish to insert into the target, then the incident ions will collide with them and project them into the target. The result is one of mixing of the surface atoms with the target atoms. An example of this is the production of an alloy from appropriately chosen target material and surface atoms. This technique is called "ion beam mixing".

10.4.4. Mechanism of physical pulverization (sputtering)

The pulverization is characterized by the level of sputtering (Y) which is defined by the equation

$$Y = Y = \left(\frac{\text{average number of atoms ejected from target}}{\text{number of incident ions}}\right)_{\text{for a given } \Delta t},$$

so that $Y = f(E, J, \theta, M_1, M_2)$, with E, J, θ (see Figure 10.13a) being the energy of the incident ions, the density of the incident ion current, and the angle of incidence of the ion beam to the normal of the target, respectively.

The condition for ejecting an atom (B) from the target is given by $T_B \geq E_b$ where

- T_B is the energy transmitted to atom B during its sequence of collisions; and
- E_b is the bonding energy of the surface (reasonably estimated to be equal to the energy of sublimation of the target, typically from 5 to 10 eV).

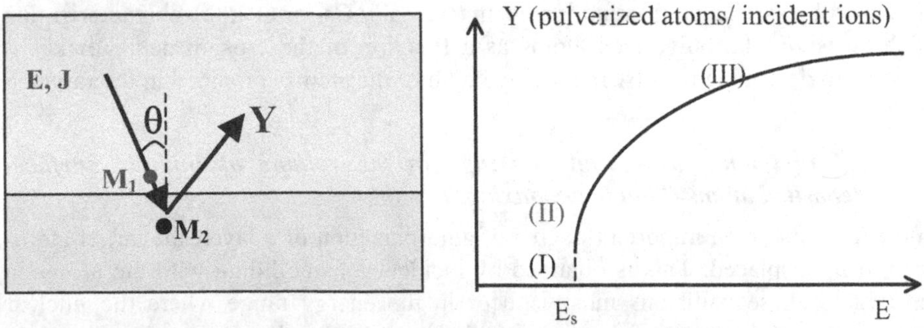

Figure 10.13. *(a) Pulverization of a target atom of mass M_2 by an incident atom of mass M_1 and (b) pulverization curve.*

In general, the pulverization curve presents three zones (Figure 10.13b):

- zone I. below the threshold ($T_B \leq E_b$);
- zone II. $E \approx E_s$ where this threshold energy is set by the type of collision sequences encountered, which can be of two types depending on $\theta = 0$ or $M_1 \geq M_2$; and
- zone III. when $E > E_s$ there is a rapid increase in Y which then reaches a saturation stage [rapidly reached if the masses M_1 (incident ion) and M_2 (target atom) are very different].

The analytical forms of the levels of pulverization were established by Thomson and Sigmund who, respectively, obtained (N is the atomic density of the target):

$$Y \propto \frac{N^{2/3}}{E_b} \frac{1}{1+\dfrac{M_2}{M_1}} \frac{1}{\cos\theta} \quad \text{and} \quad Y(E, E_b) \propto \frac{S_N(E)}{E_b}, \text{ so that :}$$

$$Y(\theta) \propto \frac{1}{\cos^n \theta}.$$

The result is that if M_1 increases, then Y also increases. Similarly, if θ increases, Y also increases. This law remains true for $\theta < 60\,°$ (see Figure 10.14).

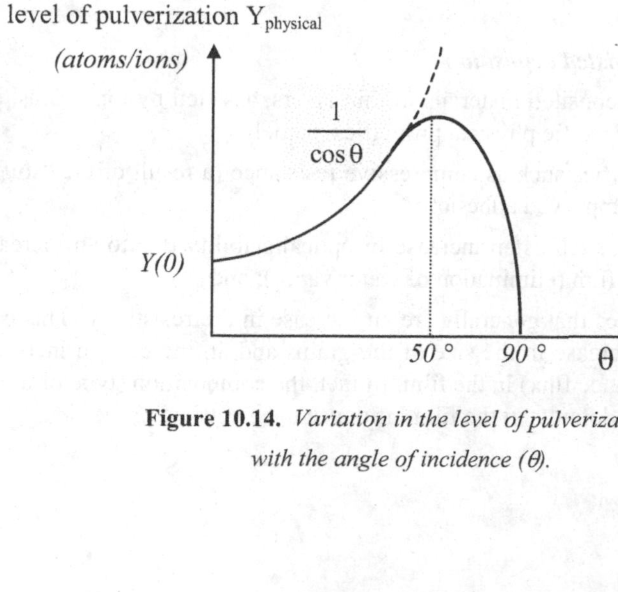

Figure 10.14. *Variation in the level of pulverization with the angle of incidence (θ).*

Finally, pulverization can be used in lithography to engrave the target. This is a widely used technique in microelectronics. In this case, the target is covered with a resin and then a mask is duplicated on the resin by insolation. The pulverization of the target results in an engraving of the target that follows the motif of the mask, as shown in Figure 10.14, which shows an engraved polyimide used for an optical guide.

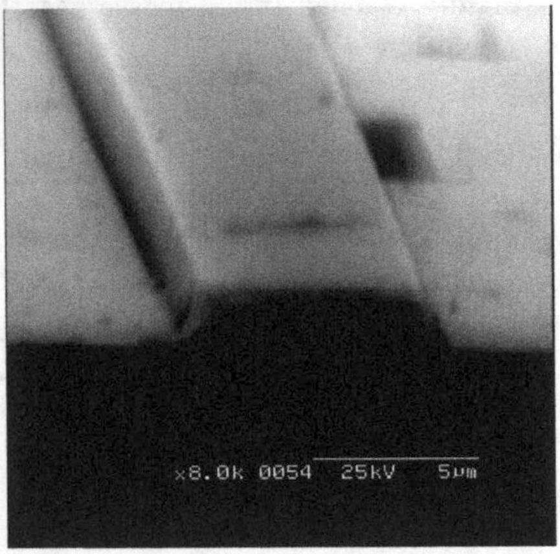

Figure 10.14. *Ion engraving to produce an optical guide*

It is equally possible to pulverize a target to generate a deposition of atoms, which make up the target, on a substrate to cover it with a thin layer. This technique is an alternative method to that of evaporation under vacuum.

10.4.5. Ion-beam–assisted deposition

The densification of deposited materials in thin layers, assisted by ion beams, gives rise to materials with specific physical properties, namely:

- mechanical properties such as compressive resistance (a result of the "stuffing" of ions), with an improved adhesion;
- optical properties, such as an increase in optical stability due to an increase in the density of the film (elimination of water vapor); and
- electrical properties that generally are an increase in the resistivity. This can be attributed to a decrease in the size of the grains and an increase in inert atoms (due to the assistance flux) in the film. In fact, the composition (type of ion) and the density of the defaults in the film influence the electrical properties.

In general terms, ion bombardment (with energies of the order of 100 eV) during a deposition can lead to an increase in the spatial density of growing islets that form on the surface of the layer. This is simultaneous to a decrease in the size of the deposited grains that tends to favor a contact between the increasing small-deposed domains and induce a densification of the layer. One of the consequences of the decrease in grain size while using beam assistance is a decrease in the roughness of the surface, which can be beneficial to the quality of interfaces in components based on multilayers. The smoothing of the surface thus can result in the increase in the mobility of adatoms (added atoms) that are under the effect of the energy of an ion bombardment. The local rearrangement of atoms is favored so that atoms relax into minimum energy sites. Dynamic molecular simulations can be used to model such arrangements. There are two main types of geometrical configurations:

- ion beam assisted deposition (IBAD) shown in Figure 10.15. Here a conventional deposition obtained by evaporation under vacuum or by the Joule effect (heating a crucible containing the atoms to be deposited) is assisted by an ion beam effect. Typically argon ions are used with energies of the order of 100 eV or a keV; or

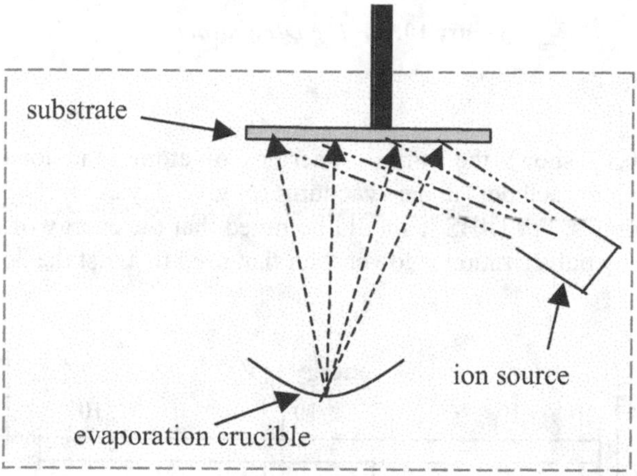

Figure 10.15. *The IBAD setup.*

- the dual ion beam sputtering (DIBS) configuration which is setup in a similar manner to the IBAD system with an assisting ion beam, but here the material is not generated by a Joule effect but rather by pulverization of a target containing the atoms to be deposited in a thin layer (Figure 10.16).

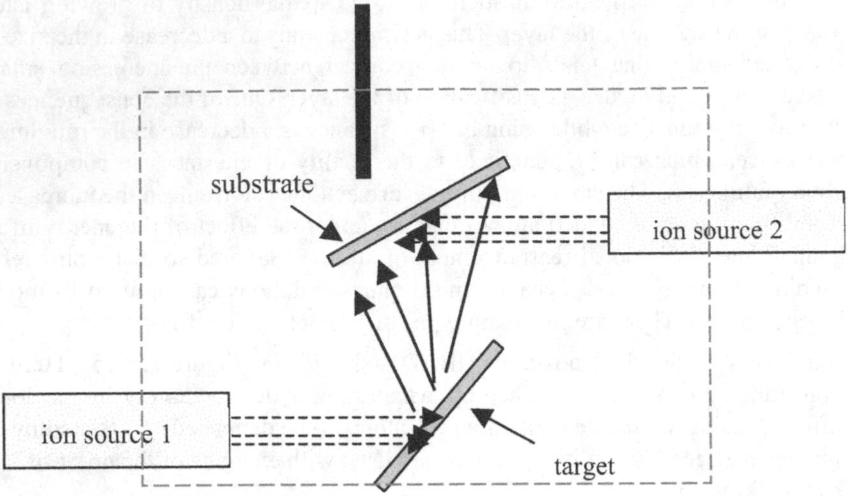

Figure 10.16. *The DIBS layout.*

Figure 10.17 shows the various energies of atoms and ions used in the different deposition methods under vacuum to give layers for applications in electronics and optics. For DIBS it should be noted that the energy of the ions used to obtain a layer by pulverization is lower than that used to assist the deposition.

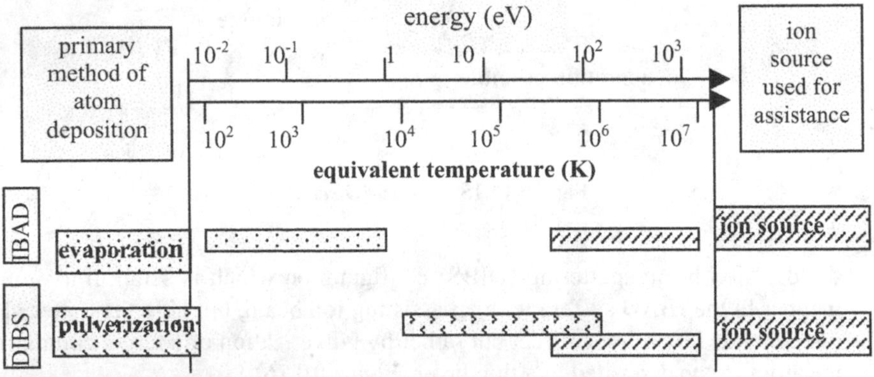

Figure 10.17. *Energies of ions and atoms used in IBAD and DIBS*

10.5. Additional information on various ion sources and their functioning

This section details the main types of ion sources that use an electromagnetic source, the setup of which varies with the source type.

10.5.1. The electron cyclotron resonance (ECR) source

Sources without filaments (used to generate electrons by thermoelectronic effects), such as the ECR, are used to generate ions from corrosive elements such as oxygen which would otherwise considerably reduce the lifetime of the filament and thus the source. Chapter 9 gives details on this type of source. In addition, we can state that the source shown in Figure 10.18 is based on a combination of the production of plasma by cyclotronic resonance (also detailed in Chapter 9, but without the divergent magnetic field produced by an electromagnet) with a magnetic confinement constructed with permanent magnets. The advantage of such a configuration is the separation of regions generating the plasma from multipolar regions. The multipolar region limits radial losses of plasma at the walls, while the interior region is not influenced by the magnetic field. A source of small dimensions with respect to the plasma in multipolar zones permits a uniform distribution. The inconvenience is that the density of the plasma is reduced to the confines of the multipolar zone.

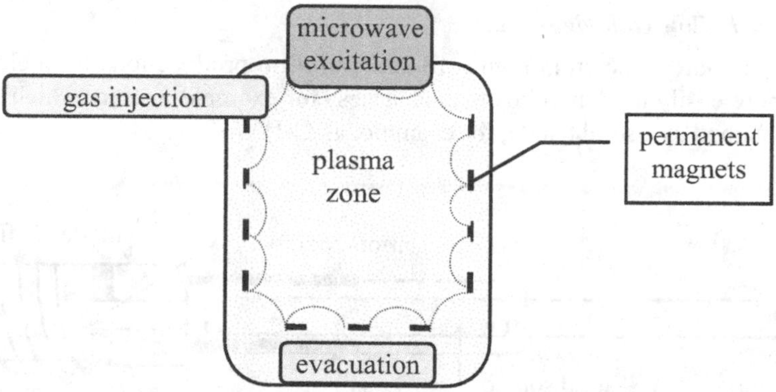

Figure 10.18. *ECR source with confinement due to permanent magnets.*

10.5.2. Basic elements in an ion source: the Penning source

The arc chamber shown in Figure 10.19 is limited by a cylindrical anode and two disks at the extremities which act as cathodes. The gas flux traverses the chamber and a low-voltage discharge results in the formation of a plasma in the chamber. An axial magnetic field imposes helical trajectories on the electrons, which undergo increased pathways so as to induce a more efficient ionization. The ions generated in the cathode then hit the cathode to produce secondary electrons (which also can be obtained by heating the cathode to yield a thermoelectronic emission). The addition of a heating filament makes it possible to more finely control the level of ionization. The ions then are extracted through the opening in the cathode with an energy closely matched to that of the discharge voltage.

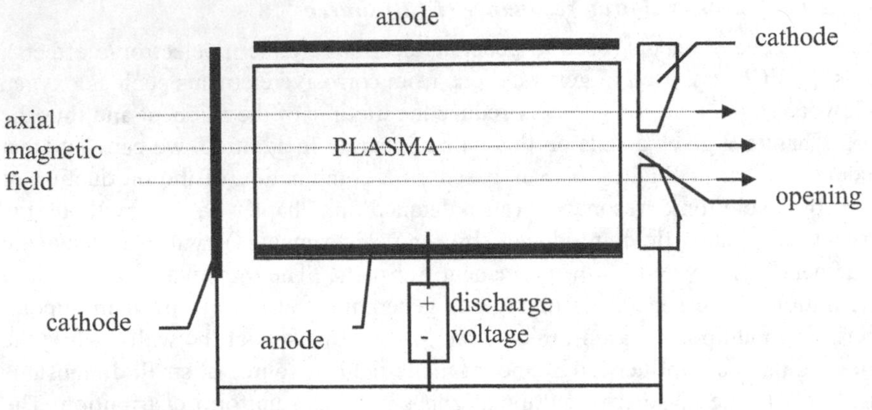

Figure 10.19. *Schematic of a Penning-type source.*

10.5.3. The hollow cathode source

This type of source, shown in Figure 10.20, is used to produce ions from elements that are more easily used as solids than as gases (for example, cesium which can be facilely obtained as a solid halide, for example, as CsI).

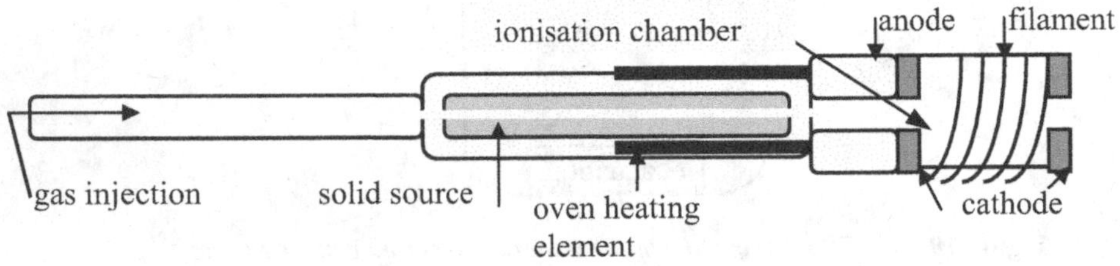

Figure 10.20. *The hollow cathode source.*

A vector gas, which carries the furnace-vaporized solid, is introduced into the ionization chamber. The cathode, made from tungsten with 2% rhenium, produces electrons by the thermoelectronic effect. The filament is encapsulated between two tantalum capsules that make up the cathode and are separated by an insulator. The tungsten exhibits a high mechanical rigidity up to 2000°C, while the rhenium increases the filament lifetime by improving its resistance to corrosion. The anode (the other discharge electrode) makes up the body of the ionization chamber. The arc is generally set off with a voltage of around 250 V, although this value can be reduced once the arc is obtained. The oven is made from a cylinder of boron nitride around which is wound a tantalum wire. Once the plasma is formed, the ions are extracted through a circular opening by way of an electrode at a negative potential (–10, –20, –30 kV) with respect to the plasma.

Outside of the internal anode–cathode zone and in a concentric manner, a solenoid can be used to produce a magnetic field which generates a helical trajectory for electrons leaving the filament, which then go on to collide with and ionize the vaporized solid.

10.5.4. Grid sources and broad beams

These apparatuses use a continuous current discharge in an initially neutral gas such as argon. The discharge chamber and the ion optics make up two essential elements in this source, as shown in Figure 10.21. These setups are used largely for assisted depositions or as ion sources for cleaning surface substrates.

10.5.4.1. Discharge chamber

The primary electrons are emitted from a hot cathode (thermoelectric emission) placed at the center of the chamber.

An axial magnetic field (also divergent is possible so as to increase the electron pathway and the generation of ions through atom–electron collisions) was used in the original setups. Permanent magnets distributed in a multipolar configuration make possible a more homogeneous ion density around the ion optics.

A coaxial cylindrical anode within the chamber is used to precisely fix the potential of the formed ions (beam current denoted Ib) that are practically accelerated with the potential of the anode (Vb) towards the extraction point (grid).

The discharge produced between the cathode and the anode depends on the pressure of the introduced gases. At low pressures, the ion current (Ib) decreases with respect to the given discharge current (Ib) and voltage (Vd). Below a minimum pressure the ion current tends to zero. Also, this current increases slowly with increasing pressures.

The ion current (Ib) appears at an optimal pressure for the given discharge conditions (Id, Vd), as shown in Figure 10.22.

The ratio between Id and Ib typically is of the order of 10 to 20.

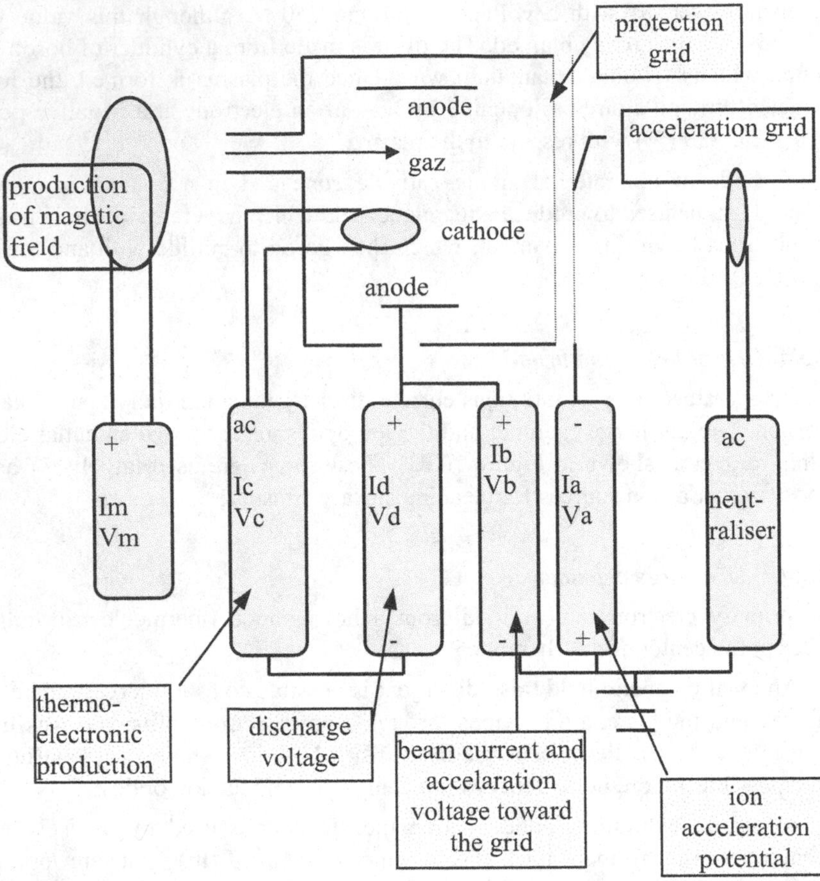

Figure 20.21. *Schematization of a Kaufman-type grid source.*

The discharge voltage must be of the order (and just below) of the sum of the first and second ionization potentials of the atoms to be ionized when looking to prepare doubly charged ions. For argon, these are 15.8 and 27.6 eV, respectively, their sum being around 43 eV. To limit secondary effects (doubly charged ions with energies twice that of the monocharged ions), a discharge tension of between 35 to 40 eV is used. Helium has a first ionization potential at 24.6 eV and a second around 55 eV, so the discharge tension should be around 80 eV.

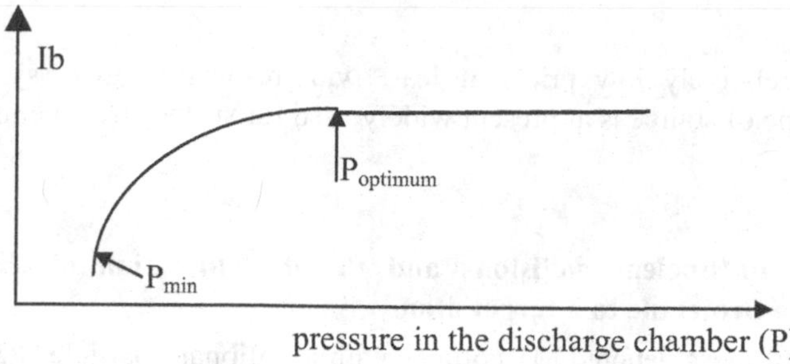

Figure 10.22. *Ion current as a function of the pressure in the chamber.*

10.5.4.2. Ion optics and the neutralizer

Ions that approach the ion optics (through being driven back by the positive tension of the anode) are accelerated by a grid carrying a negative potential Va of the order of 1000 V. To limit the premature degradation of the grid and the consequent contamination of the beam, a protection grid is placed in front. This is made from a material with a low pulverization level such as carbon.

In addition, a neutralization of the ion beam is required given the high levels of currents generated using this type of source. The main energy in the beam is carried by the ions, whereas the electrons in the ion beam can be used to facilitate the transport by neutralization of both the charge and the current. Thus:

- charge neutralization which involves the compensation of the positive charge of the ion beam by the electrons. If this does not happen, then the source will work badly giving rise to a dispersed beam due to the mutual electrostatic repulsion of the ions. The neutralization simply signifies that an elementary volume contains a closely equivalent number of positive and negative charges. It does not mean that the electrons recombine with the source ions to form neutral atoms or molecules. In principle, the level of recombinations is low under usual functioning conditions due to the differing velocities of the electrons and the ions. If the electrons are not produced by a filament neutralizer, they can come from discharges or arcs with earthed parts of the apparatus.

- neutralizing current; if the neutralization electrons have a velocity which approaches that of the ions, then the ion current can be neutralized. If there are excess electrons, then they will run off easily to earth.

A neutralization current is only really necessary if the target is an insulator (whether earthed or not), which would otherwise become positively charged and risk electrical breakdown. In practical terms, a hot neutralizer filament (tantalum or tungsten) is placed in the zone of maximum ion current density, resulting in the maximum possible coupling between electrons and ions.

Given its relatively low price, at least with respect to ECR systems, for example, this type of source is at present widely used for surface treatments.

10. 6. Problem

Detailed study of nuclear collisions and the determination of the energy transferred by a projectile to a target atom

A particle (P1) of mass denoted M1 collides with a stationary particle (P2) of mass M2. Following the collision, P2 recoils and absorbs some of the energy of P1. The latter sees its velocity fall and its trajectory modified. At any instant t, the velocities of P1 and P2 are denoted $\vec{v}(P_1)$ and $\vec{v}(P_2)$, respectively.

At any instant (t) the velocities of P1 and P2 are denoted $\vec{v}(P_1)$ and $\vec{v}(P_2)$, respectively. The coordinates also are given; for example, if we use laboratory co-ordinates (R_L), the velocities are denoted by $\vec{v}(P_1/R_L)$ and $\vec{v}(P_2/R_L)$ such that

$$\vec{v}(P_1/R_L) = \frac{d\vec{OP_1}}{dt}, \ \vec{v}(P_2/R_L) = \frac{d\vec{OP_2}}{dt},$$

where O is the origin of R_L.

In addition, the superscript ' indicated that the velocity is at an initial moment in time before the collision and in an asymptotic position. Similarly, the superscript '' indicates a final state following the collision at an asymptotic position.

1. Study the movement within laboratory coordinates. Using the notations given in Figure 10.23, determine the final velocities and directions of the particles.

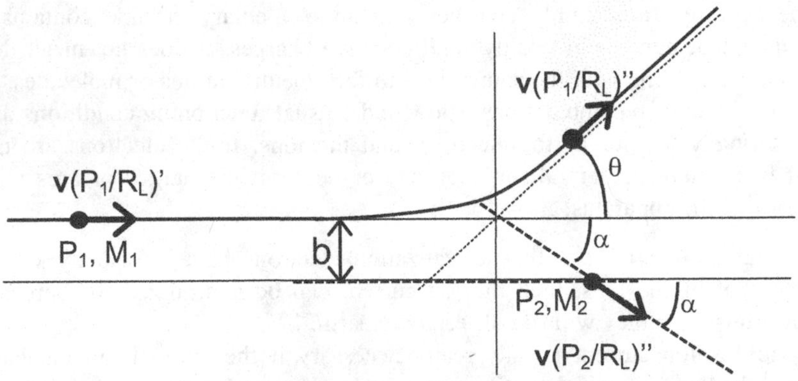

Figure 10.23. *Deviation of particles within laboratory coordinates, where θ is the angle of deviation of the projectile and α is the angle of deviation of the target particle.*

2. Study the movement within barycentric coordinates using the notations given in Figure 10.24.

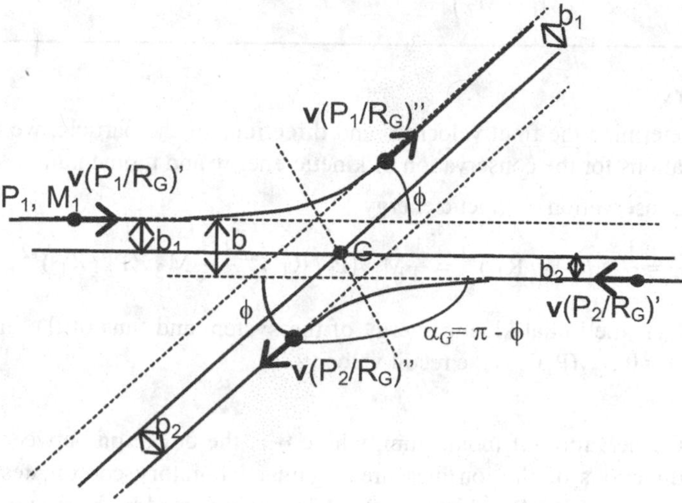

Figure 10.24. *Deviation of particles with barycentric coordinates*

In particular, show that given the definition of G that the total momentum within barycentric coordinates is zero.

3. Establish the relationships for the particles pathways for velocities using co-ordinates of the centers of mass of the particles (R_G) and R_L.

4. Establish the following equations (noting the conservation of the normal velocities):

$$\begin{cases} v(P_1/R_G)'' = v(P_1/R_L)' - v_G \\ v(P_2/R_G)'' = -v_G. \end{cases}$$

Also show that with $M^* = \dfrac{M_1 M_2}{M_1 + M_2}$, we have $v_G = \dfrac{M^*}{M_2} v(P_1/R_L)'$.

5. Geometrically represent the pathway of the center of mass of P2 with respect to R_L. In particular, show that $\alpha_G = 2\,\alpha$.

6. Show that the energy transferred from the projectile to the target particle can be written as $T = \dfrac{4E_0 M_1 M_2}{(M_1 + M_2)^2}\sin^2\dfrac{\phi}{2}$.

Answers

1. To determine the final velocities and directions of the particle, we can start with the equations for the conservation of kinetic energy and momentum.

- Conservation of kinetic energy

$$E_0 = \frac{1}{2}M_1 v(P_1/R_L)'^2 = \frac{1}{2}M_1 v(P_1/R_L)''^2 + \frac{1}{2}M_2 v(P_2/R_L)''^2, \qquad (1)$$

where E_0 is the initial kinetic energy of the system and thus of the incident ion, as $v(P_2/R_L)' = 0$; $v(P_2)''$ is the recoil velocity.

- Conservation of momentum, where θ is the deviation between the final and initial directions of the ion measured against laboratory coordinates and α is the angle between $\vec{v}(P_2/R_L)''$ and $\vec{v}(P_1/R_L)'$ as indicated in the figure.

By projection on the horizontal axis,

$$M_1 v(P_1/R_L)' = M_1 v(P_1/R_L)''\cos\theta + M_2 v(P_2/R_L)''\cos\alpha\,; \qquad (2)$$

by projection along the vertical axis,

$$0 = M_1 v(P_1/R_L)''\sin\theta + M_2 v(P_2/R_L)''\sin\alpha\,. \qquad (3)$$

2. The force that exists between two particles during the collision depends only on the direction of the line between the two particles. There are in effect no transversal forces. The movement of the two particles thus is reduced to that of the movement of a single particle in a central potential. Here this is the interatomic potential at the center on the origin G of the coordinates of R_G. The origin, the center of mass, is the barycenter for the point P_1 and P_2. It is defined in R_L by

$$\vec{OG} = \frac{M_1\vec{OP_1} + M_2\vec{OP_2}}{M_1 + M_2}\,.$$

By derivation we obtain:

$$\frac{d\vec{OG}}{dt} = \frac{M_1 \frac{d\vec{OP_1}}{dt} + M_2 \frac{d\vec{OP_2}}{dt}}{M_1 + M_2} \Leftrightarrow \vec{v}_G = \frac{M_1 \vec{v}(P_1/R_L) + M_2 \vec{v}(P_2/R_L)}{M_1 + M_2},$$

$\Rightarrow \vec{v}_G = \vec{C}^{ste}$ as the numerator is a constant taking into account the conservation of momentum in an elastic shock. The $\vec{v}_G$ is the velocity vector of the point G in R_L and it is constant which indicates that the coordinates of R_G move uniformly within R_L.

This is true at all instants and in particular those prior to the collision. We can thus write:

$$\vec{v}_G = \vec{C}^{ste} = \frac{M_1 \vec{v}(P_1/R_L)'}{M_1 + M_2} \qquad \forall t;$$

$\vec{v}_G$ is collinear to $\vec{v}(P_1/R_L)'$, and the vector $\vec{v}_G$ thus is horizontal (see figure).

In the coordinates of the center of mass, the barycentric equation also can be written as:

$$M_1 \vec{GP_1} + M_2 \vec{GP_2} = (M_1 + M_2)\vec{GG} = \vec{0}.$$

By derivation we obtain:

$$M_1 \frac{d\vec{GP_1}}{dt} + M_2 \frac{d\vec{GP_2}}{dt} = \vec{0},$$

or again $M_1 \vec{v}(P_1/R_G) + M_2 \vec{v}(P_2/R_G) = \vec{0}$, (4)

where $\vec{v}(P_1/R_G)$ and $\vec{v}(P_2/R_G)$ are the velocities of the particles P_1 and P_2 in the center mass system.

The amount $M_1 \vec{v}(P_1/R_G) + M_2 \vec{v}(P_2/R_G)$ represents the total momentum of the system. To conclude, the total momentum in the barycentric coordinate is zero, taking into account the definition of G.

3. For the equation of the change in coordinates of the center of mass (R_G) with respect to the laboratory coordinates (R_L), we have

$$\vec{GP_1} = \vec{OP_1} - \vec{OG}, \qquad (5)$$

$$\Leftrightarrow \frac{d\vec{GP_1}}{dt} = \frac{d\vec{OP_1}}{dt} - \frac{d\vec{OG}}{dt}$$

$$\Leftrightarrow \vec{v}(P_1/R_G) = \vec{v}(P_1/R_L) - \vec{v}_G$$

$$\Leftrightarrow \begin{cases} \vec{v}(P_1/R_G)' = \vec{v}(P_1/R_L)' - \vec{v}_G \\ \vec{v}(P_1/R_G)'' = \vec{v}(P_1/R_L)'' - \vec{v}_G \end{cases} \qquad (5 \text{ bis})$$

and

$$\vec{GP_2} = \vec{OP_2} - \vec{OG} \qquad (6)$$

$$\Leftrightarrow \frac{d\vec{GP_2}}{dt} = \frac{d\vec{OP_2}}{dt} - \frac{d\vec{OG}}{dt}$$

$$\Leftrightarrow \vec{v}(P_2/R_G) = \vec{v}(P_2/R_L) - \vec{v}_G$$

$$\Leftrightarrow \begin{cases} \vec{v}(P_2/R_G)' = -\vec{v}_G \\ \vec{v}(P_2/R_G)'' = \vec{v}(P_2/R_L)'' - \vec{v}_G . \end{cases} \qquad (6 \text{ bis})$$

4. The velocities of the two particles, with respect to the center of mass, conserve their normals prior to and after the reaction.

Equation (4) derived from the equation defining the barycenter is written prior to the collision (within the center of mass coordinates) in the form:

$$M_1\vec{v}(P_1/R_G)' + M_2\vec{v}(P_2/R_G)' = \vec{0} .$$

With Eqs. (5 bis) and (6 bis), we can determine that:

$$M_1\left(\vec{v}(P_1/R_L)' - \vec{v}_G\right) - M_2\vec{v}_G = \vec{0} .$$

By projecting along the horizontal axis, we obtain:

$$M_1\left(v(P_1/R_L)' - v_G\right) - M_2 v_G = 0 . \qquad (7)$$

Following the collision, the same Eq. (4) gives $M_1\vec{v}(P_1/R_G)'' + M_2\vec{v}(P_2/R_G)'' = \vec{0}$,

from which by projecting on the final asymptotic direction we have

$$M_1 v(P_1/R_G)'' + M_2 v(P_2/R_G)'' = 0 . \qquad (8)$$

By identification of Eqs. (7) and (8) we obtain:

$$\begin{cases} v(P_1/R_G)'' = v(P_1/R_L)' - v_G \\ v(P_2/R_G)'' = -v_G. \end{cases} \quad (9)$$

In fact this result can be generalized to all instants. By using an appropriate axis for each instant it is possible to show that the normal of the velocity vectors is constant at all times.

A consequence of Eq. (7) is that it is possible to write:

$$M_1 v(P_1/R_L)' = (M_1 + M_2)v_G,$$

so that:

$$v_G = \frac{M_1}{M_1 + M_2} v(P_1/R_L)' \overset{M^* = \frac{M_1 M_2}{M_1 + M_2}}{=} \frac{M^*}{M_2} v(P_1/R_L)'. \quad (10)$$

5. Geometric representation and angular relations

Before calculating the transferred energy (T), we will give the equations that make it possible to go from diffusion angles (defined in the laboratory system R_L) to diffusion angles defined within a system of the mass center R_G (see figures).

Following Eqs. (5 bis) and (6 bis), we can go from one system to the other by a simple vector translation ($\vec{v}_G$, so that, for example):

$$\vec{v}(P_2/R_L)'' = \vec{v}_G + \vec{v}(P_2/R_G)'';$$

We thus can construct Figure 10.25 that defines an isosceles triangle as, according to Eq. (9): $\|\vec{v}(P_2/R_G)''\| = \|-\vec{v}_G\| = v_G$

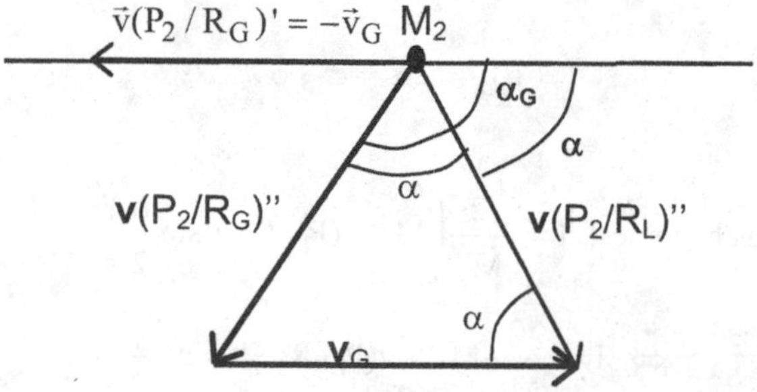

Figure 10. 25. *Change from center of mass to laboratory coordinates.*

We can see that α_G and α are bound by the equation

$$\alpha_G = 2\,\alpha. \qquad (11)$$

We also have $\cos\alpha = \dfrac{v(P_2/R_L)''/2}{v_G}$, so that

$$v(P_2/R_L)'' = 2v_G\cos\alpha\,. \qquad (12)$$

6. Calculation of the energy transferred (T) to a target atom

During the collision, the energy transferred to the target can be written as: $T = \dfrac{1}{2}M_2\left(v(P_2/R_L)''\right)^2$, from which with Eq. (12) :

$$T = \frac{M_2}{2}\left(2v_G\cos\alpha\right)^2.$$

By substituting Eq. (10), we obtain: $T = \dfrac{M_2}{2}\left(\dfrac{2v(P_1/R_L)'M^*\cos\alpha}{M_2}\right)^2$, so that also:

$$T = \frac{2}{M_2}\left(v(P_1/R_L)'M^*\cos\alpha\right)^2. \qquad (13)$$

We can see in the above figure that $\alpha_G = \pi - \phi$, so that with Eq. (11),

$$\alpha_G = 2\alpha = \pi - \phi\,.$$

Equation (13) thus becomes: $T = \dfrac{2}{M_2}\left(M^*\,v(P_1/R_L)'\sin\dfrac{\phi}{2}\right)^2$

$$\Leftrightarrow\ T = \frac{2}{M_2}M^{*2}\,v(P_1/R_L)'^2\sin^2\frac{\phi}{2},$$

and thus $T = \dfrac{4E_0M^{*2}}{M_1M_2}\sin^2\dfrac{\phi}{2} = \dfrac{4E_0}{M_1M_2}\left(\dfrac{M_1M_2}{M_1+M_2}\right)^2\sin^2\dfrac{\phi}{2}$, where $E_0 = \frac{1}{2} M_1 v(P_1 / R_L)'^2$ and is the kinetic energy of the incident ion defined in Eq. (1).

Finally:

$$T = \frac{4E_0 M_1 M_2}{(M_1 + M_2)^2} \sin^2 \frac{\phi}{2}. \qquad (14)$$

This equation is particularly fundamental to calculations for nuclear stopping powers. It represents, in effect, the energy lost by the projectile during the collision.

References

ANDERSON J.C., *Diélectriques*, Dunod, Paris, 1966.

BÖTTCHER C.J.F., *Theory of electric polarization*, 2nd édition, Elsevier, Paris, vol. 1, 1973; and vol. 2, 1978.

COELHO R., ALADENIZE B., *Les diélectriques*, Hermès, Paris, 1993.

COELHO R., *Physics of dielectrics*, Elsevier, Paris, 1979.

COMBES P.F., *Micro-ondes*, tomes 1 et 2, Dunod, Paris, 1996.

FEYNMAN R., « Le cours de physique de Feynman », *Electromagnétisme 2*, Inter Editions, Paris, 1992.

GARDIOL F., *Electromagnétisme*, Presses Polytechniques Universitaires Romandes, 2002.

HILL N., VAUGHAN W.E., PRICE A.H., DAVIES M., *Dielectrics properties and molecular behaviour*, Van Nostrand Reinhold Company, London, 1969.

JACKSON J.D., *Electrodynamique classique*, Dunod, 2001.

Original edition : JACKSON J.D., *Classical electrodynamics*, 3rd edition, John Wiley & Sons, 1999

JONSCHER A.K., *Dielectric relaxation in solids*, Chelsea Dielectrics Press, London, 1983.

LORRAIN P., CORSON Dale R., LORRAIN F., *Les phénomènes électromagnétiques*, Dunod, 2006,

Original edition : LORRAIN P., CORSON Dale R., LORRAIN F., *Fundamentals of Electromagnetic phenomena,* W.H. Freeman and Company (2000).

MOLITON A., *Optoélectronics of molecules and polymers*, Springer, 2006.

PAUL Clayton R., Electromagnetics for Engineers, Wiley Edition, 2004

PEREZ J.P., CARLES R., FLECKINGER R., *Electromagnétisme : vide et milieux matériels*, Masson, Paris, 1991 et 1996.

SALEH B.E., TEICH M.C., *Fundamentals of photonics*, Wiley, New York, 1991.

SMITH R. , *Atomic and ion collisions in solids and at surfaces*, Cambridge University Press, 1997.

Index